Methods in Bioengineering

Cell Transplantation

The Artech House Methods in Bioengineering Series

Series Editors
Martin L. Yarmush, M.D., Ph.D.
Robert Langer, Sc.D.

Methods in Bioengineering: 3D Tissue Engineering,
François Berthiaume and Jeffrey Morgan, editors

Methods in Bioengineering: Biomicrofabrication and Biomicrofluidics,
Jeffrey D. Zahn and Luke P. Lee, editors

Methods in Bioengineering: Microdevices in Biology and Medicine,
Yaakov Nahmias and Sangeeta N. Bhatia, editors

Methods in Bioengineering: Nanoscale Bioengineering and Nanomedicine,
Kaushal Rege and Igor Medintz, editors

Methods in Bioengineering: Stem Cell Bioengineering,
Biju Parekkadan and Martin L. Yarmush, editors

Methods in Bioengineering: Systems Analysis of Biological Networks,
Arul Jayaraman and Juergen Hahn, editors

Methods in Bioengineering

Cell Transplantation

Alejandro Soto-Gutierrez
University of Pittsburgh

Nalu Navarro-Alvarez
Columbia University

Ira J. Fox
University of Pittsburgh

Editors

ARTECH
HOUSE

BOSTON | LONDON
artechhouse.com

Library of Congress Cataloging-in-Publication Data
A catalog record for this book is available from the U.S. Library of Congress.

British Library Cataloguing in Publication Data
A catalogue record for this book is available from the British Library.

ISBN-13: 978-1-60807-015-2

Cover design by Vicki Kane

685 Canton Street
Norwood MA 02062

10 9 8 7 6 5 4 3 2 1

To all those individuals who have dedicated themselves
to looking for cures through new and innovative therapeutics.

To Mary and Javier—thank you for your love and support
and showing me the world through hope,
persistence, and questioning eyes.
–Alejandro

To Lupita, for all your love and for teaching me by example
that everything is possible in life. To Emmanuel for changing my life.
–Nalu

To Marta, and our sons Alex and Michael.
–Ira

Contents

Preface

The primary purpose of this book is to describe in detail the basic techniques and protocols in regenerative medicine. This methodology book is meant to be a primary resource for scientists, clinicians, and students intended in cell and tissue transplantation. Regenerative medicine cell-based therapeutics may incorporate grown tissues, cells, laboratory-made compounds, and engineered organs, bringing together experts in medicine, biology, chemistry, computer science, bioengineering, genetics, and other fields.

Successful experimentation requires attention to detail, and this guidebook makes it possible to understand the many intricate steps involved in tissue engineering, cell isolation, progenitors and stem cell differentiation, and in vivo models for assessing new strategies. Each chapter includes detailed methodology, as well as assistance with troubleshooting so that readers can establish these techniques in their own laboratories.

The future is bright and we hope that this book will assist readers to better understand the field of regenerative medicine and therapeutics. We are especially indebt to Martin L. Yarmush; our publisher, Artech House, without whose trust and guidance this work would not have begun; and our editorial assistant, Maria Bond, without whose hard work it would not have been completed.

Alejandro Soto-Gutierrez
University of Pittsburgh
Pittsburgh, Pennsylvania

Nalu Navarro-Alvarez
Columbia University
New York, New York

Ira J. Fox
University of Pittsburgh
Pittsburgh, Pennsylvania

May 2011

1

Extracellular Matrices for Hepatic Tissue Engineering

Chong-Su Cho[1], Hu-Lin Jiang[2], Takashi Hoshiba[3], Sang-Hyon Kim[4], and Toshihiro Akaike[5*]

[1]Department of Agricultural Biotechnology and Research Institute for Agriculture and Life Sciences, Seoul National University, Seoul 151-921, Korea
[2]College of Veterinary Medicine, Seoul National University, Seoul 151-742, Korea
[3]Biomaterials Center, National Institute for Materials Science, Tsukuba, Japan
[4]Biomaterials Research Center, Korea Institute of Science and Technology, Seoul 136-791, Korea
[5]Graduate School of Bioscience and Biotechnology, Tokyo Institute of Technology, Yokohama 226-850, Japan

*Correspondence: Graduate School of Bioscience and Biotechnology, Tokyo Institute of Technology, Yokohama 226-8501, Japan
E-mail: takaike@bio.titech.ac.jp
Phone: +81-45-924-5790; Fax: +81-45-924-5815

Abstract

Recently, the development of biomaterials for hepatic tissue engineering has focused on the design of galactose-carrying artificial extracellular matrix (ECM) that can perform specific hepatocyte behaviors and direct liver regeneration through molecular recognition. Specifically, the bioartificial liver-assist device or regeneration of the liver–tissue substitutes for hepatic tissue engineering requires a suitable ECM for hepatocyte culture because the hepatocytes are anchorage-dependent cells and are highly sensitive to the ECM environment for maintaining their viability and differentiated functions. This chapter will review mechanisms of specific interaction between galactose moieties of the galactose-carrying ECMs and asialoglycoprotein receptors (ASGPR) on the hepatocytes. Hepatic tissue engineering typically employs artificial three-dimensional ECMs as the scaffold of hepatocytes to induce long-term and stable differentiated hepatocyte functions. The artificial ECMs can be fabricated from two classes of biomaterials: synthetic polymers and natural ones. The design criteria of artificial ECMs such as density of galactose, chemical structure of galactose, deposition of ECM, coculture system, and cell source will be explained together with detailed descriptions of the relevant experimental approaches.

Key terms extracellular matrix, hepatocyte, galactose, tissue engineering, scaffold, spheroid, liver function

1.1 Introduction

The extracellular matrix (ECM) is very important in tissue engineering because it plays an important role in regulating cellular behaviors such as cell adhesion, cell proliferation, migration, differentiation, and survival and it also serves many functions including the provision of structural support and tensile strength, attachment sites for cell surface receptors, and as a reservoir for signaling factors [1]. It became clear that the ECM is much more than a passive supporter in the events of tissue and organ development because the concept of a close relationship between the ECM and intracellular cytoskeletal and nuclear elements has become widely accepted [2]. Therefore, the development of an artificial ECM has been a big concern in tissue engineering because cellular growth and differentiation, in the two-dimensional cell culture as well as in the three-dimensional space of the developing organism, require ECMs with which the cells can interact. The bioartificial liver-assist devices or the regeneration of liver–tissue substitutes for hepatic tissue engineering require a suitable ECM for hepatocyte culture because the hepatocytes are anchorage-dependent cells and are highly sensitive to the ECM environment for maintaining their viability and differentiated functions. The development of galactose-carrying artificial ECMs to replace natural ECMs is increasingly essential and promising in hepatic tissue engineering because they play important roles in providing an alternative to organ and tissue transplantation, both of which suffer from limitation of supply [3].

This chapter will describe methods for checking specific interaction between galactose moieties of the galactose-carrying artificial ECMs and asialoglycoprotein receptors (ASGPRs) on the hepatocytes. The artificial ECMs classified by synthetic polymers and natural ones among the biomaterials will be covered. The design criteria of artificial ECMs such as density of galactose, chemical structure of galactose, deposition of ECM, coculture system, and cell source will be explained together with detailed descriptions of the relevant experimental approaches. Natural ECMs will not be covered in this review owing to space limitation; however, recent reviews on hepatocyte behaviors by natural ECMs are available [4].

1.2 Experimental Design

Specific interaction between galactose moieties of galactose-carrying ECMs and ASGPR on hepatocytes was checked because the ASGPR on hepatocytes was first confirmed by Pricer et al. [5], who found that circulating asialoglycoproteins (ASGPs) bound to and were degraded by hepatocytes. The hepatocytes recognize and bind molecules carrying galactose, N-acetylgalactosamine, or glucose residues through specific interaction between cell receptors and ligands. The ASGPs are internalized via coated pits and coated vesicles and subsequently appear in a complex network of tubules and uncoated vesicles before reaching the lysosomes where they are degraded after binding receptors [6]. The hepatic plasma membrane receptors mediate the specific binding and uptake of partially deglycosylated glycoproteins through receptor-mediated endocytosis. To investigate specific interactions between ECM and hepatocytes, cell adhesion, cell detachment, inhibition assay, Western blot analysis, flow cytometric analysis, and lectin-enzyme assay were used.

1.3 Materials

- NaOH (Sigma);
- Sodium dodecyl sulfate (Sigma);
- Williams' Medium E (WE) medium (pH 7.4, 10 mM HEPES, 10 mM $NaHCO_3$, 50 μg/ml penicillin and streptomycin, and 100 μg/ml neomycin) (Sigma);
- Polystyrene dish (Becton Dickinson);
- Phosphate buffered saline (PBS, Invitrogen);
- 3-(4,5-dimethylthiazol-2-yl)-2,5-diphenyl tetrazolium bromide (MTT, Sigma);
- Calcium chloride (Sigma);
- Ethylene glycol-bis[β-amino ethyl ether]-N,N,N´N´´-tetraacetic acid (EGTA, Sigma);
- Magnesium chloride (Sigma);
- Asialofetuin (ASF, Sigma);
- Galactose (Wako Pure Chemical);
- Lactose (Wako Pure Chemical);
- Fluorescein isothiocyanate (FITC, Wako Pure Chemical);
- Fetuin (Sigma);
- *p*-Vinyl benzylamine (Wako Pure Chemical);
- Lactobionic acid (Wako Pure Chemical);
- Aminohexyl-D-galactopyranoside (Wako Pure Chemical);
- 1-amino-1-deoxy-β-D-lactose (Wako Pure Chemical);
- Bicinchoninic acid (BCA, Pierce);
- Poly-(L-lysine) (PLL, Sigma);
- Alginate (Sigma);
- Heparin (Sigma);
- Fibronectin (FN, Sigma);
- Anti-phospho-FAK (New England Biolabs);
- Concanavalin A (ConA, Sigma);
- Lens culinaris (LcH, Sigma);
- Arachis hypogaea (Sigma);
- α-D-mannose (Wako Pure Chemical);
- α-D-glucose (Wako Pure Chemical).

1.4 Methods

1.4.1 Cell Adhesion Assay

The percentage of cell adhesion was accurately quantified by protein assay method. Briefly, protein was extracted from adhered hepatocytes with lysis solution (0.25% NaOH and 0.1% SDS) after detached cells were removed, and measured by the Lowry method of the protein assay kit. All values are normalized to the percentage of adhesion of the maximum, which was taken as 100%.

1.4.2 Cell Detachment Assay

Each well of 96 well polystyrene (PS) microplate was coated with ECM solution. Isolated hepatocytes (1×10^4 cells/well) in WE base, 100 μl, were seeded to each well and allowed to be adhered at 37°C for indicated times. The final volume in each well was brought to 420 μl with cold PBS. The plate was carefully sealed by rolling on an adhesive plate sealer without air bubbles. Subsequently, sealed plates were inverted and centrifuged for 10 min at 50g or 500g force. After centrifuge, nonadherent cells were carefully removed and wiped in an inverted position on tissue paper. The remaining adhered cells were quantified by MTT assay.

1.4.3 Inhibition Assay

1.4.3.1 The Use of Chelators

Ca^{2+} is essential to interactions between hepatocytes and galactose-carrying ECMs. To determine affinity of ASGPR for the ECM surfaces, the effects of Ca^{2+} on hepatocyte adhesion to the ECM were investigated. Figure 1.1 shows hepatocyte adhesion to the different kinds of carbohydrate-carrying glycopolymers (Figure 1.2) as the ECMs against concentration of Ca^{2+} [7]. The results indicated that hepatocyte adhesion occurred in the presence of 50, 100, and 300 μM $CaCl_2$ on surfaces coated with less than 0.5 μg/ml of galactose-carrying glycopolymer. On the other hand, hepatocytes adhered to surfaces coated with glucose-carrying glycopolymers in the presence of more than 200 μM Ca^{2+} as shown in Figure 1.1. Also, the requirement for Ca^{2+} was increased when hepatocytes adhered to glycoplymers at low coating density.

Removal of Ca^{2+} was accomplished with use of chelators such as EGTA and EDTA. Therefore, adhered hepatocytes were almost detached after treatment with 2.5 mM EGTA. Also, effective adhesion of hepatocytes could not be induced by Mg^{2+} ions alone as shown in Table 1.1 [8], suggesting that the interaction between ASGPRs and galactoses requires Ca^{2+} ions.

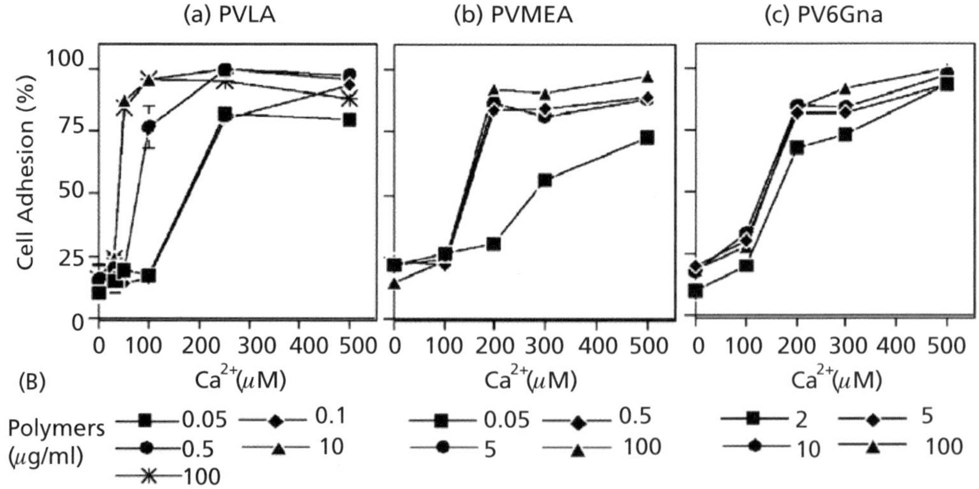

Figure 1.1 Effect of Ca^{2+} on hepatocyte adhesion for PVLA (a), PVMEA (b), and PV6Gna (c) surfaces. Hepatocytes were allowed to be adhered in the presence of 0–500 μm of Ca^{2+} in Hanks solution for 30 min. All values were normalized to the percentage of adhesion of the maximum, which was taken as 100%. (From [7].)

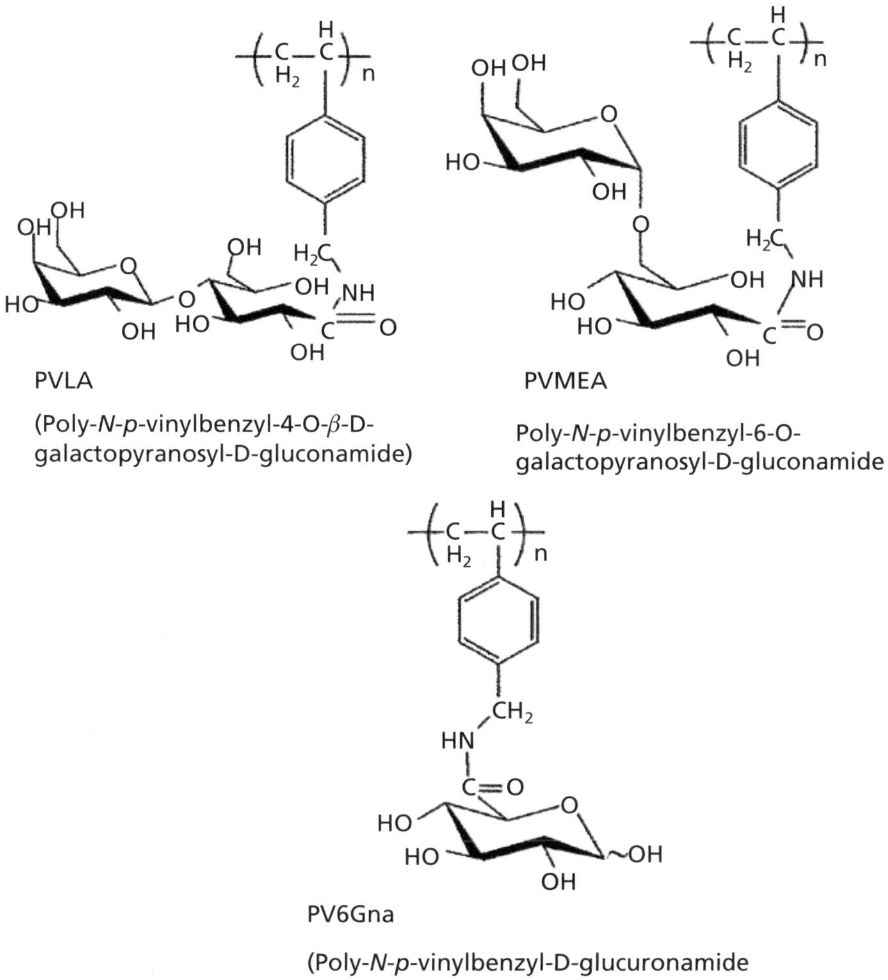

Figure 1.2 Molecular structure of synthetic glycopolymers. (From [7].)

Table 1.1 The Effect of Divalent Cations (Ca^{2+} and Mg^{2+} ions, 1.8 mM) on Hepatocyte Attachment to PVLA-Coated Dishes with Various Coating Densities

	PVLA-Coating Density ($\mu g\ cm^{-2}$)				
	Cell Attachment (%)				
Conditions	0	0.07	0.76	0.90	1.08
Ca^{2+}, Mg^{2+} (1.8 mM)	40 ± 6	77 ± 2	73 ± 3	75 ± 5	71 ± 5
Ca^{2+} (1.8 mM)	38 ± 5	76 ± 4	72 ± 4	71 ± 2	69 ± 5
Mg^{2+} (1.8 mM)	43 ± 6	33 ± 3	12 ± 7	10 ± 4	8 ± 7
Without Ca^{2+}, Mg^{2+}	32 ± 5	30 ± 3	11 ± 4	10 ± 5	9 ± 5

Source: [8].

1.4.3.2 The Use of Inhibitors

It is already reported that hepatocytes are specifically interacted with galactose-carrying glycopolymers mediated by a carbohydrate-binding protein such as ASGPR on the cell

Table 1.2 The Effect of Inhibitors on Hepatocyte Attachment to PVLA-Coated Dishes at Various Coating Densities

		PVLA-Coating Density ($\mu g\ cm^{-2}$)			
		Inhibition Rate of Cell Attachment (%)			
Inhibitor concentration	(mM)	0.07	0.76	0.90	1.08
Glucose	30	8 ± 4	5 ± 8	7 ± 6	9 ± 7
	240	5 ± 3	3 ± 5	1 ± 2	14 ± 4
Galactose	30	18 ± 8	14 ± 2	7 ± 3	12 ± 6
	240	45 ± 7	28 ± 5	25 ± 7	25 ± 4
Lactose	30	15 ± 5	8 ± 6	1 ± 3	6 ± 9
	240	80 ±7	88 ± 9	78 ± 9	86 ± 8
PVMA	2.1	12 ± 3	2 ± 5	7 ± 7	7 ± 4
VLA	30	95 ± 7	92 ± 6	96 ± 2	97 ± 5
PVLA	2.1	97 ± 10	97 ± 5	94 ± 7	98 ± 8
Fetuin	0.5	4 ± 5	14 ± 4	19 ± 7	28 ± 6
Asialofetuin	0.5	96 ± 8	94 ± 9	96 ± 5	97 ± 2

Source: [8].

membrane surface. Therefore, inhibition of hepatocyte adhesion to the galactose-carrying glycopolymers by galactose-type carbohydrates can be expected. Table 1.2 shows the inhibitory effects of various carbohydrates on the hepatocyte adhesion onto the PVLA surface [8].

The results indicated that the adhesion was remarkably inhibited by quite small amounts of PVLA and asialofetuin as the natural ligand for ASGPR whereas only small adhesion was inhibited by glucose, glucose-carrying glycopolymer (PVMA), and fetuin. Also, galactose, lactose, and VLA (the monomeric substance of PVLA) were found to inhibit adhesion although rather large amounts of these low-molecular-weight carbohydrates were required for effective inhibition. It is suggested from the effect of inhibitors that adhesion of hepatocytes to PVLA-coated surfaces having low and high coating densities is mediated by ASGPR.

1.4.4 Western Blot Assay

It has been already reported that focal adhesion kinase (FAK) is a key molecule involved in the formation of local contact that regulates cellular behaviors such as cell growth, survival, spreading, and migration through integrin signaling [9]. Integrin-mediated adhesion of cells to the ECM triggers autophosphorylation at the Tyr-397 residue of FAK [9]. Therefore, to check whether integrin-mediated signals occur in hepatocytes cultured on galactose-carrying ECMs, the phosphorylation of FAK in hepatocytes was evaluated compared with FN as the natural ECM by Western blot analysis with anti-pFAK (Tyr-397) antibody [10]. As shown in Figure 1.3(a), phosphorylated FAK was not detected in hepatocytes cultured on PVLA even 12 hours after seeding, whereas the phosphorylation level of FAK was rapidly increased in hepatocytes cultured on FN with time, as shown in Figure 1.3(b), suggesting a different signal transduction pathway between receptor-mediated adhesion to galactose-carrying artificial ECM and integrin-mediated adhesion to natural ECM.

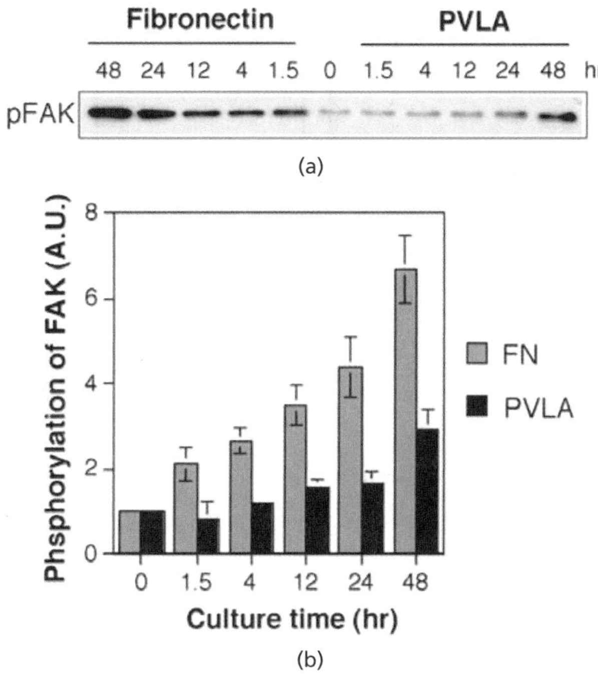

Figure 1.3 FAK is not phosphorylated in hepatocytes adhered onto a PVLA matrix. (a) Western blot analysis of FAK phosphorylation: hepatocytes were allowed to adhere for an indicated time. (b) The mean intensity of the band was quantified as described in Section 1.4.4; the values were determined from experiments performed separately at least three times. All values were normalized to the ratio of phosphorylated FAK to total FAK, which was 1. (*From:* [10].)

1.4.5 Flow Cytometry Analysis

Flow cytometry analysis is carried out using FITC-labeled glycopolymers to investigate ASGPR-mediated interactions between the hepatocyte cell receptors and ligands in the glycopolymers. As shown in Figure 1.4(a), (b), and (c) [11], the surface-labeled hepatocyte fraction (Region 1) was determined from the PI-unlabeled fraction (Region 1 + Region 3). Hepatocytes were labeled to 25% of the PI-unlabeled fraction in the presence of 0.1 mg/ml FITC-PV6Gna (Figure 1.4(b)). The FITC-PV6Gna binding to hepatocytes increased with an increase of concentration of added FITC-PV6Gna as shown by a hyperbolic plot in Figure 1.4(d). However, the FITC-PV6Gna binding to hepatocytes was remarkably decreased by the addition of EGTA, GalNAc, and ASF but not fetuin as shown in Figure 1.4(c) and (e), suggesting ASF or GalNAc as competitive inhibitors. Also, the soluble FITC-PV6Gna binding to hepatocytes was inhibited 5×10^4-fold more effectively than by GalNAc, indicating multivalent effect on the inhibition of the interaction between PV6Gna and hepatocytes. Furthermore, it is suggested that ASGPRs could bind to the glucose-derivatized polymer substituted at the hydroxyl group at the C-6 position of glucose [11].

1.4.6 Lectin Binding Assay

It is well known that lectins are responsible for cell surface sugar recognition in bacteria, animals, and plants [12]. Therefore, a direct lectin-enzyme assay can be used to determine the type of the carbohydrate moiety of glycopolymers. Dose-dependent binding of

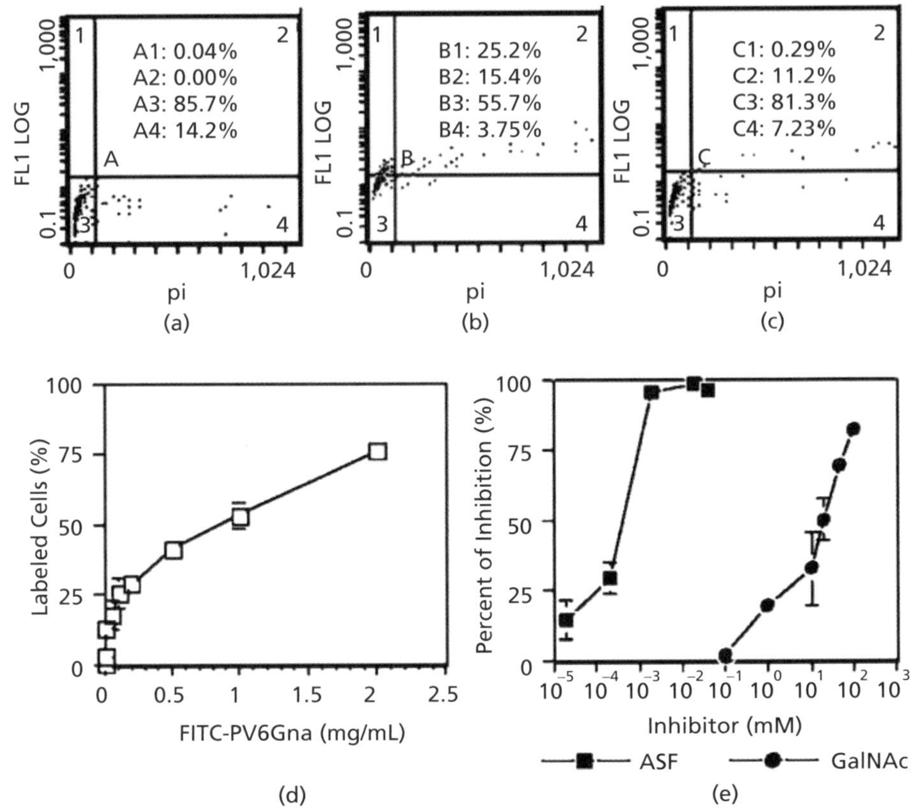

Figure 1.4 Interaction between soluble FITC-conjugated PV6Gna and hepatocytes. Freshly isolated hepatocytes were incubated with unlabeled PV6Gna (a) and with 0.1 mg/ml FITC-conjugated PV6Gna (b) as controls. (c) Inhibitory effect of EGTA. (d) Dose-dependent increase of FITC-conjugated PV6Gna binding to hepatocytes. (e) Inhibitory effect of GalNAc and ASF. FL1 (*x*-axis) and pi (*y*-axis) are the mean intensity of FITC and PI, respectively. PI-labeled regions indicate a membrane-damaged cell fraction. Regions 1, 2, 3, and 4 in individual panels (a, b, and c) represent the undamaged FITC-labeled cell fraction, damaged FITC-labeled cell fraction, undamaged FITC-unlabeled cell fraction, and damaged FITC-unlabeled cell fraction, respectively. The percentage of FITC-labeled cells was determined from the PI-unlabeled cell fraction. (From [11].)

lectins to the several glycopolymers is shown in Figure 1.5 [11]. ConA and LcH showed a higher selectivity in binding to PVMA, PVG, and PV6Gna substituted at C-1, C-3, and C-6 of glucose, respectively, than to PVLA and PVMEA, which have galactose moieties [7]. Allo A bound only to PVLA whereas Arachis hypogaea lectin (PNA) bound to both PVLA and PVMEA, and neither lectin bound to the glycopolymers having glucose moieties. Weis et al. also reported that ConA and LcH had lectin specificity to α-D-mannose or α-D-glucose whereas Allo A and PNA had lectin specificity to β-D-galactose [12]. In particular, ConA bound to PVMA with a higher affinity than to PV6Gna and PVG [7].

1.4.7 Coupling of Galactose Groups with ECM Bulk and ECM Surface

Weigel at al. first introduced galactose groups to polyacrylamide (PAAm) as the bulk state [13]. He also reported that rat hepatocytes were bound to galactosylated PAAm in a calcium-dependent manner via a patch of ASGPR [14]. Weisz et al. reported that the rat hepatic lectins on the hepatocyte cell surface mediate adhesison of isolated primary rat hepatocytes to galactosylated PAAm gel [15].

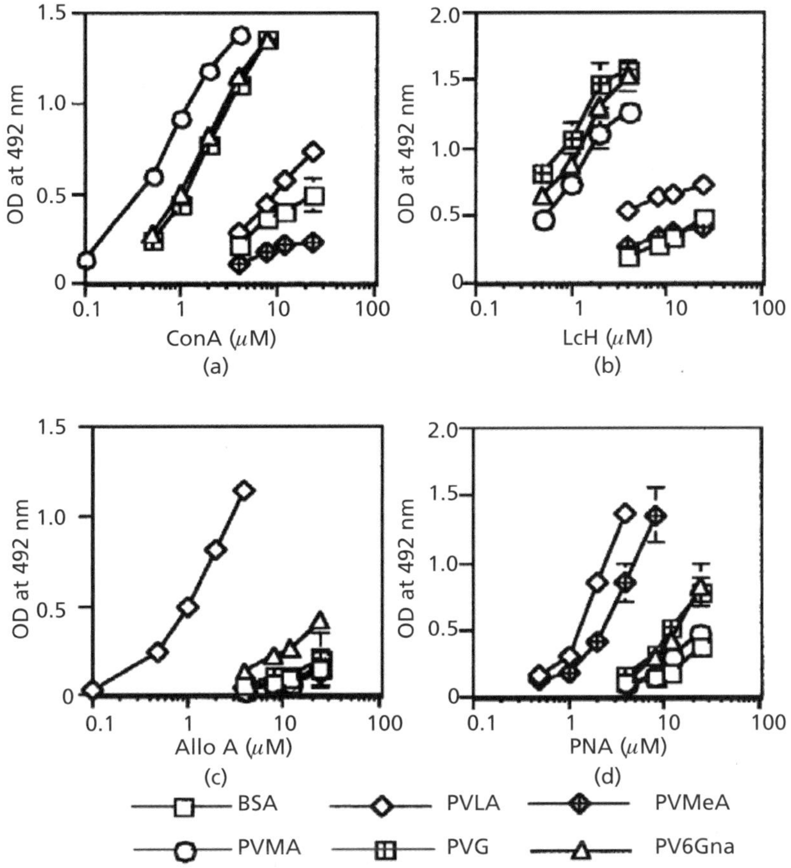

Figure 1.5 Direct lectin-enzyme assay for glycopolymers. ConA (a) or LcH (b) and Allo A (c) or PNA (d) were used as lectins specific for glucose and galactose, respectively. (*From:* [11].)

Akaike and his coworkers reported that galactose-derivatized PS, poly(*N*-*p*-vinylbenzyl-4-O-*β*-D-galactopyranosyl-D-gluconamide)(PVLA), as a representative galactose-carrying synthetic polymer, has been studied as an excellent artificial ECM to guide hepatocyte adhesion through the unique ASGPR-galactose interaction between ASGPR of hepatocytes and highly concentrated galactose moieties along the polymer chains, although ASGPR is a nonadhesion cell surface receptor [16]. Galactose groups were covalently coupled with poly(ethylene oxide) [17], poly(styrene-co-maleic acid) [18], poly(allyl amine) [19], and phosphorylcholine polymer [20] to obtain artificial ECMs capable of selectively recognizing hepatocytes and preserving their functions.

Cell-ECM interaction is a surface phenomenon although bulk properties of the ECM dictate the mechanical properties of the ECM. Therefore, galactose groups are only introduced to the ECM surface for hepatic tissue engineering. Galactose moieties were covalently coupled with the surface of poly(D, L-lactic-co-glycolic acid) [21], acrylic acid-grafted poly(ethylene terephthalate) (PET) [22], PS [23], and PET [24]. The results indicated that hepatocytes cultured on those galactosylated substrata exhibited good attachment and promoted spheroid formation of the attached cells although the galactose groups only existed on the ECM surface.

There are several chemical structures of galactose such as lactose monohydrate [16], lactobionic acid [25], aminohexyl-D-galactopyranoside [22], 1-amino-1-deoxy-*β*-D-lactose [17] and 1-amino-1-deoxy-β-D-galactose [18], which were selected according to the functional group in the artificial ECMs. The chemical reaction will be explained briefly.

Kobayashi et al. coupled lactose with *p*-vinyl benzylamine through imine reaction between the amino group of *p*-vinyl benzylamine and the ketone group of lactose lactone obtained by oxidation of lactone as the monomer, as shown in Figure 1.6 [16]. The galactose-carrying polymer (PVLA) was prepared by free radical polymerization of lactose lactone-coupled *p*-vinyl benzylamine.

Galactose-carrying lactobionic acid was coupled with amino groups of chitosan as the natural polymer using EDC and NHS as activating agents as shown in Figure 1.7 [25].

The amino group in aminohexyl-D-galactopyranoside reacted with carboxylic acid in acrylic acid-grafted PET through amide linkage, as shown in Figure 1.8 [22].

The amino group in 1-amino-1-deoxy-β-D-lactose was coupled with carboxylic acid in poly(styrene-co-maleic acid) dissolved in a mixture of solvent of DMSO and water [18].

Amino group in 1-amino-1-deoxy-β-D-lactose was coupled with poly(ethylene oxide) hydrogel after activiation with tresyl chiloride [17].

A terminal end group of carboxylic acid in poly(D,L-lactic-co-glycolic acid)(PLGA) was conjugated with the amino group of lactose obtained by reductive amination reaction as shown in Figure 1.9 [21].

Kim et al. [26] compared attachment and morphology of hepatocytes between monosaccharide and disaccharide of the terminal galactose, and the position of

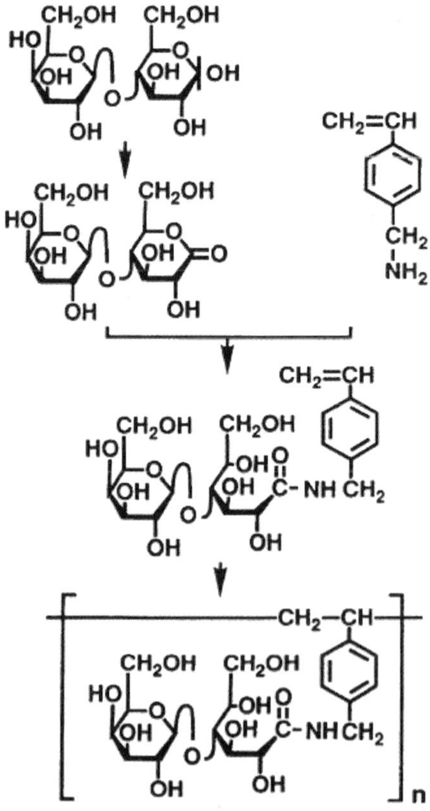

Figure 1.6 Synthetic scheme of galactose-carrying polystyrene. (From [16].)

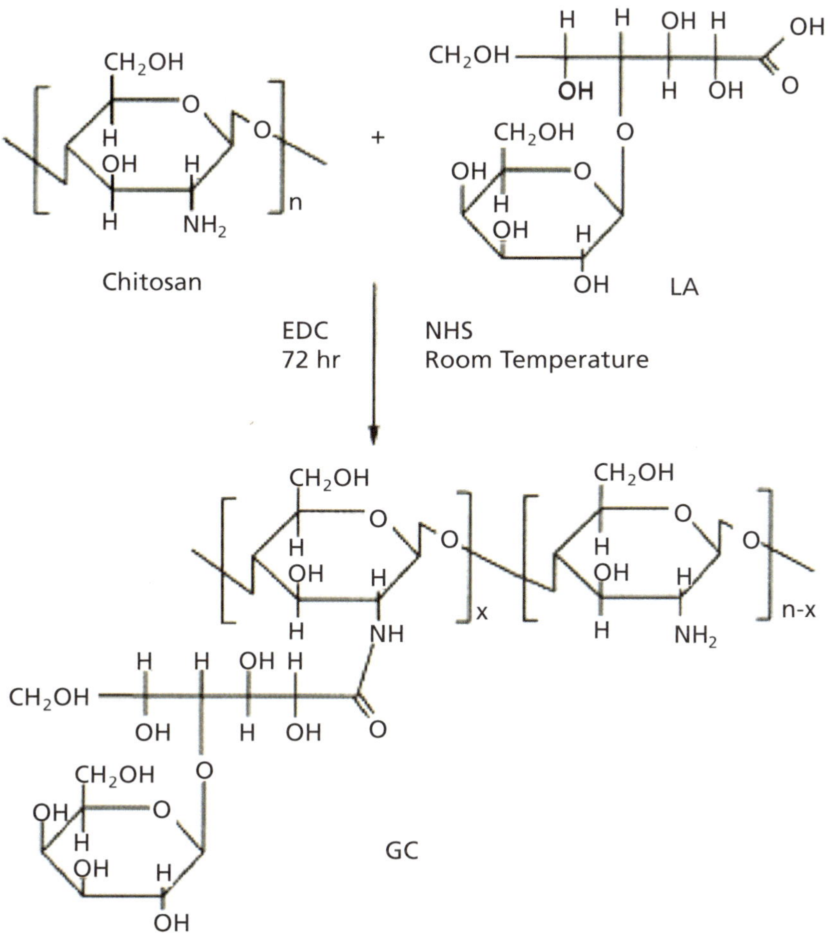

Figure 1.7 Synthesis scheme of GC. (From [25].)

galactose residue. The results indicated that monosaccharide of terminal galactose and β-galactose residue in the disaccharide had a higher affinity to hepatocytes compared to disaccharide of the terminal galactose and α-galactose residue in the disaccharide. The chemical structure of galactose is shown in Figure 1.10.

1.4.8 Regulation of Density of Galactose and Spatial Orientation of Galactose

Density of galactose in the galactose-carrying artificial ECMs is a very important parameter in the regulation of hepatocellular behaviors because triantennary molecules with three terminal galactose residues bind to the lectins with the higher affinity than biantennary ones with one or two terminal galactose residues [27] and multisubunit receptors of hepatocytes are responsible for binding galactose residues on desialylated glycoproteins [28].

Akaike and his coworkers reported that hepatocytes attached at low concentrations of PVLA exhibited high ^{3}H-thymidine uptake activity whereas hepatocytes attached at high concentrations of PVLA exhibited low ^{3}H-thymidine uptake activity, an indication of low differentiation of hepatocytes [8]. They also reported that hepatocytes attached

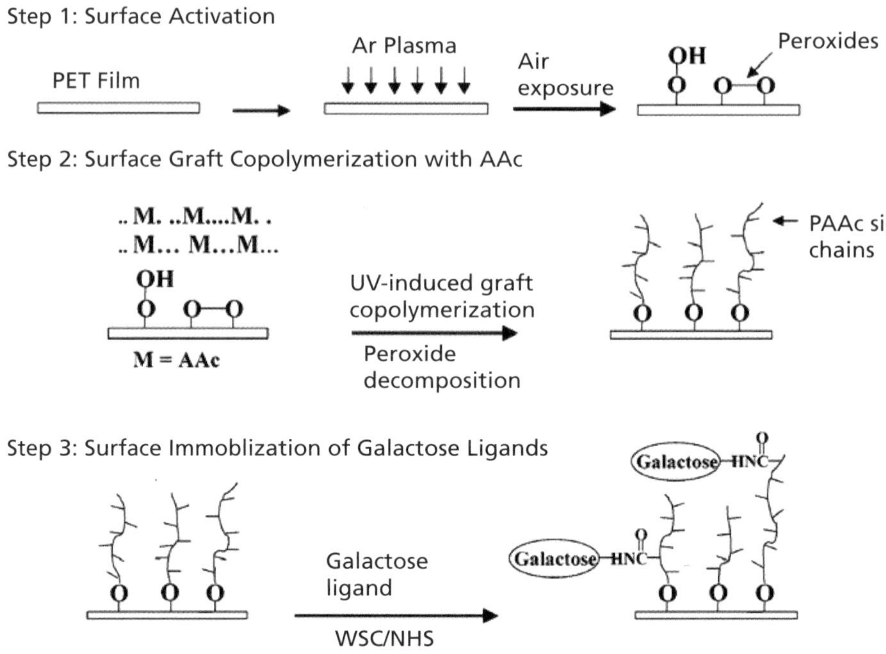

Figure 1.8 Surface modification scheme for the UV-induced graft copolymerization of AAc with the argon-plasma-pretreated PET and the subsequent galactose ligand immobilization. (From [22].)

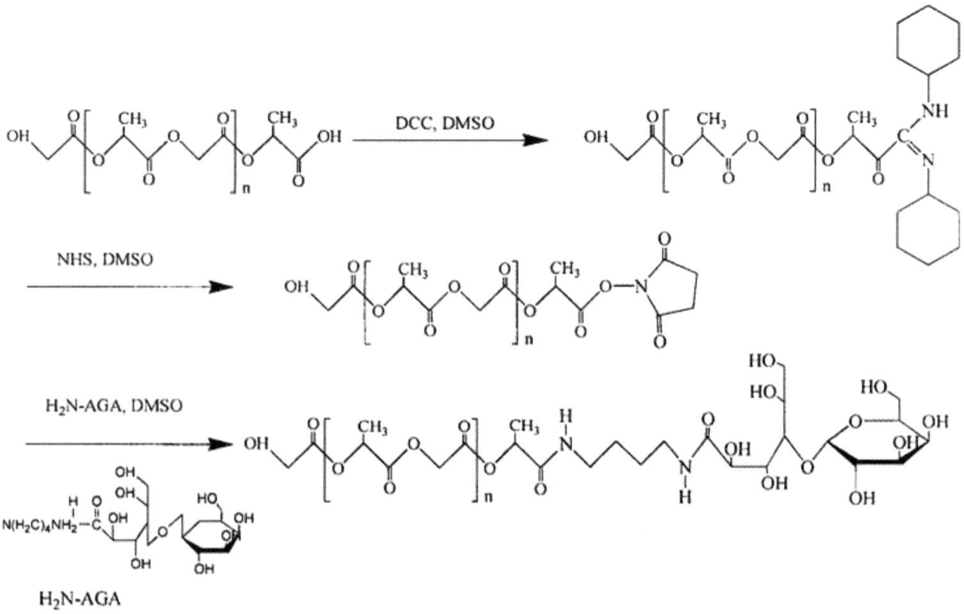

Figure 1.9 Synthetic scheme of PLGA-AGA. (From 21].)

onto PVLA below 20 ng/ml coating density expressed low levels of ASGPR with higher proliferative capacities than above 50 ng/ml coating density [29]. Furthermore, they reported that hepatocytes adhered at 100 μg/ml PVLA appeared to be round whereas hepatocytes adhered at 0.5 μg/ml PVLA appeared to be spread [10] as shown in Figure 1.11, suggesting that hepatocyte morphologies were also dependent on galactose densi-

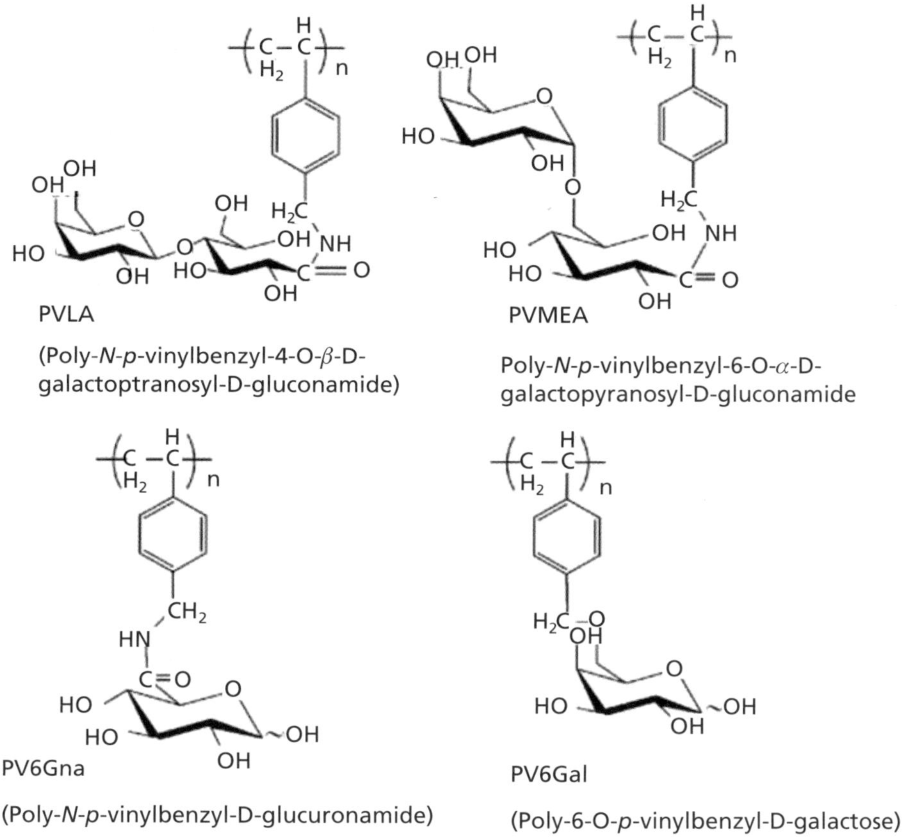

Figure 1.10 Molecular structure of PV-sugars. (From 26].)

ties of the PVLA. They explained the mechanism on the relationship between morphology of hepatocytes and density of galactose as followed [10]. ASGPRs on hepatocyte membranes are clustered within sites of focal adhesion as a large patch and result in the prevention of integrin receptors from the adhesion processes during culture at high galactose density. On the other hand, hepatocytes adhered at low galactose density allow integrin receptors to participate in adhesion processes within the space where ECM proteins are secreted during culture, indicating that the integrin receptors play an important role in changing round shapes of hepatocytes to spread ones, and in FAK phosphorylation. The illustration of hepatocyte behaviors on PVLA surfaces against galactose density in PVLA is shown in Figure 1.12 [10]. With similar tendency Mao's group reported that hepatocytes adhered at a high galactose density in the galactosylated PET [30] and galactosylated polyvinylidene difluoride (PVDF) [31] formed spheroid formation with higher albumin synthesis.

The spatial orientation or spatial microdistribution of the galactose in the galactose-carrying ECMs affects galactose ligand-ASGPR interaction. Cho et al. reported that the hepatocytes adhered on the PVLA Langmuir-Blodgett (LB) surface were observed as round shapes even at the very low galactose density due to the spatial orientation of galactose in the PVLA [32]. Also, Griffith et al. reported that spatial microdistribution of galactose in the PEO regulated morphologies and functions of the hepatocytes by the accessibility of ligand to be clustered in spatial microdomains [33].

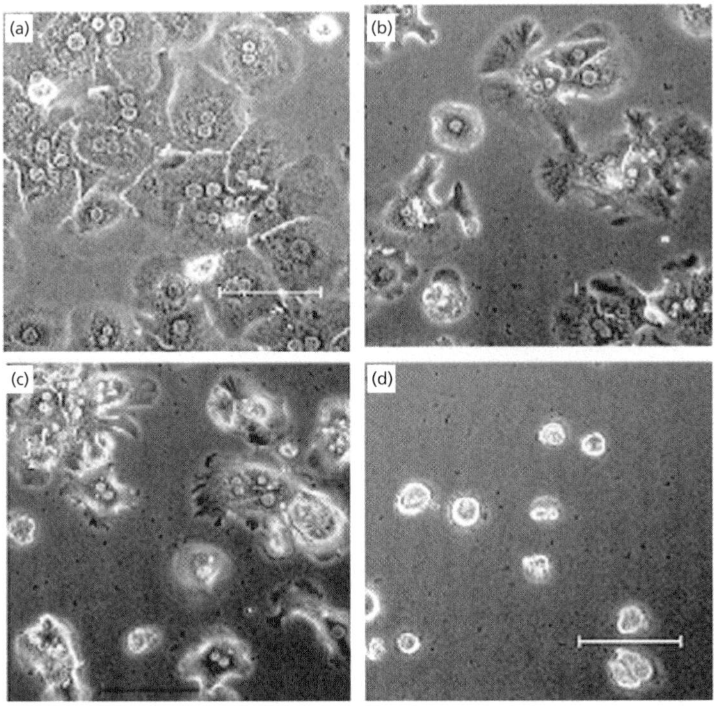

Figure 1.11 Hepatocyte morphology depending on adhesion matrix. Hepatocytes were cultured for 16 h in WE base on PS dishes coated with (a) 10 μg/ml collagen type I, (b) 5 mg/ml PLL, (c) 0.5 μg/ml PVLA, or (d) 100 μg/ml PVLA. Scale bar = 100 μm. (From [10].)

1.4.9 Regulation of Topology of ECM

Topology of ECM affects cell morphology, function, and physiological responsiveness because three-dimensional ECMs induce differentiated hepatocyte functions more effectively than two-dimensional ones. Many researchers are more interested in the design of topology of ECM for the spheroid formation of hepatocytes because this structure is similar to the tight cell-cell contact in native liver and tight junctions to mimic ultrastructure of the native lobule. Therefore, the design of 3-D scaffolds to form spheroids of hepatocytes is a powerful method to enhance liver-specific functions in hepatic tissue engineering although the hepatocytes in large spheroids encounter mass transfer limitations by the absence of a vascular network. Packed-bed modules from a multicapillary polyurethane form [34], electrospun nanofibers from galactosylated poly(epsilon-caprolactone-co-ethyl ethylene phosphate) [35] and hollow fiber bioreactor from galactosylated PVDF [36] were used as the scaffolds to get the spheroid formation of the hepatocytes.

1.4.10 Regulation of Deposition of Natural ECM

It was already reported that the natural ECMs such as FN, collagen vitronectin, and laminin are deposited onto the surface of artificial ECMs during the cell culture and results in affecting cellular behaviors [37] although the type of natural ECM molecules deposited onto the artificial ECM surface has not been fully understood. Hoshiba et al. checked deposited natural ECMs by PVLA as a hepatocyte-specific ECM and PLL as a

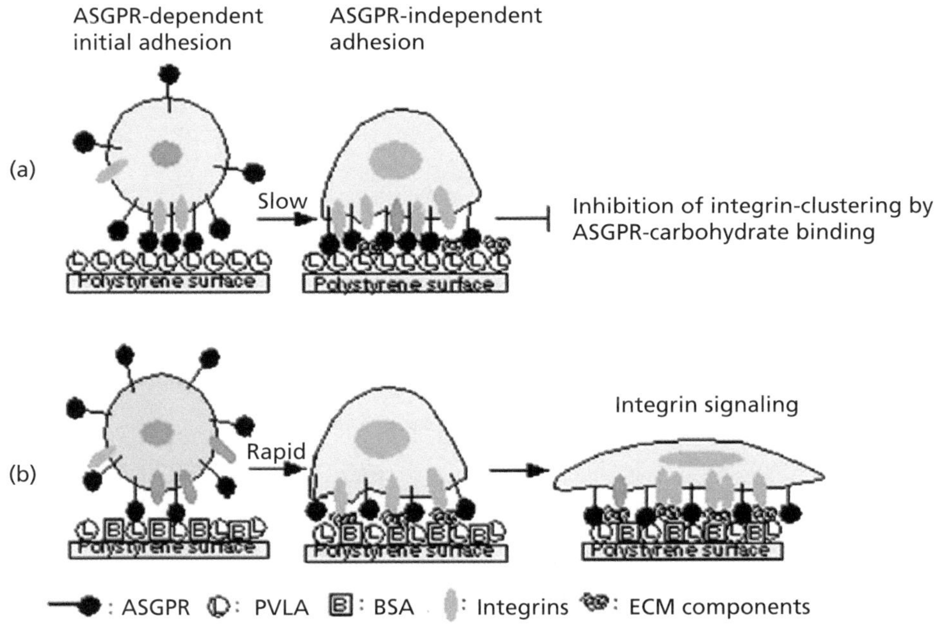

Figure 1.12 (a, b) Illustration of hepatocyte behavior on PVLA surfaces. (From [10].)

nonspecific one [38]. The results showed that type I collagen was dominantly deposited on a PVLA-coated surface whereas FN was dominantly deposited on a PLL-coated one, indicating that the adsorption of natural ECMs secreted from hepatocytes during culture on biomaterial surface is dependent on the type of biomaterial.

1.4.11 Design of Coculture System

It is already reported that coculture systems of hepatocyes with other types of cells enhanced liver-specific functions of hepatocytes and regulated cellular behaviors although the precise mechanisms remain unclear. This might involve the cell-cell interactions between hepatocytes and other cells as the feeder cells, and the soluble growth factors from the feeder cells. Most commonly used feeder cells are epithelial- and fibroblast-like cells. Cho and his coworkers compared liver-specific functions of hepatocytes between monoculture with hepatocytes and coculture with hepatocytes and 3T3 fibroblasts in the alginate/GC scaffold [39]. The results indicated that the liver-specific functions such as albumin synthesis, ammonia eliminaton, and cytochrome P450 activity of hepatocytes cocultured with hepatocytes and 3T3 fibroblasts were enhanced compared with those monocultured with hepatocytes as shown in Figure 1.13. They also reported that coculture of hepatocytes in alginate/GC/heparin scaffolds with NIH3T3 in a transwell insert resulted in significant increase of liver-specific functions compared with those in hepatocyte monoculture [40], suggesting that coculture systems enhance liver-specific functions of hepatocytes. Also, the liver-specific functions of the hepatocytes were improved under the coculture with sinusoidal cells, mouse embryonic fibroblasts, liver epithelial cells, pancreatic islet cells, bone marrow cells, Kupffer cells, fetal hepatocytes, and hepatic stellate cells.

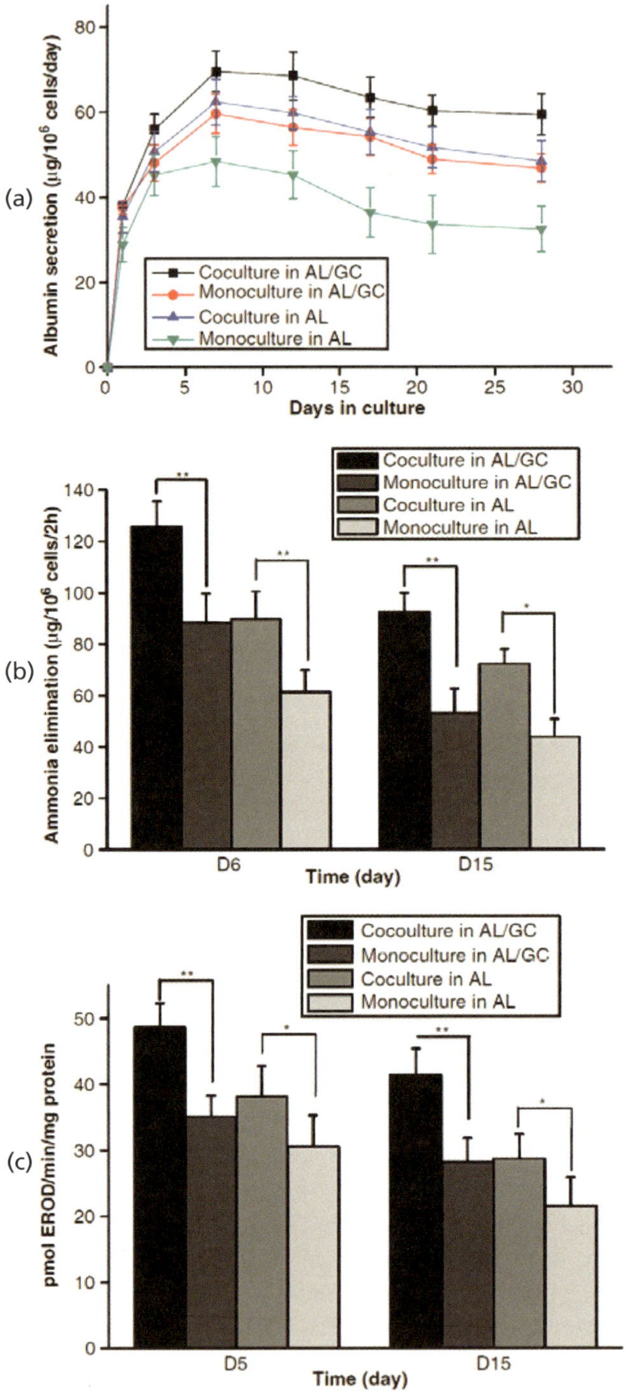

Figure 1.13 Comparision of liver-specific functions such as albumin secretion (a), ammonia elimination (b), and EROD activities (c) in AL and AL/ GC sponges for monoculture and coculture of hepatocytes with NIH3T3. (a) Albumin secretion rates for both systems were measured over 28 days in culture. ELISA specific for mouse albumin were used to quantify protein in 24h conditioned medium. (b) Ammonia elimination was compared between both systems on day 5 and 16 of culture. (c) EROD deethylase activity by cytochrome P4501A1 was compared between both systems on day 6 and 16 of culture. EROD activity was measured using the highly fluorescent resorufin in both systems. Data is represented as mean ± $SD_{(n-1)}$ (n = 6). Statistics (two-tailed paired Student's *t*-test) compared monoculture with coculture; *P <0.05 and **P <0.01. (From [39].)

1.4.12 Choice of Cell Sources

The choice of cell type and cell source is very important in hepatic tissue engineering because it is necessary to select cells that demonstrate the particular phenotype of interest [41]. The various potential cell sources for liver tissue engineering are shown in Table 1.3 and discussed below.

Human hepatocytes appear to be the natural choice for liver tissue engineering although the supply is limited due to the competing demand of liver transplantation. On the other hand, adult human hepatocyte is not easy to replicate in vitro and the daughter cells express low levels of liver-specific functions although they are very biocompatible to the host.

A promising new possibility is the discovery of liver stem cells. Recent studies suggest that there are stem cells present within [42] as well as outside the liver [43], which can differentiate into fully mature hepatocytes. Tateno et al. reported the presence of a subpopulation of small hepatocytes in rat liver with a high proliferative potential in in vitro study [44]. A major hepatic source is stem cells of the bone marrow that may take part in normal tissue renewal as well as in liver regeneration after severe induced hepatic injury [45]. Hepatic progenitor cells (oval or Ito cells) display a distinct phenotype and have been shown to be a bipotential progenitor of two types of epithelial cells found in the liver, hepatocytes, and bile ductular cells [46]. Recently, commercially available human mesenchymal stem cells [47] and human umbilical cord blood-derived cells [48] with hepatic differentiation were reported.

Immortalized human cell lines can be also used as one source of cells. Kobayashi et al. reported a reversible immortalization system with the SV40 T antigen and Cre-Lox recombination targeting human hepatocytes in their final differentiated state [49]. This system allowed them to establish a safe human hepatocyte line that is capable of differentiation at low cost and on a large scale although immortalized C3A cells have lower levels of P450A1 activity, ammonia removal, and amino acid metabolism [50] and there are concerns of the possibility of transmission of tumorigenic products into the patient [51].

Xenotransplantation is very interesting because the xenogenetic hepatocytes offer no risk of transmitting malignancies to the patient. Porcine hepatocytes exhibit P-450 function, synthesis of urea, albumin, and other proteins, and are activated by growth

Table 1.3 Characteristics of Various Cell Sources Available for Hepatic Tissue Engineering

Cell Source	Supply	Liver-Specific Function	Host Compatibility	Risks to Patient	Reference
Primary human hepatocytes	Limited	Unstable, but limited experience with cell culture	Excellent	Small	[41]
Stem-cell derived human hepatocytes	Unlimited	Low efficiency of differentiation to hepatocytes	Excellent	Small	[42–48]
Immortalized human hepatocytes	Unlimited	Function poorly characterized	Excellent	Transmission of tumorigenic products	[49–51]
Human hepatoma cell line	Unlimited	Reduced level compared to primary cells	Excellent	Transmission of tumorigenic products	[49–51]
Primary xenogenic hepatocytes	Unlimited	Express high level but generally decreases over time	Excellent	Transmission of zoonosis	[52,53]

From: [51].

factors similar to human cells [52]. However, there are several disadvantages such as the risk of xenozoonosis and immune rejection by the patient although the cells are easily obtained in large quantities and the cells demonstrate the same qualities and therapeutic effects of human hepatocytes [53].

1.5 Anticipated Results

By using galactose-carrying synthetic and natural polymers among the artificial ECMS, hepatocellular behaviors such as adhesion, morphology, proliferation, migration, differentiation, and survival can be regulated through specific interaction between galactose-carrying ECMs and ASGPR of hepatocytes. Through cell adhesion assay, cell detachment assay, inhibition assay, Western blot assay, flow cytometry analysis, and lectin binding assay, specific interaction between galactose moieties of galactose-carrying ECMs and ASGPR on hepatocytes has been proven. Several chemical structures of galactose such as lactose monohydrate, lactobionic acid, aminohexyl-D-galactopyranoside, 1-amino-1-deoxy-β-D-lactose, and 1-amino-1-deixt-β-D-lactose can be used to couple with artificial ECMs as the bulk or surface state. The cellular behaviors can be regulated by density of galactose and spatial orientation of galactose, topology of ECM, deposition of natural ECM, coculture system, and cell sources.

1.6 Discussion and Commentary

Bioartificial liver-assist devices or regeneration of liver–tissue substitutes for hepatic tissue engineering require a suitable ECM for hepatocyte culture because the hepatocytes are anchorage-dependent cells and are highly sensitive to the ECM environment for maintaining their viability and differentiated function. In previous results, several chemical structures of galactose moieties were coupled with synthetic and natural polymers as the artificial ECMs to guide hepatocellular behaviors because ASGPR of the hepatocytes recognize and bind ECMs having galactose through specific interaction between cells and ligands. Of coupling reaction, ketone, amino, or carboxylic acid groups in the galactose moieties were used for coupling with ECMs through amide reaction or reductive amination as the bulk or surface state.

As highlighted in this chapter, cell adhesion assay, cell detachment assay, inhibition assay, Western blot assay, flow cytometry analysis, and lectin binding assay were used to confirm the specific interaction between galactose moieties of galactose-carrying ECMs and ASGPR on hepatocytes. Density of galactose and spatial orientation or spatial microdistribution of the galactose in the galactose-carrying artificial ECM affect galactose ligand-ASGPR interaction and regulate hepatocellular behaviors.

Also, topology of three-dimensional ECMs induces differentiated hepatocyte functions more effectively than two-dimensional ones because 3-D scaffolds form more spheroids of hepatocytes and enhance liver-specific functions. Furthermore, deposition of natural ECMs during the cell culture, coculture of hepatocytes with other types of cells, and choice of cell sources should be regulated because they affect hepatocellular behaviors. Overall, galactose-carrying synthetic and natural polymers for enhancing specific interaction between cells and ligand continue to be explored, and these

Troubleshooting Table

Problem	Explanation	Potential Solutions
Lack of cell attachment.	Incoupling of galactose with ECM. Highly hydrated ECM. Different sugar moiety.	Adequate coupling chemistry. Increase of galactose moiety. Regulation of hydration. Adequate sugar moiety.
Nonspecific interaction.	Nonspecific interaction between hepatocytes and ECM.	Treatment of albumin.
Irregulation of galactose density.	Inadequate galactose concentration in ECM.	Different differentiation of hepatocytes according to galactose density.
Irregulation of pore size.	Inadequate pore size in 3-D scaffold.	Different differentiation of hepatocytes according to pore size in 3-D scaffold.

approaches should ultimately aid in providing an alternative to organ and tissue transplantation.

1.7 Application Notes

The design of ECM for hepatic tissue engineering covered in this chapter explains introduction of galactose moiety to the artificial ECMs because ASGPR of the hepatocytes recognize galactose in the ECMs as a specific ligand. This coupling chemistry is applicable for other tissue engineering. Several methods can be used to confirm the specific interaction between cells and ligands. Among them, cell adhesion and inhibition assays are mostly used whereas other methods are time-consuming and expensive. Although galactose moieties can be immobilized to the bulk state of the ECM, they can be coupled to the surface of the ECM. Also, fabrication of 3-D scaffolds for hepatic tissue engineering can be broadly utilized to regulate cellular behavior. Furthermore, the discovery of liver stem cells will likely be a key extension of regenerative medicine fields.

1.8 Summary Points

In this chapter, we have presented a summary of the published works on the design of ECMs for hepatic tissue engineering. The galactose-carrying synthetic and natural polymers among the artificial ECMs can guide hepatocellular behaviors such as adhesion, morphology, proliferation, migration, differentiation, and survival through specific interaction between galactose-carrying ECMs and ASGPR of hepatocytes.

Several methods have been explained to investigate specific interaction between ECM and hepatocytes. The density of galactose and spatial orientation of galactose, chemical structure of galactose, coupling of galactose with bulk ECM and surface of ECM, topology of ECM, deposition of ECM, coculture system, as well as cell sources that affected hepatocyte behaviors were also described. Precise design of galactose-carrying ECMs as artificial ECMs will contribute to hepatic tissue engineering as a new form of bioartificial liver-assist devices or regeneration of liver-tissue substitutes.

Acknowledgments

This work was supported by a grant from the 21st Century Centers of Excellence Program and a grant-in-aid for Scientific Research from the Ministry of Education, Culture, Sports, Science and Technology of Japan.

References

[1] Badylak, S. F., "The Extracellular Matrix as a Scaffold for Tissue Reconstruction," *Semin. Cell Dev. Biol.,* Vol.13, 2002, pp. 377–83.

[2] Ingber, D., "Extracellular Matrix and Cell Shape: Potential Control Points for Inhibition of Angiogenesis," *J. Cell. Biochem.,* Vol. 47, 1991, pp. 236–41.

[3] LeCluyse, E. L., P. L. Bullock, and A. Parkinson, "Strategies for Restoration and Maintenance of Normal Hepatic Structure and Function in Long-Term Cultures of Rat Hepatocytes," *Adv. Drug Deliv. Rev.,* Vol. 22, 1996, pp. 133–86.

[4] Hansen, L. K., J. Wilhelm, and J. T. Fassett," Regulation of Hepatocyte Cell Cycle Progression and Diffrentiation by Type I Collagen Structure," *Curr. Top Dev. Biol.,* Vol. 72, 2006, pp. 205–36.

[5] Pricer, W. E.. and G. Ashwell, "The Binding of Desialylated Glycoproteins by Plasma Membranes of Rat Liver," *J. Biol. Chem.,* Vol. 246, 1971, pp. 4825–33.

[6] Wall, D. A., G. Wilson, and A. L. Hubbard, "The Galactose-Specific Recognition System of Mammalian Liver: The Route of Ligand Internalization in Rat Hepatocytes," *Cell,* Vol. 21, 1980, pp. 79–93.

[7] Kim, S. H., T. Hoshiba, and T. Akaike, "Hepatocyte Behavior on Synthetic Glycopolymer Matrix: Inhibitory Effect of Receptor-Ligand Binding on Hepatocyte Spreading," *Biomaterials,* Vol. 25, 2004, pp.1813–23.

[8] Kobayashi, A., et al., "Receptor-Mediated Regulation of Differentiation and Proliferation of Hepatocytes by Synthetic Polymer Model of Asialoglycoprotein," *J. Biomater. Sci. Polymer Edn.,* Vol. 6, 1994, pp. 325–42.

[9] Richardson, A., and T. Parsons, "A Mechanism for Regulation of the Adhesion-Associated Proteintyrosine Kinase pp125 FAK," *Nature,* Vol. 380, 1996, pp.538–40.

[10] Kim, S. H., J. H. Kim, and T. Akaike, "Regulation of Cell Adhesion Signaling by Synthetic Glycopolymer Matrix in Primary Cultured Hepatocytes," *FEBS. Lett.,* Vol. 553, 2003, pp. 433–39.

[11] Kim, S. H., M. Goto, and T. Akaike, "Specific binding oOf Glucose-Derivatized Polymers to the Asialoglycoprotein Receptor of Mouse Primary Hepatocytes," *J. Biol. Chem.,* Vol. 276, 2001, pp. 35312–319.

[12] Weis, W. I., and K. Drickamer, "Structural Basis of Lectin-Carbohydrate Recognition," *Ann. Rev. Biochem.,* Vol. 65, 1996, pp. 441–73.

[13] Weigel, P. H., et al., "Specific Adhesion of Rat Hepatocytes to Beta-Galactosides Linked to Polyacrylamide Gels," *J. Biol. Chem.,* Vol. 253, 1978, pp. 330–33.

[14] Weigel, P. H., "Rat Hepatocytes Bind to Synthetic Galactoside Surfaces Via a Patch of Asialoglycoprotein Receptors," *J. Cell Biol.,* Vol. 87, 1980, pp. 855–61.

[15] Weisz, O. A., and R. L. Schnaar, "Hepatocyte Adhesion to Carbohydrate-Derivatized Surfaces I. Surface Topography of the Rat Hepatic Lectin," *J. Cell Biol.,* Vol. 115, 1991, pp. 485–93.

[16] Kobayashi, K., A. Kobayashi, and T. Akaike, "Culturing Hepatocytes on Lactose-Carrying Polystyrene Layer Via Asialoglycoprotein Receptor-Mediated Interactions," in *Methods in Enzymology: Neoglycoconjugates. Part B: Biomedical Applications,* Y. C. Lee, and R. T. Lee, (eds.), Vol. 247, San Diego, CA: Academic Press, 1994, pp. 409–18.

[17] Lopina, S. T., et al., "Hepatocyte Culture on Carbohydrate-Modified Star Polyethylene Oxide Hydrogels," *Biomaterials,* Vol. 17, 1996, pp. 559–69.

[18] Donati, I., et al., "Synthesis, Characterization, and Preliminary Biological Study of Glycoconjugates of Poly(Styrene-Co-Maleic Acid)," *Biomacromolecules,* Vol. 3, 2002, pp. 805–12.

[19] Kang, I. K., et al., "Morphology and Metabolism of Ba-Alginate Encapsulated Hepatocytes with Galactosylated Poly (Allyl Amine) and Poly (Vinyl Alcohol) as Extracellular Matrices," *J. Mater. Sci. Mater. Med.,* Vol. 16, 2005, pp. 533–39.

[20] Iwasaki, Y., et al., "Selective Biorecognition and Preservation of Cell Function on Carbohydrate-Immobilized Phosphorylcholine Polymers," *Biomacromolecules,* Vol. 8, 2007, pp. 2788–94.

[21] Yoon, J. J., et al., "Surface Immobilization of Galactose onto Aliphatic Biodegradable Polymers for Hepatocyte Culture," *Biotechnol. Bioeng.,* Vol. 78, 2002, pp. 1–10.

[22] Ying, L., et al., "Immobilization of Galactose Ligands on Acrylic Acid-Graft-Copolymerized Poly(Ethylene Terephthalate) Film snd Its Application to Hepatocyte Culture," *Biomacromolecules,* Vol. 4, 2003, pp. 157–65.

[23] Kang, I., et al., "Grafting of Lactose-Carrying Styrene onto Polystyrene Dishes Using Plasma Glow Discharge snd Their Interaction with Hepatocytes," *J. Mater Sci. Mater. Med.,* Vol. 14, 2003, pp. 611–16.

[24] Du, Y., et al., "Identification and Characterization of a Novel Prespheroid 3-Dimensional Hepatocyte Monolayer on Galactosylated Substratum," *Tissue Eng.,* Vol. 13, 2007, pp. 1455–68.

[25] Park, I. K., et al., "Galactosylated Chitosan As A Synthetic Extracellular Matrix For Hepatocyte Attachment," *Biomaterials,* Vol. 24, 2003, pp. 2331–37.

[26] Kim, S. H., T. Hoshiba, and T. Akaike, "Effect of Carbohydrates Attached to Polystyrene on Hepatocyte Morphology on Sugar-Derivatized Polystyrene Matrices," *J. Biomed. Mater. Res.,* Vol. 67A, 2003, pp. 1351–59.

[27] Baenziger, J. U., and D. Fiete, "Galactose and N-Acetylgalactosamine-Specific Endocytosis of Glycopeptides by Isolated Rat Hepatocytes," *Cell,* Vol. 22, 1980, pp. 611–20.

[28] Ashwell, G., and J. Harford, "Carbohydrate-Specific Receptors of the Liver," *Annu. Rev. Biochem.,* Vol. 51, 1982, pp. 531–54.

[29] Ise, H., et al., "Low Asialoglycoprotein Receptor Expression as Markers for Highly Proliferative Potential Hepatocytes," *Biochem. Biophys. Res. Commun.,* Vol. 285. 2001, pp. 172–82.

[30] Yin, C., et al., "High Density of Immobilized Galactose Ligand Enhances Hepatocyte Attachment and Function," *J. Biomed. Mater. Res.,* Vol. 67A, 2003, pp. 1093–1104.

[31] Lu, H., et al., "Galactosylated PVDF Membrane Promotes Hepatocyte Attachment and Functional Maintenance," *Biomaterials,* Vol. 24, 2003, pp. 4893–4903.

[32] Cho, C. S., et al., "Effect of Ligand Orientation on Hepatocyte Attachment onto the Poly (N-P-Vinylbenzxyl-O-β-D-Galactopyranosyl-D-Gluconamide) as a Model Ligand of Asialoglycoprotein," *J. Biomater. Sci. Polym Ed.,* Vol. 7, 1996, pp. 1097–1104.

[33] Griffith. L. G., and S. Lopina, "Microdistribution of Substratum-Bound Ligands Affects Cell Function: Hepatocyte Spreading on PEO-Tethered Galactose," *Biomaterials,* Vol. 19, 1998, pp. 976–86.

[34] Mizumoto, H., and K. Funatsu, "Liver Regeneration Using a Hybrid Artificial Liver Support System," *Artif. Organs,* Vol. 28, 2004, pp. 53–7.

[35] Chua, K. N., et al., "Stable Immobilization of Rat Hepatocyte Spheroids on Galactosylated Nanofiber Scaffold," *Biomaterials,* Vol. 26, 2005, pp. 2537–47.

[36] Lu, H. F., et al., "Galactosylated Poly(Vinylidene Difluoride) Hollow Fiber Bioreator For Hepatocyte Culture," *Tissue Eng.,* Vol. 11, 2005, pp. 1167–77.

[37] Kato, S., S. Kidoaki, and T. Matsuda, "Substrate-Dependent Cellular Behavior of Swiss 3T3 Fibroblasts and Activation of Rho Family During Adhesion and Spreading Processes," *J. Biomed. Mater. Res.,* Vol. 68A, 2004, pp. 314–24.

[38] Hoshiba, T., et al., "The Effect of Natural Extracellular Matrix Deposited on Synthetic Polymers on Cultured Primary Hepatocytes," *Biomaterials,* Vol. 27, 2006, pp. 4519–28.

[39] Seo, S. J., et al., "Enhanced Liver Functions of Hepatocytes Cocultured with NIH3T3 in the Alginate/Galactosylated Chitosan Scaffold," *Biomaterials,* Vol. 27, 2006, pp. 1487–95.

[40] Seo, S. J., et al., "Alginate/Galactosylated Chitosan/Heparin Scaffold as a New Synthetic Extracellular Matrix for Hepatocytes," *Tissue Eng.,* Vol. 12, 2006, pp. 33–44.

[41] Kulig, K. M., and J. P. Vacanti, "Hepatic Tissue Engineering," *Transpl. Immnuol.* Vol. 12, 2004, pp. 303–10.

[42] Thorgeirsson. S. S., "Hepatic Stem Cells in Liver Regeneration," *FASEB J.,* Vol. 10, 1996, pp. 1249–56.

[43] Theise, N. D., et al., "Liver from Bone Marrow in Humans," *Hepatology,* Vol. 32, 2000, pp. 11–16.

[44] Tateno, C., et al., "Heterogeneity of Growth Potential of Adult Rat Hepatocytes In Vitro," *Hepatology,* Vol. 31, 2000, pp. 65–74.

[45] Petersen, B. E., et al., "Bone Marrow as a Potential Source of Hepatic Oval Cells," *Science,* Vol. 284, 1999, pp. 1168–70.

[46] Oh, S. H., H. M. Hatch, and B. E. Petersen, "Hepatic Oval Stem Cell in Liver Regeneration," *Semin. Cell Dev. Biol.,* Vol. 13, 2002, pp. 405–9.

[47] Ong, W. Y., H. Dai, and K. W. Leong, "Hepatic Differentiation Potential of Commercially Available Human Mesenchymal Stem Cells," *Tissue Eng.,* Vol. 12, 2006, pp. 3477–85.

[48] Sun, Y., et al., "Formation of Human Heptocyte-Like Cell with Different Cellular Phenotypes by Human Umbilical Cord Blood-Derived Cells in the Human-Rat Chimeras," *Biochem. Biophys. Res. Comun.,* Vol. 357, 2007, pp. 1160–65.

[49] Kobayashi, N., et al. "Hybrid Bioartificial Liver: Establishing a Reversibly Immortalized Human Hepatocyte Line and Developing a Bioartificial Liver for Practical Use," *J. Artif. Organs,* Vol. 6, 2003, pp. 236–44.

[50] Wang, L, et al., "Comparison of Porcine Hepatocytes with Human Hepatoma (C3A) Cells for Use in a Bioartificial Liver Support System," *Cell Tranpl.,* Vol. 7, 1998, pp. 459–68.

[51] Chan, C. B., et al., "Hepatic Tissue Engineering for Adjunct and Temporary Liver Support: Critical Technologies," *Liver Transpl.,* Vol. 10, 2004, pp. 1331–42.

[52] Lorenti, A., et al., "Culture of Porcine Hepatocytes: The Dogma of Exogeneous Matrix Revisted," *Artif. Organs,* Vol. 25, 2001, pp. 546–50.

[53] Allen, J. W., and S. N. Bhatia, "Engineering Liver Therapies for the Future," *Tissue Eng.,* Vol. 8, 2002, pp. 725–37.

2

Tissue Engineering Using a Temperature-Responsive Cell Culture Dish

Takanori Iwata, Masayuki Yamato*, and Teruo Okano

Institute of Advanced Biomedical Engineering and Science, Tokyo Women's Medical University, 8-1 Kawada-cho, Shinjuku-ku, Tokyo 162-8666, Japan

*Correspondence: Institute of Advanced Biomedical Engineering and Science, Tokyo Women's Medical University, 8-1 Kawada-cho, Shinjuku-ku, Tokyo 162-8666, Japan
Phone: +81 3 5367 9945 ext. 6211; Fax: +81 3 3359 6046
E-mail: myamato@abmes.twmu.ac.jp

Abstract

Temperature-responsive culture dishes, an inventive integration of polymer chemistry and cell biology, are now utilized to create a "cell sheet" for scaffold-free transplantation and successful outcomes are accumulating both in animal and human studies. Just by reducing the temperature below 32°C, a contiguous cell sheet, which preserves intact extracellular matrix proteins and cell-cell interactions, can be harvested and ready for transplantation without scaffolds. As cell sheets contain extracellular adhesive proteins, cell sheets spontaneously attach to the recipient sites or other cell sheets. It is possible to transplant cell sheets to the recipient sites without sutures and to deposit multilayered cell sheets to create thick tissues in vitro. Therefore, this method can be an alternative method for cell transplantation. In this chapter, we will introduce how to generate cell sheets using temperature-responsive culture dishes, and review the potentials of cell sheets for regenerative medicine and cell biology.

Key terms *N*-isopropylacrylamide (PIPAAm), temperature-responsive culture dishes, UpCell™ Surface, cell sheet, multilayering, scaffold-free transplantation

2.1 Introduction

Tissue engineering, a concept proposed by Vacanti and Langer [1], has been widely accepted in the field of regenerative medicine to create several kinds of tissues by the combined use of cells, scaffolds, and growth factors. However, the limitations of this concept have been exposed due to graft rejections induced by inflammation during degradation of scaffolds, necrosis, or low stability of transplanted cells [2]. To overcome these problems, we have developed a temperature-responsive polymer named *N*-isopropylacrylamide (PIPAAm) for the surface of cell culture dishes [3]. At temperatures lower than 32°C, PIPAAm is fully hydrated with an extended-chain conformation; however, at temperatures higher than 32°C, PIPAAm is extensively dehydrated and compact. Cells generally adhere to hydrophobic surfaces, but not to highly hydrated hydrophilic surfaces [4]. Our laboratory grafted PIPAAm to develop temperature-responsive culture dishes by irradiation with an electron beam [5]. The advantage of a temperature-responsive dish is the possibility of harvesting intact cells and proteins with just low-temperature treatment. Just by reducing the temperature, cells spontaneously detach from the surface of dishes. Compared to retrieving cells with enzymes such as trypsin or dispase, loss of cell viability and degradation of surface proteins are minimized. Recent reports clearly demonstrated that various kinds of single cells harvested from temperature-responsive dishes have highly preserved functions as well as intact proteins [6–8]. In addition, the preserved subcellular matrix proteins of harvested cell sheets provide the adhesive properties for stacking. Therefore, cell sheets can be stabilized at the recipient sites without any glues or sutures in a corneal reconstruction model [9]. It is also possible to create thick tissues by layering multiple cell sheets.

In this chapter, we will introduce how to create cell sheets from single suspension cells and the practical applications toward regenerative medicine.

2.2 Experimental Design

The adhesive ability of cultureware with UpCell™ Surface is usually lower than that of conventional polystyrene cultureware (Figure 2.1(a)). Thus we usually precoat the UpCell Surface with adhesive proteins and incubate at 37°C for several hours to allow the adhesion of these proteins to the UpCell Surface at first (Figure 2.1(b)). Then cells are plated on the surface of these proteins. (Figure. 2.1(c)). After cells reached confluency, the temperature is reduced to less than 32°C. The characteristics of PIPAAm dramatically changes from hydrophobic to hydrophilic and PIPAAm absorbs water. Cells with extracellular matrices and adhesive proteins can be harvested as a unit that we call the cell sheet (Figure 2.1(d)). Enzyme-free harvest permits the cell sheets to be readily manipulated, transferred, layered, or fabricated, because they adhere rapidly to other surfaces, such as traditional culture plastics, other cell sheets, and tissues in vivo [10].

2.3 Materials

In addition to the conventional cell culture devices, such as an incubator and a water bath, you will need the following items:

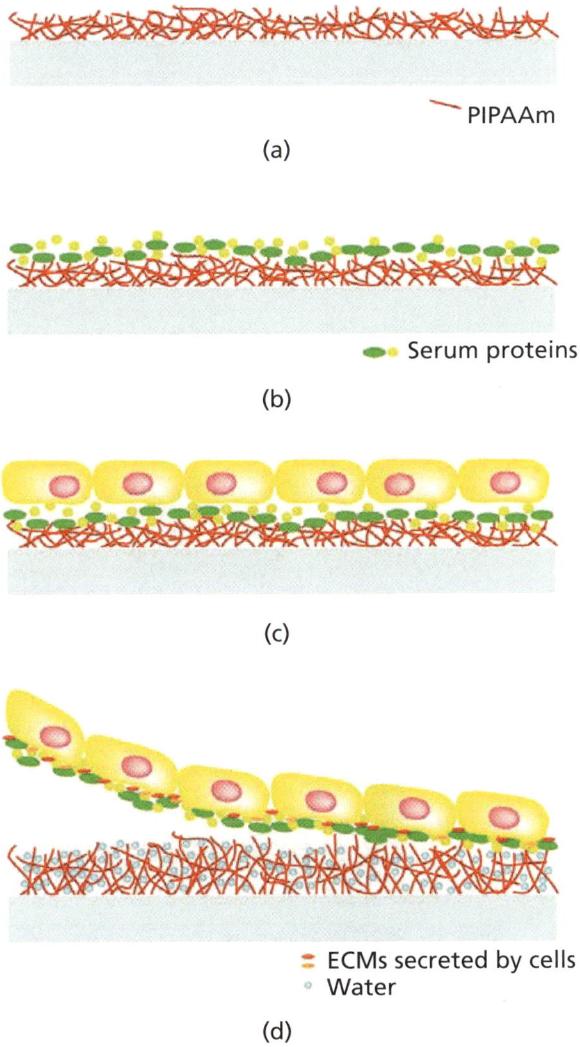

Figure 2.1 (a) UpCell™ Surface above 32°C. UpCell™ Surface contains a thin (~ 20 nm) PIPAAm layer on the tissue culture-grade polystyrene dish. PIPAAm is extensively dehydrated and compact above 32°C. (b) Precoating adhesive proteins above 32°C. To enhance the initial cell attachment to the UpCell™ Surface, we recommend precoating the UpCell™ Surface with adhesive proteins, such as serum proteins and collagens. (c) Spreading cells on the adhesive proteins. Cells are seeded on the UpCell™ Surface at an appropriate cell density. (d) Harvesting cells as a sheet. An appropriate culture period allows cell harvest as a contiguous cell sheet. Cells secrete extracellular matrices (ECMs) around themselves, and cell–cell and cell–matrix interactions are formed. At temperatures lower than 32°C, PIPAAm is fully hydrated with an extended-chain conformation. Cell sheets preserved intact cells and proteins can be harvested at below 32°C.

- Cultureware with UpCell Surface (Nunc, Naperville, IL. http://www.nuncbrand.com) (Figure 2.2(a)).
- Precoating reagents: we usually use fetal bovine serum (FBS).
- Thermo plate (for example, MATS-55W or MATS-55WL, Tokai Hit, Shizuoka, Japan) (Figure 2.2(b) and (c)).
- Transfer sheets for manipulation and transplantation (for example, DURAPORE® membrane filter (SVLP, Millipore, Bedford, MA), or Cellshifter™ (Cat. No. CSD001, CellSeed, Tokyo, Japan)).

(a)

(b)

(c)

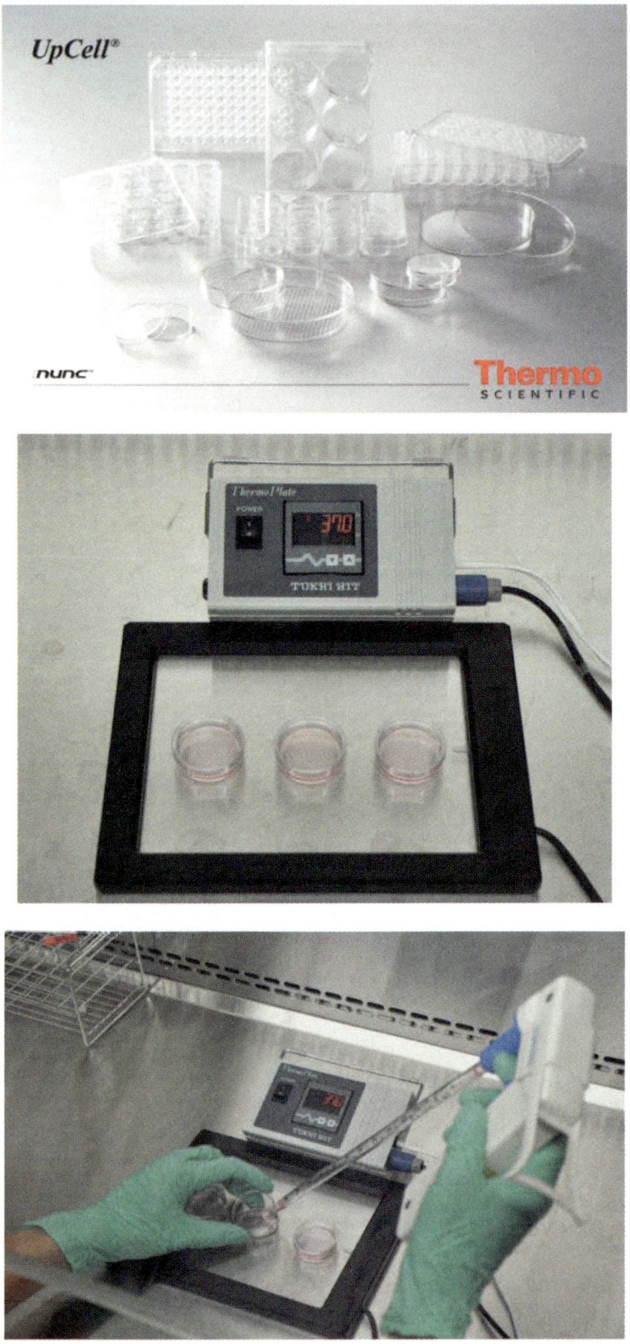

Figure 2.2 (a) Products of UpCell™ Surface. Various sizes (100, 60, and 35 mm) and multidishes (96, 48, 24, 12, and 6 wells/dish) are commercially available all over the world. (b) A thermo plate set at 37°C is usually used to prevent the temperature change of UpCell™ Surface culture dishes. Culture media should be also warmed with a conventional water bath to prevent the detachment of cells or proteins. (c) Medium change on a thermo plate. Set a thermo plate into the clean bench. Change the media, which has been prewarmed, and quickly return the dishes to the incubator so as not to change the temperature of the culture dishes.

• (If applicable), L-ascorbic acid phosphate magnesium salt n-hydrate (Wako Pure Chemical, Tokyo, Japan) to strengthen the stability of the cell sheet.

2.4 Methods

2.4.1 Seeding Cells

2.4.1.1 Precoating UpCell Surface

The adhesive ability of cultureware with UpCell Surface is usually lower than that of conventional polystyrene cultureware. Thus, we strongly recommend precoating the UpCell Surface with adhesive proteins such as FBS, collagen, and so forth (Figure 2.1(b)). The type of adhesive protein depends on the type of cell seeding; therefore you have to optimize the initial adhesive condition with your cells. We usually use a growth medium that contains 10% FBS or 100% FBS itself to enhance the adhesion of cells because FBS contains several adhesive molecules, such as fibronectin and vitronectin. We put 2 ml of FBS solution on the 35 mm UpCell Surface, and incubate for several hours at 37°C in a humidified atmosphere with 5% CO_2. If the cells do not adhere strongly to the UpCell Surface, try precoating with 100% FBS for longer incubation time (for example, overnight).

2.4.1.2 Design the Seeding Density of Cells

It is important to determine the initial seeding density of cells based on your experimental design. Generally, the longer you culture cells on the dishes, the stronger the cell sheets will be because cells secrete extracellular matrices around themselves. However, the long culture period induces the differentiation of cells. In cases with cardiomyocytes, hepatocytes, and islet cells, the short culture period can prevent the loss of cell function. Thus, you have to seed cells at high density initially. In contrast, you can seed cells at low density if you want to induce differentiation of cells during cultivation.

In the case of mesenchymal stromal cells (MSCs), cells can be spread at the density of 1×10^4 cells/cm^2 for long-term culture (up to 2 weeks) and 5×10^4 cells/cm^2 for short-term culture (3 to 4 days).

2.4.1.3 Spreading Cells

After the enzymatic digestion, cells are ready for seeding to the precoated UpCell Surface. To avoid the change of surface characteristics, you have to keep the temperature of cell suspension, as well as the UpCell Surface, above 32°C, or precoated proteins will detach from the UpCell Surface. We usually warm the cell suspension in a water bath at a temperature of 37°C and use the thermo plate on a clean bench (Figure 2.2(c)). Following the count of the number of cells, discard the precoating solution and seed the cells at the appropriate cell density you designed. To avoid a drop of the medium temperature, return the dishes into the incubator quickly.

2.4.2 Change Medium

If you do not want to culture cells for a long time, you do not have to change the medium. You can harvest cells 3 to 4 days after the seeding.

Cells tend to detach from the UpCell Surface during changing of the medium. Thus, both careful handling and strict control of temperature are needed. We always use a water bath to warm the medium and a thermo plate to maintain the temperature of the dishes while changing the medium. If you cannot obtain a thermo plate, change medium and return the culture dish to the incubator in a short time.

For MSCs with osteoinduction, we usually change medium twice a week for 2 weeks with great care of temperature and pipetting. Otherwise cells spontaneously detach from the UpCell. Longer cultivation (such as 3 weeks or 4 weeks) may also induce the detachment of cell sheet.

2.4.3 Harvest of Cell Sheets

Upon temperature reduction below the polymer's lower critical solution temperature (LCST) of 32°C, the polymer surface becomes hydrophilic and swells, forming a hydration layer between the dish surface and the cultured cells (Figure 2.1(d)), allowing for their spontaneous detachment without the need for enzymatic treatments such as trypsinization. We usually put the dishes inside a clean bench at room temperature, and wait for the cell sheets to start detaching from the side edges of the dishes. If cell sheets do not detach spontaneously, tear the most outside cell sheet with forceps or a disposable fine tip. The contiguous sheet of cells usually floats within 30 min. Pipetting the medium and stirring the culture dish can accelerate the detachment speed; however strong handling may break the cell sheet structure. Incubation at a low temperature (20°C) may be helpful if the cell sheets stick on the surface strongly. Detached sheets shrink spontaneously by the tension of cell–cell interaction and cell–matrix interaction, thus the size of cell sheet decreases with time.

2.4.4 Transplantation of Cell Sheets

Transfer membranes are used for harvesting and transporting the cell sheets to the recipient site. PVDF membranes are commercially available and they enable the multilayering of cell sheets if the sheet is strong. Biological or polymer membranes are also available for these purposes. If you do not want the cell sheets to shrink during the procedure, a ring-shaped membrane can allow you to harvest cell sheets without shrinkage [9].

For MSCs, the culture dish is incubated at room temperature for several minutes. Then, culture medium is removed and a wet membrane soaked in saline is set onto the cell sheet. At this point, the membrane must be set from the edge of the cell sheet to prevent the bubble formation between the membrane and the cell sheet. Wait for several minutes to prompt the membrane to stick to the cell sheet, and peel off the membrane-cell sheet complex from the edge in small steps.

2.4.5 Multilayering of Cell Sheets

Multilayering of cell sheets is essential to create thick tissues. However, you have to pay attention to the nutrient supply of multiple cell sheets, or the cell sheets undergo necrosis. From the study using cardiomyocyte transplantation, the thickness limit for layered cell sheets in subcutaneous tissue was approximately 80 μm (3 layers) [11]. The thickness limit varies according to the type of cells and the recipient sites; therefore, researchers need to consider the suitable number of cell sheet layers for each experiment.

Generally, there are three ways of multilayering of cell sheets.

- Multilayering with a supporting membrane [12, 13];
- Multilayering with a tip of a pipette [11, 14];
- Multilayering with a manipulator [15–17].

Confluent cells on the UpCell Surface are harvested by low-temperature treatment, and transfer of the cell sheet to another cell sheet. Repeat this procedure as designed. For details, refer to the literature listed in Section 2.8.

2.5 Data Acquisition, Anticipated Results, and Interpretation

To evaluate the effectiveness of cell sheet transplantation, you need to analyze the parameters you designed. It varies by the type of cells and experiments, thus we do not discuss here. Generally, we search the longevity of cell sheets in the following graft models.

2.5.1 Autograft/Allograft Models

Labeling the cells with green fluorescent protein (GFP), bromodeoxyuridine (BrdU), or dextran coated superparamagnetic iron oxide nanoparticles (SPION) are used to chase the transplanted cells. However, the uptake of these molecules by host tissue macrophages with a fixed probability was recently reported [18]. Thus, additional experiments may be needed to distinguish whether labeled cells exist or not. This problem has to be taken into consideration in graft models using labeling molecules.

2.5.2 Xenograft Models

Severe immunodeficient animals are usually used for xenograft models to minimize the effect of immune rejections. SCID mice and nude rats are often used for cell sheet transplantation to evaluate the effectiveness of xenograft transplantation. In these models, donor species or tissue-specific antibodies are available to detect the transplanted cells in immunohistochemical studies.

2.6 Discussion and Commentary

The limitations of tissue engineering using biodegradable scaffolds and injection of cell suspension are known and alternative strategies have been proposed recently. The aspect of "cell sheet engineering" is the application of temperature-responsive polymer to the surface of cell culture dishes to overcome the problems. UpCell Surface, which is commercially available now, enables the harvesting of cells without enzymes, and it permits the cell sheets to be readily manipulated, transferred, layered, or fabricated, because they adhere rapidly to other surfaces, cell sheets, and recipient sites. Cell sheet engineering has already been tested in clinical settings such as corneal reconstruction [9], endoscopic treatment of esophageal ulceration [19], and improvement of left ventricular function in patients with dilated and ischemic cardiomyopathy [20].

In this chapter, the procedure of making cell sheets is described and the technical points, which are not mentioned in the literature, are explained. To know the nature of UpCell Surface, temperature-responsive and slightly lower adhesive property compared to polystyrene can assist troubleshooting. Strict control of temperature prevents unexpected detachment of cells. As the adhesive property of the UpCell Surface is usually inferior to that of normal culture dish surfaces, initial enhancement of cell adhesion may be needed, especially with cells of low adhesion. Cells should be confluent before transplantation, or the cell sheet tends to be fragile. Initial cell density and culture period have to be optimized. Even though the procedure may have been performed properly, it is possible that the cell sheet is too fragile for transplantation. In such a case, we strongly recommend adding ascorbic acid (AA) if your cells can response to AA. AA enhances both synthesis and posttranslational modification of collagen, thus rich extracellular matrices strengthen the stability of cell sheets [21].

2.7 Applications Notes

2.7.1 Single Cells

By using the UpCell Surface and low temperature treatments at low cell density, it is possible to harvest viable cells such as macrophages and mature osteoclasts, which are considered to be declined in function by enzymatic digestion. It is also possible to analyze native cell surface proteins of adherent cells, which are conventionally analyzed after enzymatic treatments.

2.7.2 Cell Sheets

Culturing cells over the confluency enables the harvest of cells-matrices complexes as a contiguous cell sheet. It permits the cell sheets to be readily manipulated, transferred,

Troubleshooting Table

Problem	Explanation	Potential Solutions
Cells do not attach to UpCell Surface.	Cell adhesion to UpCell Surface is low.	Precoat UpCell Surface with adhesive proteins.
Cell sheet does not detach from UpCell Surface.	Temperature is too high.	Low temperature treatment.
	Cells stick to the side surface.	Tear cells with forceps or yellow tips.
Cell sheet is fragile.	Low cell density.	Increase initial seeding density.
	Low extracellular matrix.	Add ascorbic acid (if applicable).
Cell sheet shrinks.	Naturally occurring over time.	Shrinkage happens spontaneously. Ring-shaped membrane can prevent shrinkage.
Cannot transfer cells to membrane.	Too wet.	Remove culture medium thoroughly then put pre-soaked semi-dry membrane onto cell sheet.
Membrane does not release cell sheet.	Cell sheet does not attach to the recipient.	Wait 5 to 30 min before removing membrane.

layered, or fabricated, because they adhere rapidly to other surfaces, cell sheets, and recipient sites. Therefore, transplantation without sutures and multilayering of cell sheets are feasible and they remain stable at the recipient sites. To create cell sheets, the initial cell density is critical for each type of cells. Table 2.1 may be useful to design both the initial cell density and the culture period of your experiments.

2.8 Summary Points

- Grafting a temperature-responsive polymer, PIPAAm, to the surface of the cell culture dish allows harvesting of cells with a simple low temperature treatment.
- The aspect of harvested cells exhibits as a contiguous sheet. The cell sheet preserves both cell–cell interactions and cell–matrix interactions, which are damaged by conventional enzymatic treatments. Therefore, it is possible to transplant cells as a high functioning tissue.
- As the cell sheets preserve extracellular adhesive proteins, they adhere to the recipient sites or to other cell sheets without sutures.
- Multilayering of cell sheets enables the creation of thick tissues.
- Because the UpCell Surface has lower adhesive ability compared to a polystyrene surface, precoating with adhesive molecules, such as serum proteins or collagens, enhances the initial cell adhesion to the surface.
- To maximize the potential of cells, the initial seeding density and the culture period before the transplantation should be optimized.
- Strict temperature control can prevent unexpected cell detachment.

Acknowledgments

We thank Ryo Takagi, Katsuhisa Sakaguchi, and Makoto Kondo for their valuable comments on cell culture. This research was supported by Formation of Innovation Center for Fusion of Advanced Technologies in the Special Coordination Funds for Promoting Science and Technology from the Ministry of Education, Culture, Sports, Science and Technology (MEXT) of Japan.

Table 2.1

Type of Cell	Approximate Initial Density (cells/cm^2)	Culture Period	References
Epithelial cells	$2–5\times10^5$	8–15 days	[9, 22–24]
oCardiomyocytes	$3–6\times10^5$	4–5 days	[25, 26]
Chondrocytes	1×10^4	7 days	[27]
Hepatocytes	1×10^4	3 days	[28]
MSCs	1×10^5	3 days	[29]
Smooth muscle cells	1×10^5	7 days	[30]
Urothelial cells	2×10^4	21 days	[31, 32]
Islet cells	5×10^5	2 days	[33]
Thyroid cells	1×10^5	7 days	[34]

References

[1] Langer, R.,and J. P.Vacanti, "Tissue Engineering," *Science,* Vol. 260, No. 5110, 1993, pp. 920–26.

[2] Yang, J., et al., "Cell Sheet Engineering: Recreating Tissues without Biodegradable Scaffolds," *Biomaterials,* Vol. 26, No. 33, 2005, pp. 6415–22.

[3] Yamada, N., et al., "Thermo-Responsive Polymeric Surfaces; Control of Attachment and Detachment of Cultured Cells," *Die Makromolekulare Chemie, Rapid Communications,* Vol. 11, No. 11, 1990, pp. 571–76. Authors establish that temperature-responsive polymeric surfaces can control both cell attachment and detachment.

[4] Ratner, B. D., et al., "Cell Adhesion to Polymeric Materials: Implications with Respect to Biocompatibility," *J. Biomed. Mater. Res.,* Vol. 9, No. 5, 1975, pp. 407–22.

[5] Okano, T., et al., "A Novel Recovery System for Cultured Cells Using Plasma-Treated Polystyrene Dishes Grafted with Poly(N-Isopropylacrylamide)," *J. Biomed. Mater. Res.,* Vol. 27, No. 10, 1993, pp. 1243–51.

[6] Nakajima, K., et al., "Intact Microglia are Cultured and Non-Invasively Harvested without Pathological Activation Using a Novel Cultured Cell Recovery Method," *Biomaterials,* Vol. 22, No. 11, 2001, pp. 1213–23.

[7] Ishii, K. A., et al., "Coordination of PGC-1 Beta and Iron Uptake in Mitochondrial Biogenesis and Osteoclast Activation," *Nat. Med.,* Vol. 15, No. 3, 2009, pp. 259–66.

[8] Collier, T. O., et al., "Adhesion Behavior of Monocytes, Macrophages, snd Foreign Body Giant Cells on Poly (N-Isopropylacrylamide) Temperature-Responsive Surfaces," *J. Biomed. Mater. Res.,* Vol. 59, No. 1, 2002, pp. 136–43.

[9] Nishida, K., et al., "Corneal Reconstruction with Tissue-Engineered Cell Sheets Composed of Autologous Oral Mucosal Epithelium," *N. Eng.l J. Med.,* Vol. 351, No. 12, 2004, pp. 1187–96. The first clinical trial using cell sheets to treat a human disease.

[10] Nishida, K., et al., "Functional Bioengineered Corneal Epithelial Sheet Grafts from Corneal Stem Cells Expanded Ex Vivo on a Temperature-Responsive Cell Culture Surface," *Transplantation,* Vol. 77, No. 3, 2004, pp. 379–85.

[11] Shimizu, T., et al. "Polysurgery of Cell Sheet Grafts Overcomes Diffusion Limits to Produce Thick, Vascularized Myocardial Tissues," *FASEB J.,* Vol. 20, No. 6, 2006, pp. 708–10.

[12] Shimizu, T., et al., "Electrically Communicating Three-Dimensional Cardiac Tissue Mimic Fabricated by Layered Cultured Cardiomyocyte Sheets," *J. Biomed. Mater. Res.,* Vol. 60, No. 1, 2002, pp. 110–17.

[13] Iwata, T., et al., "Periodontal Regeneration with Multi-Layered Periodontal Ligament-Derived Cell Sheets in a Canine Model," *Biomaterials,* Vol. 30, No. 14, 2009, pp. 2716–23.

[14] Shimizu, T., et al.,"Fabrication of Pulsatile Cardiac Tissue Grafts Using a Novel 3-Dimensional Cell Sheet Manipulation Technique and Temperature-Responsive Cell Culture Surfaces," *Circ. Res.,* Vol. 90, No. 3, 2002, pp. e40. This is the first study to create a functional thick tissue by multilayering of cell sheets.

[15] Tsuda, Y., et al., "Cellular Control of Tissue Architectures Using a Three-Dimensional Tissue Fabrication Technique," *Biomaterials,* Vol. 28, No. 33, 2007, pp. 4939–46.

[16] Elloumi Hannachi, I., et al., "Fabrication of Transferable Micropatterned-Co-Cultured Cell Sheets with Microcontact Printing," *Biomaterials,* Vol. 30, No. 29, 2009, pp. 5427–32.

[17] Sasagawa, T., et al., "Design of Prevascularized Three-Dimensional Cell-Dense Tissues Using a Cell Sheet Stacking Manipulation Technology," *Biomaterials,* Vol. 31, No. 7, 2010, pp. 1646–54.

[18] Pawelczyk, E.,et al., "In Vivo Transfer of Intracellular Labels from Locally Implanted Bone Marrow Stromal Cells to Resident Tissue Macrophages," *PLoS ONE,* Vol. 4, No. 8, 2009, pp. e6712.

[19] Ohki, T., Yet al., "Endoscopic Transplantation of Human Oral Mucosal Epithelial Cell Sheets—World's First Case of Regenerative Medicine Applied to Endoscopic Treatment," *Gastrointest. Endosc.,* Vol. 69, No. 5 2009, pp. AB253–AB254. This study is the first human reconstruction of endoscopic treatment with cell sheets.

[20] Hoashi, T., et al., "Skeletal Myoblast Sheet Transplantation Improves the Diastolic Function of a Pressure-Overloaded Right Heart,"*J. Thorac. Cardiovasc. Surg.,* Vol. 138, No. 2, 2009, pp. 460–67.

[21] L'Heureux, N., et al., "A Completely Biological Tissue-Engineered Human Blood Vessel," *FASEB J.,* Vol. 12, No. 1, 1998, pp. 47–56.

[22] Watanabe, K., et al., "Development of Transplantable Genetically Modified Corneal Epithelial Cell Sheets for Gene Therapy," *Biomaterials,* Vol. 28, No. 4, 2007, pp. 745–49.

[23] Obokata, H., et al., "Subcutaneous Transplantation of Autologous Oral Mucosal Epithelial Cell Sheets Fabricated on Temperature-Responsive Culture Dishes," *J. Biomed. Mater. Res. A,* Vol. 86, No. 4, 2008, pp. 1088–96.

[24] Ohki, T., et al., "Treatment of Oesophageal Ulcerations Using Endoscopic Transplantation of Tissue-Engineered Autologous Oral Mucosal Epithelial Cell Sheets in a Canine Model," *Gut,* Vol. 55, No. 12,, 2006, pp. 1704–10.

[25] Shimizu, T.,et al., "Long-Term Survival and Growth of Pulsatile Myocardial Tissue Grafts Engineered by the Layering of Cardiomyocyte Sheets," *Tissue Eng.,* Vol. 12, No. 3, 2006, pp. 499–507.

[26] Haraguchi, Y., et al., "Electrical Coupling of Cardiomyocyte Sheets Occurs Rapidly Via Functional Gap Junction Formation," *Biomaterials,* Vol. 27, No. 27, 2006, pp. 4765–74.

[27] Kaneshiro, N.,et al., "Bioengineered Chondrocyte Sheets May Be Potentially Useful for the Treatment of Partial Thickness Defects of Articular Cartilage," *Biochem. Biophys. Res. Commun.,* Vol. 349, No. 2, 2006, pp. 723–31.

[28] Ohashi, K., et al., "Engineering Functional Two- and Three-Dimensional Liver Systems In Vivo Using Hepatic Tissue Sheets," *Nat. Med.,* Vol. 13, No. 7, 2007, pp. 880–85. This report provided the first evidence that ectopic cell sheet transplantation rescues the systemic condition in mice.

[29] Miyahara, Y., et al., "Monolayered Mesenchymal Stem Cells Repair Scarred Myocardium after Myocardial Infarction," *Nat. Med.,* Vol. 12, No. 4, 2006, pp. 459–65.

[30] Hobo, K., et al., "Therapeutic Angiogenesis Using Tissue Engineered Human Smooth Muscle Cell Sheets," *Arterioscler. Thromb. Vasc. Biol.,* Vol. 28, No. 4, 2008, pp. 637–43.

[31] Shiroyanagi, Y., et al., "Urothelium Regeneration Using Viable Cultured Urothelial Cell Sheets Grafted on Demucosalized Gastric Flaps," *BJU Int.,* Vol. 93, No. 7, 2004, pp. 1069–75.

[32] Shiroyanagi, Y., et al., "Transplantable Urothelial Cell Sheets Harvested Noninvasively from Temperature-Responsive Culture Surfaces by Reducing Temperature," *Tissue Eng.,* Vol. 9, No. 5, 2003, pp. 1005–12.

[33] Shimizu, H., et al., "Bioengineering of a Functional Sheet of Islet Cells for the Treatment of Diabetes Mellitus," *Biomaterials,* Vol. 30, No. 30, 2009, pp. 5943–49.

[34] Arauchi, A., et al., "Tissue-Engineered Thyroid Cell Sheet Rescued Hypothyroidism in Rat Models after Receiving Total Thyroidectomy Comparing with Nontransplantation Models," *Tissue Eng. Part A,* Vol. 15, No. 12 2009, pp. 3943-3949.

[35] Hasegawa, M., Yamato, M., Kikuchi, A., Okano, T.and Ishikawa, I., "Human periodontal ligament cell sheets can regenerate periodontal ligament tissue in an athymic rat model," *Tissue Eng.,* Vol. 11, No. 3–4, 2005, pp. 469–78.

[36] Kanzaki, M., et al., "Dynamic Sealing of Lung Air Leaks by the Transplantation of Tissue Engineered Cell Sheets," *Biomaterials,* Vol. 28, No. 29, 2007, pp. 4294–4302.

[37] Kanzaki, M., et al., "Functional Closure of Visceral Pleural Defects by Autologous Tissue Engineered Cell Sheets," *Eur. J. Cardiothorac. Surg.,* Vol. 34, No. 4, 2008, pp. 864–69.

3

In-Vivo Functional Assessment of Engineered Human Insulin-Producing Cells

Holger A. Russ and Shimon Efrat*

Department of Human Molecular Genetics and Biochemistry, Sackler School of Medicine, Tel Aviv University, Ramat Aviv, 69978 Tel Aviv, Israel

*Correspondence: Department of Human Molecular Genetics and Biochemistry, Sackler School of Medicine, Tel Aviv University, Ramat Aviv, 69978 Tel Aviv, Israel
Phone: +972-3-640 7701; Fax: +972-3-640 9950
E-mail: sefrat@post.tau.ac.il

Abstract

Function in vivo represents the ultimate test for insulin-producing cells engineered in vitro, since it can evaluate cell capacity to respond to an array of physiological signals and maintain phenotypic stability over time. In addition, the in vivo milieu can further promote cell differentiation following manipulations in vitro. The most widely used model for assaying the function of insulin-producing cells in vivo is transplantation under the renal capsule of immunodeficient mice made hyperglycemic by injection of a beta-cell toxin. This chapter describes methods for generation of recipient mice, cell transplantation, and mouse and graft phenotyping following transplantation.

Key terms: immunodeficient mice, hyperglycemia, cell transplantation, serum C-peptide, glucose responsiveness

3.1 Introduction

Type 1 diabetes is caused by autoimmune destruction of the pancreatic islet insulin-producing beta cells. Insulin administration does not constitute a cure, due to the difficulty of adjusting the precise amounts of administered insulin to changing physiological needs. As a result, episodes of hypo- and hyperglycemia lead in the long-run to accumulation of microvascular damage and diabetes complications. Replacement of damaged beta cells through regeneration or transplantation represents the most promising approach for restoration of a regulated insulin supply. This approach faces a dual challenge: generating a sufficient mass of cells which produce adequate amounts of insulin and release it in response to normal physiological signals, and protecting these cells from recurring autoimmunity.

Islet transplantation is successful in restoring euglycemia in most recipients during the initial 12 to 24 months, but fails later due to poorly characterized causes, likely including allogeneic rejection and recurring autoimmunity, toxicity of immunosuppression, and limited cell longevity coupled with insufficient cell renewal [1]. In addition, application of islet transplantation on a large scale is limited by the number of human organ donors. Three major avenues have been explored for dealing with the tissue shortage: adult human islet expansion in vitro, induced differentiation of human embryonic stem (ES) cells into insulin-producing cells, and differentiation of cells from a variety of fetal and adult tissues into betalike cells by nuclear reprogramming. To be clinically useful, cell-based therapies would have to represent a significant advantage in maintaining euglycemia over the current treatment of insulin administration, while matching its safety record.

Ex-vivo expansion of functional adult human beta cells has proven to be difficult (see [2] for a recent review). Recent lineage-tracing studies demonstrated the ability of cells derived from human beta cells to survive and significantly replicate in vitro [3]. However, the expanded cells lose their capacity to produce insulin, and so far an efficient and reproducible way for restoring it has not been identified. Much progress has been made in developing protocols for differentiation of human ES cells into insulin-producing cells [4]; however, incomplete differentiation and the risk of teratoma formation remain significant concerns. The third approach utilizes stem/progenitor cells and/or differentiated cells from a number of fetal or adult tissues for ex-vivo generation of insulin-producing cells for allotransplantation, or for inducing differentiation of the patient's own tissues into betalike cells in vivo. Nuclear reprogramming in vivo or ex vivo using transfer of genes encoding dominant transcription factors [5–7], as well as treatment with soluble factors [8, 9], resulted in generation of functional insulin-producing cells from liver cells, pancreatic acinar cells, and bone marrow mesenchymal stem cells, among others.

While insulin-producing cells generated in vivo or ex vivo from autologous cells are not expected to elicit allograft rejection, recurring autoimmunity against autologous insulin-producing cells may represent a serious obstacle, making in-vitro engineered allogeneic cells potentially more attractive for transplantation. Engineering of insulin-producing cells in vitro allows cell banking and detailed cell characterization prior to transplantation. These analyses may include screening for infective pathogens, assessment of chromosomal integrity, assays for gene expression, and functional evaluation of insulin storage and regulated release. In addition, cell engineering in vitro provides an opportunity for equipping the cells with protective devices against immune responses (e.g., antiapoptosis genes [10] and cell encapsulation [11]). However, reconstitution of the normal cellular environment in vitro is difficult. Thus, engineered insulin-producing cells are likely to differentiate only partially under these conditions, and realize their full maturation potential in vivo following transplantation. In addition, the ultimate functional assessment of cell function must be performed in vivo, where the cells are exposed to changing physiological conditions and a wide spectrum of signals. This chapter focuses on methods for phenotypic and functional characterization of insulin-producing cells, generated in vitro from different cell sources and by various approaches, using transplantation into hyperglycemic animals.

3.2 Experimental Design

Mice represent the most widely used experimental animal in transplantation studies. Their biochemical and physiological characteristics are sufficiently similar to those of humans to provide clinically relevant data, while their small size and resilience to surgical manipulations allow relatively easy and reproducible experimentation with a limited number of cells. Testing human cells requires the use of immunodeficient mice to avoid xenograft rejection. Severe combined immunodeficient (SCID) mice, carrying a recessive mutation in a gene encoding a DNA recombination protein, are frequently used in such studies. The mutation prevents rearrangement of genes coding for antigen-specific receptors, resulting in impaired maturation of T and B lymphocytes and failure to mount immune responses. By crossing SCID mice with mice carrying mutations in other genes needed for normal immune responses, further immunocompromised strains can be created [12]. SCID mice may spontaneously develop "leakiness" in a strain-dependent manner. This leakiness is minimal on the nonobese diabetic (NOD) genetic background. However, in our experience even on the C3H/HeJ background the SCID mutation alone provides sufficient tolerance to allografts and xenografts, making these mice a good model for cell transplantation experiments.

To generate a model for hyperglycemia, the mouse beta cells are eliminated by treatment with alloxan or streptozotocin (STZ), the latter being more commonly used. STZ is a glucosamine-nitrosourea alkylating agent, which causes damage to the DNA, although other mechanisms may also be involved. STZ resembles glucose in its structure and is recognized by the glucose transporter GLUT2, but not by the other glucose transporters, making it specifically toxic to cells expressing high levels of GLUT2, such as pancreatic islet beta cells, liver, and kidney [13]. However, beta cells are by far more susceptible, probably due to their low ability to inactivate the drug.

The cells can be transplanted in one of several sites; however, the renal subcapsular site has particular advantages. This location is rich in vasculature, and the transplanted cells are likely to become vascularized and well-supplied with oxygen, nutrients, and differentiation factors. Exposure to circulating insulin secretagogues is also adequate. Usually cells are transplanted into one kidney, while the second kidney remains fully functional. Survival one-side nephrectomy can be performed at the end of the experiment, to demonstrate the glycemic effect of the transplanted cells by restoration of hyperglycemia following their removal.

Following transplantation the mice are continuously monitored for the desired period of time, usually from several weeks to several months, to determine the effect of the transplanted cells on their glycemia. Control mice transplanted with unmanipulated cells are monitored in parallel. Once blood glucose levels approach the normal range, serum samples from fed mice are analyzed for the presence of human C-peptide. Depending on the STZ treatment and the duration of the follow-up period, the endogenous mouse beta cells may partially regenerate. Thus, it is crucial to demonstrate that the glycemic changes are due to the transplanted human insulin-producing cells. Given the similarity between mouse and human mature insulin, the available antibodies cannot distinguish them. In contrast, human and mouse C-peptides are sufficiently different to allow specific detection by RIA or ELISA. In addition to analyzing fed serum glucose and human C-peptide levels, glucose-induced insulin release is determined in fasting conditions in a glucose-tolerance test (GTT). The kinetics of glucose clearance and insulin release is determined following a glucose bolus.

At the termination of the desired follow-up period, survival one-side nephrectomy is performed. The mice are followed for 1 to 2 additional days, to determine glycemia changes. The kidney is subjected to RNA and immunostaining analyses to characterize the phenotype of the transplanted cells following the period in vivo. The untransplanted kidney is used as control.

3.3 Materials

- Sodium citrate (Sigma-Aldrich, United States);
- Streptozotocin (Acros Organics, United States);
- Glucometer and strips: Accu-Check (Roche, Germany);
- Tribromoethanol (Fluka, Germany);
- Tert-amyl-alcohol (Merck, Germany);
- Transplantation syringe: Insumed.0.5-ml syringe, prefitted with a 30-gauge needle (0.3×8 mm) (Grandate, Italy);
- Suture, silk braided, 75 cm, 4-0 USP with needle (Assut sutures, Switzerland);
- Wound clips (MikRon Precision Inc., United States);
- C-peptide ELISA kits: human (Millipore, United States) and mouse (Mercodia, Sweden);
- D-glucose (Sigma-Aldrich, United States);
- 4% paraformaldehyde (Electron Microscopy Sciences, United States);
- Trizol (Sigma-Aldrich, United States).

3.4 Methods

3.4.1 Generation of Hyperglycemic Immunodeficient Recipient Mice

Follow guidelines of your Institutional Animal Care and Use Committee regarding animal housing and manipulation. Mice are preferably males 2 to 4 months of age.

1. Prepare 0.1M sodium citrate buffer for STZ solution. The buffer is adjusted to pH 4.5 with 1M HCl, filtered using a 0.2-μm filter, and stored for long term at 4°C.
2. Fast mice overnight.

3. Prepare fresh 10 μg/ml STZ solution in sodium citrate buffer and use it within 15 min.
4. Inject 20 μl of STZ solution (200 μg) per gram mouse weight intraperitoneally (i.p.) using a 1-ml syringe and a 25-gauge needle.
5. Continue fast for 1 more hour after injection.
6. Monitor blood glucose levels each day using glucometer strips by taking blood samples from the tail vein of fed mice. Mice are considered hyperglycemic when fed blood glucose levels are >300 mg/dl for 2 consecutive days. Mice become hyperglycemic within 2 to 3 days. The STZ dose and time before development of hyperglycemia vary somewhat among different strains.

3.4.2 Cell Transplantation Under the Renal Capsule

1. Preparation of anesthetic: Prepare avertin stock solution by adding 5g tribromoethanol to 5 ml tert-amyl-alcohol. Rotate light-protected for 30 min at room temperature. Filter using a 0.2-μm filter. Avertin stock solution can be kept light-protected at 4°C for up to 4 months.
2. Prepare fresh avertin working solution by diluting the stock solution to 2.5% (v/v) in phosphate-buffered saline (PBS). Rotate light-protected for 30 min at room temperature.
3. To anesthetize mice inject i.p. 15 μl of 2.5% avertin per gram mouse body weight using a 1-ml syringe and a 25-gauge needle. Pinch the hind leg to test if anesthesia is complete.
4. Preparation of cells: Wash trypsinized cells in PBS and pellet them is a 1.5-ml Eppendorf tube. Remove most of the supernatant, leaving behind only a few μl. Remove the plunger from the transplantation syringe. Using forceps bend the needle slightly at a point close to the syringe, with the beveled side of the needle facing the angle (Figure 3.1(a)). Resuspend the cell pellet with a pipette and load the cell slurry from the back into the syringe. Deposit the cells only on one side of the wall to allow free air flow. Insert the plunger and push down carefully until the plunger touches the cell slurry, then move the plunger slightly back and forth without creating air bubbles, until the cell slurry covers the entire plunger tip. Press the plunger down until a small drop of cell slurry appears at the tip of the needle. Place the syringe on ice until the injection.
5. Place mouse on its right side in a Petri dish and disinfect the left flank with 70% ethanol. Using dissection scissors make a 13- to 18-mm-long transverse incision running from the spine lateral above the left kidney. Move the mouse under a dissection stereomicroscope. Slide the skin around until the kidney is visible through the body wall. Pull up the body wall with fine blunt forceps and make a small incision just above the kidney. Enlarge the incision parallel to the ribcage for a final length of 7 to 9 mm. Hold the body wall open with the left hand using fine straight blunt forceps. Use curved blunt forceps to gently grab the kidney and pull it out of the abdominal cavity. The kidney should remain outside the body wall without sliding back. Pierce the renal capsule with the needle of the transplantation syringe on the right top of the convex side and carefully insert the needle between the capsule and the kidney, in parallel to the surface (Figure 3.1(a)). Advance the needle as far as possible along the top of the kidney (but not beyond the needle bend). Increase the angle between

the needle and the kidney slightly so to avoid piercing the capsule. Hold the syringe with the left hand and push the plunger with the right hand to inject up to 30 μl of cell slurry. Wait a few seconds before withdrawing the needle. The kidney capsule should hold the cell slurry in place without leakage (Figure 3.1(b)), since the hole generated by the needle is at some distance. Gently push the kidney back into the abdominal cavity. Stitch the body wall with silk suture and close the skin with wound clips.

3.4.3 Monitoring Blood Glucose, Insulin, and C-Peptide Levels

Monitor blood glucose levels twice a week (same days and time) in blood samples obtained from the tail vein of fed mice. To determine serum levels of mouse and human C-peptide, blood samples are withdrawn from the retro-orbital plexus of fed mice into 1.5-ml Eppendorf tubes. Serum is separated by spinning the clotted blood for 1 min at 14,000 rpm and stored at –80°C. C-peptide levels are determined using ELISA kits, following the manufacturer's protocol.

3.4.4 Assessment of Glucose Responsiveness Using GTT

Fast mice overnight. Measure fasting blood glucose levels using glucometer strips. Inject mice i.p.with 100 mg/ml D-glucose in saline at 1 mg (10 μl) per gram body weight. Monitor blood glucose levels 15, 30, 60, 90, and 120 min after glucose injection in blood samples obtained from the tail vein.

3.4.5 Survival Left-Side Nephrectomy and Tissue Removal

1. For survival nephrectomy follow the steps outlined in Section 3.4.2, items 3 and 5, to pull the left kidney out of the abdominal cavity. Grasp the renal pelvis and blood vessels with blunt straight forceps right under the kidney. Use a silk suture to tie off the renal pelvis and blood vessels under the forceps to avoid bleeding when the kidney is cut off (Figure 3.2(a)). This is best done by another person. Carefully cut between the forceps and the suture knot using fine curved dissection scissors (Figure 3.2(b)). Stitch the body wall with silk suture and close the skin with wound clips. Monitor blood glucose levels daily using glucometer strips. Hyperglycemia should ensue within 1 to 2 days. The experiment is then terminated and the mouse euthanized. The right kidney can be removed at this stage and used as a negative control for the following analyses.
2. Transfer the kidney into a Petri dish containing ice-cold PBS. Place the dish under a stereomicroscope. The cell transplant will be visible as a white-yellow area on the red background of kidney parenchyma (Figure 3.3(a)). Slice the kidney in the middle of the transplant into two halves (Figure 3.4). Place one half in 4% paraformaldehyde in PBS on ice. Fix tissue for 4 to 24 hours at 4°C, wash several times with PBS, and proceed to a standard protocol for paraffin embedding and sectioning.
3. Trim off excessive kidney tissue from the other half of the kidney and transfer into 1.5-ml Eppendorf tube containing Trizol reagent. Use a pestle to crush the tissue. Isolate RNA according to the manufacturer's protocol.

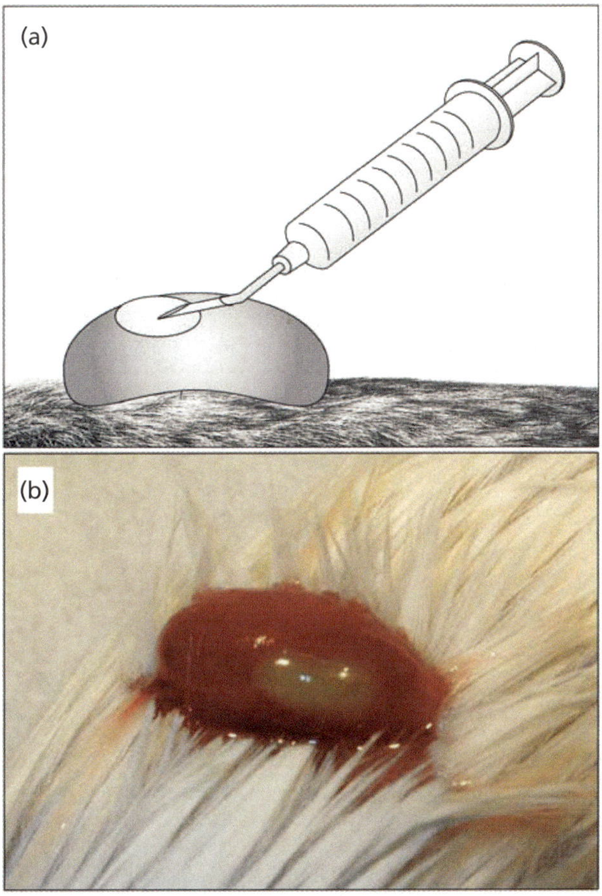

Figure 3.1 Cell transplantation under the renal capsule. (a) Schematic representation of cell injection. (b) Kidney with graft following withdrawal of injection needle.

3.4.6 Graft Histological and RNA Analyses

1. Process paraffin sections according to standard protocols for rehydration and antigen retrieval. Stain with hematoxylin and eosin (H&E) to visualize transplant area (Figure 3.3(b)). Use adjacent sections for immunofluorescence analyses of antigens of interest, such as insulin and human C-peptide (Figure 3.3(d) and (e)). If human C-peptide is not detected, staining with another human-specific antibody, such as antihuman heat shock protein (HSP) 27, or antihuman nuclear proteins, should be used to positively demonstrate the presence of human cells (Figure 3.3(f)). Analysis for cell proliferation can be performed by injecting bromodeoxyuridine (BrdU) the day before kidney removal and staining for BrdU incorporation as described [8] (Figure 3.3(c)).
2. RNA is analyzed by quantitative RT-PCR with human-specific primers, preferably spanning exon junctions, using standard protocols. If human-specific primers have a high human-mouse sequence homology, RNA extracted from the untransplanted right kidney should be employed as a negative control. Typical transcripts analyzed include those encoding for islet hormones, beta-cell transcription factors, and proinsulin-processing enzymes.

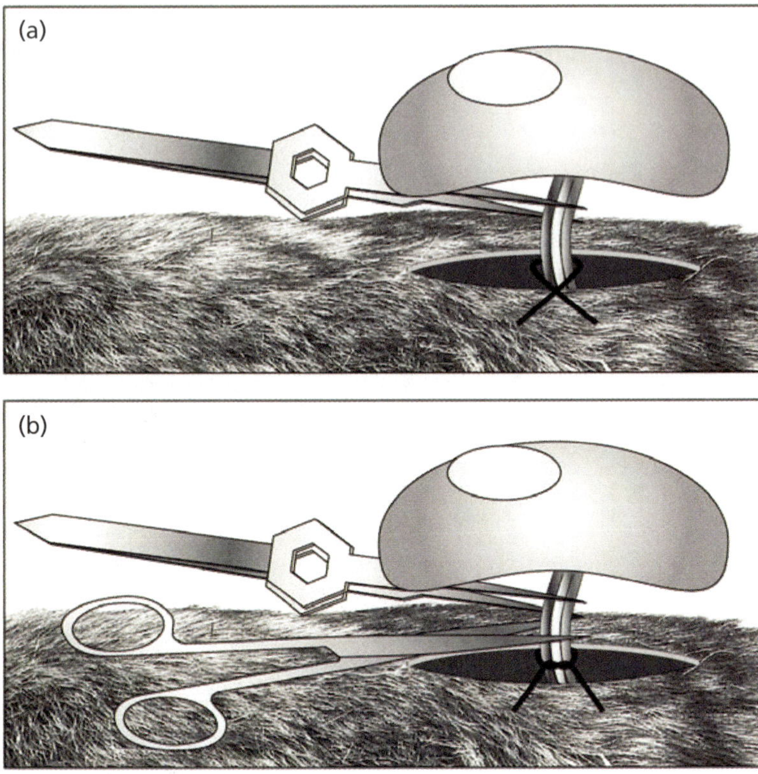

Figure 3.2 Schematic representation of nephrectomy. (a) A silk suture is used to tie off the renal blood vessels. (b) The kidney is removed by cutting between the forceps and the suture knot.

3.5 Anticipated Results

STZ-treated mice should become hyperglycemic within a few days. If untreated further, blood glucose may continue to rise and the mice will die within several weeks. Control mice transplanted with unmanipulated cells, which do not release insulin, are expected to manifest the same phenotype. In contrast, mice transplanted with a sufficient mass of insulin-producing cells should show a gradual decrease in glycemia towards the normal range (fed blood glucose values <200 mg/dl). Sufficient numbers of mice should be included in each group (5–10), to allow generation of statistically-significant data. Cells producing low amounts of insulin in vitro may further differentiate in vivo, resulting in an increase in insulin production, as well as improved regulation of its release. This process may take from several days to several weeks. Fed serum human C-peptide levels around 1 ng/ml would be indicative of adequate function of the insulin-producing cells. GTT should reveal kinetics of glucose clearance within 2 hours, which resembles that of nondiabetic mice. Phenotypic stability of the cells, as well as lack of xenograft rejection, should result in long-term maintenance of stable blood glucose levels. Mouse C-peptide should be absent, or detected at very low levels. Its presence may indicate incomplete STZ toxicity to the endogenous beta cells, or beta-cell renewal, and complicate interpretation of the results. In this case survival one-side nephrectomy will be particularly valuable. By detecting an increase in blood glucose levels following graft removal, the contribution of the transplanted cells to glycemia correction can be demonstrated. The region under the renal capsule containing the majority of the transplanted cells should

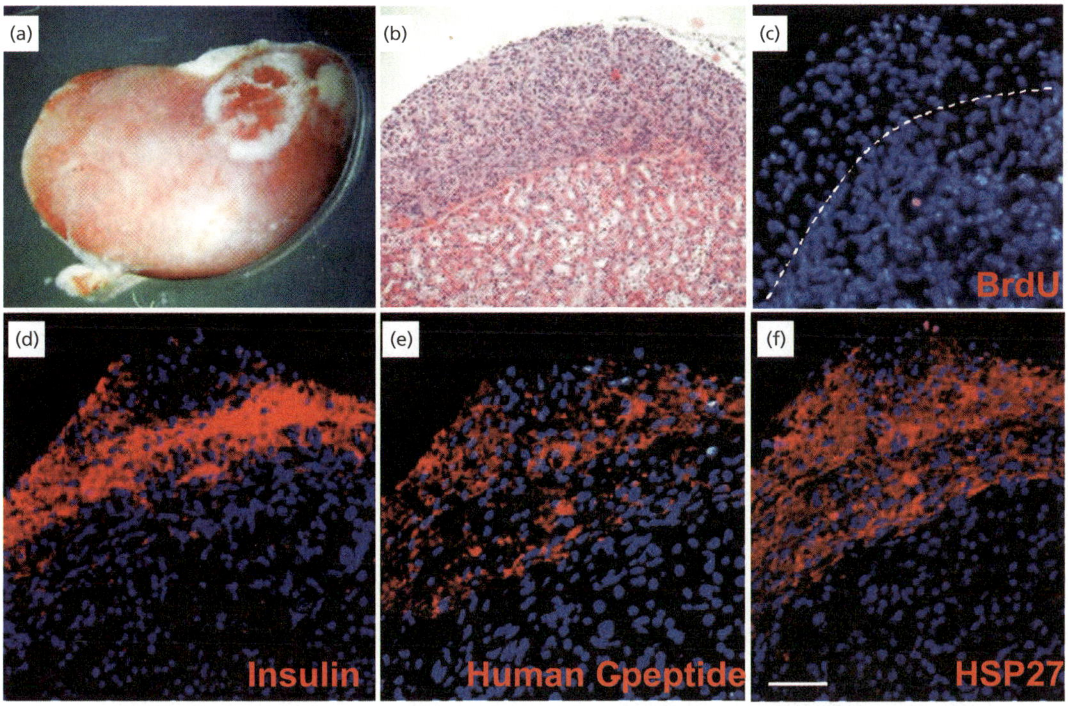

Figure 3.3 Images of cell transplant under the renal capsule. (a) Kidney with transplanted cells seen at its top right corner. (b) H&E staining of a kidney section showing the transplanted cells in the top half. (c–f) Immunofluorescence analyses of adjacent sections with the indicated antibodies, as visualized with Cy3-conjugated second antibodies. All nuclei are labeled blue with DAPI. In (c), the dashed line marks the boundary between the transplanted cells and the kidney parenchyma. A single nucleus labeled by BrdU is seen in the latter. Bar = 50 μm. (From: [8].)

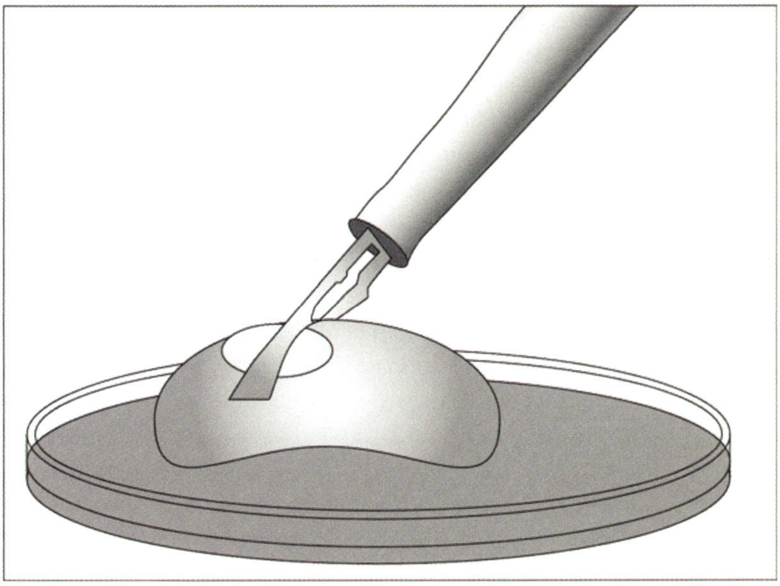

Figure 3.4 Schematic representation of graft processing for analysis. The kidney is cut into two halves in the middle of the graft area. One half is fixed for histological analyses. Excessive kidney tissue is trimmed off the graft area in the other half, and the remaining graft is used for RNA extraction.

be visible in the dissected kidney. Histological analyses of this region should show an area under the renal capsule that stains for human C-peptide and other human-specific markers. Alternatively, if the transplanted cells are prelabeled by expression of an ectopic marker not present in the endogenous mouse cells, such as green fluorescent protein, detection of this marker can positively demonstrate the presence of human cells. Analysis of RNA extracted from the transplant area by quantitative RT-PCR performed with human-specific primers is expected to reveal changes in gene expression, compared with the cultured cells, which are likely to indicate further differentiation during the period in vivo.

3.6 Discussion and Commentary

STZ-treated SCID mice are a useful model organism for functional evaluation of human insulin-producing cells. However, STZ is not strictly a beta-cell-specific toxin. Thus, there is a relatively narrow dose range in which the majority of beta cells are destroyed without compromising mouse health in other ways. This range varies depending on mouse strain and age, and requires some optimization in each experimental system. Recently, a new immunodeficient mouse strain that spontaneously develops hyperglycaemia has been described that offers an attractive alternative. This strain combines a dominant mutation in the murine insulin 2 gene, termed $Ins2^{Akita}$, which impairs normal insulin folding and induces beta-cell apoptosis, with 2 null mutations in genes required for normal immune function, *Rag1* and *Prf1*, on the NOD background [14]. Initial studies have shown it to be an excellent model organism for transplantation of human insulin-producing cells [14].

While islet transplantation under the renal capsule is a well-established method, transplantation of cultured cells in this site presents some technical problems. This site is limited in the cell volume it can accommodate, to several million cells, depending on cell size. This cell number may not be sufficient for reversing hyperglycemia if cellular insulin content is low. Moreover, injection of a large volume of loose cell suspension, as opposed to intact islets, may lead to cell leakage through the hole in the renal capsule created by the injection needle. It is recommended to form small cell aggregates or clusters (e.g., by gentle agitation in Petri dishes [15]) prior to transplantation. In addition to facilitating cell transfer and immobilization under the renal capsule, cell clustering may contribute to further differentiation of the transplanted cells [15]. The engineered cells may have the potential to release other islet hormones, which may affect the function of the insulin-producing cells and the final physiological outcome. However, insulin-producing cells alone are likely to be capable of phenotype normalization, since function of the mouse nonbeta islet cells is expected to be maintained throughout the experiment.

This experimental system is intended for the limited purpose of testing cell function in the context of an intact organism, and represents only the first step in preclinical characterization of engineered insulin-producing cells. Successful results will warrant further testing in larger animals such as pigs or nonhuman primates, as well as in models of autoimmunity such as mice with a humanized immune system.

3.7 Application Notes

This analysis is suitable for functional assessment of insulin-producing cells derived from a variety of sources and engineered by different approaches, following propagation and manipulation in vitro. Cells which manifest significant levels of expression of beta-cell genes in vitro and store detectable amounts of insulin are more likely, of course, to perform well in vivo, compared with cells with lower expression levels. However, the potential differentiation effects of the in vivo milieu justify analysis in vivo even of cells manifesting a relatively low differentiation level in vitro.

3.8 Summary Points

1. Function in vivo represents the ultimate test for insulin-producing cells engineered in vitro, since it can evaluate cell capacity to respond to an array of physiological signals and maintain phenotypic stability over time.
2. The in vivo milieu can further promote cell differentiation on top of the effects of the manipulations in vitro.

Troubleshooting Table

Problem	Possible Explanations	Potential Solutions
Mice injected with STZ do not become hyperglycemic.	STZ not fresh; insufficient STZ.	Use fresh STZ solution and optimize dosage for each mouse strain.
Cell slurry clots the transplantation syringe.	Formed cell clusters are too big; cell death during preparation.	Alter protocol to induce cluster formation; work on ice to avoid cell death; add 10 μg/ml DNase I to cell suspension to digest free DNA.
Cell slurry spills out of the injection site.	Too much supernatant left behind when creating cell slurry; needle is not inserted far enough; needle pierced the kidney capsule.	Remove more of the supernatant; try to advance the needle as far as possible along the kidney membrane without piercing it.
Blood glucose of hyperglycemic mice transplanted with control cells is normalized.	Endogenous beta-cell regeneration; control cell differentiation in vivo.	Slightly increase STZ dose; analyze serum human and mouse C-peptide levels to determine insulin source.
Hyperglycemia is not corrected following cell transplantation.	Graft cell number or cell insulin content too low.	Increase transplanted cell number; use other in vitro differentiation methods to increase cell insulin content.
Hyperglycemia is corrected initially but reappears later.	Graft cell death following transplantation may release large insulin amounts; graft cell phenotypic instability; residual immune responses.	Manipulate cells carefully during transplantation to reduce cell death; alter differentiation protocol to increase graft cell phenotypic stability; use mice with additional immune system mutations to further reduce immune responses.
Transplanted mice develop hypoglycemia.	Unregulated graft cell proliferation or insulin release.	Transplant fewer cells; alter differentiation protocol to achieve growth arrest and regulated insulin secretion.
No increase in glycemia following nephrectomy.	Endogenous beta-cell regeneration; graft leakage out of the kidney.	Slightly increase STZ dose; analyze serum human and mouse C-peptide levels to determine insulin source.
Human cells are not detected in mouse kidney sections.	Graft cell leakage out of the kidney; wrong kidney area analyzed.	Use cell clusters instead of loose cell suspension to reduce leakage; section entire kidney and stain every fifth slide with H&E to identify graft area.

3. SCID mice treated with STZ are a simple and adequate model for functional assessment of human insulin-producing cells generated by different approaches from cells derived from different tissue sources, although additional mutations in immune system genes and the use of mice on NOD background may further reduce the chances for immune responses against the graft.
4. The renal capsule is a convenient site for transplantation of insulin-producing cells since it is well-vascularized, restricts the transplanted cells to a discrete location and allows their clear visualization, and allows transplant retrieval for further analysis and demonstration of phenotypic reversibility upon graft removal.

References

[1] Ryan, E. A., et al., "Five-Year Follow-Up after Clinical Islet Transplantation," *Diabetes,*Vol. 54, 2005, pp. 2060–69.

[2] Efrat, S., "Ex-Vivo Expansion of Adult Human Pancreatic Beta Cells," *Rev. Diabet. Stud.,* Vol. 5, 2008, pp. 110–116.

[3] Russ, H. A., et al. "In Vitro Proliferation of Cells Derived from Adult Human Beta Cells Revealed by Cell-Lineage Tracing," *Diabetes,* Vol. 57, 2008, pp. 1575–83.

[4] Kroon, E., et al., "Pancreatic Endoderm Derived from Human Embryonic Stem Cells Generates Glucose-Responsive Insulin-Secreting Cells In Vivo," *Nat. Biotechnol.,* Vol. 26, 2008, pp. 443–52.

[5] Zalzman, M., et al., "Reversal of Hyperglycemia in Mice by Using Human Expandable Insulin-Producing Cells Differentiated from Fetal Liver Progenitor Cells," *Proc. Natl. Acad. Sci. USA,* Vol. 100, 2003, pp. 7253–58.

[6] Karnieli, O., et al., "Generation of insulin-producing cells from human bone marrow mesenchymal stem cells by genetic manipulation," *Stem Cells,* Vol. 25, 2007, pp. 2837–44.

[7] Zhou. Q., et al., "In Vivo Reprogramming of Adult Pancreatic Exocrine Cells to Beta-Cells," *Nature,* Vol. 455, 2008, 627–32.

[8] Zalzman, M., L. Anker-Kitai, and S. Efrat, "Differentiation of Human Liver-Derived Insulin-Producing Cells Towards the Beta-Cell Phenotype," *Diabetes,* Vol. 54, 2005, pp. 2568–75.

[9] Baeyens, L., et al., "In Vitro Generation of Insulin-Producing Beta Cells from Adult Exocrine Pancreatic Cells," *Diabetologia,* Vol. 48, 2005, pp. 49–57.

[10] Pierce MA, Chapman HD, Post CM, Svetlanov A, Efrat S, Horwitz MS, Serreze DV. Adenovirus early region 3 (E3) anti-apoptotic 10.4K, 14.5K, and 14.7K genes decrease the incidence of autoimmune diabetes in NOD mice. *Diabetes* 2003, 52:1119-1127.

[11] Lacy, P. E., et al., "Maintenance of Normoglycemia in Diabetic Mice by Subcutaneous Xenografts of Encapsulated Islets," *Science,* Vol. 254, 1991, pp. 1782–84.

[12] Ito, M., et al., "NOD/SCID/Gamma(c)(Null) Mouse: An Excellent Recipient Mouse Model for Engraftment of Human Cells," *Blood,* Vol. 100, 2002, pp. 3175–82.

[13] Schnedl, W. J., et al., "STZ Transport and Cytotoxicity. Specific Enhancement in GLUT2-Expressing Cells," *Diabetes,* Vol. 43, 1994, pp. 1326–33.

[14] Pearson, T., et al., "A New Immunodeficient Hyperglycaemic Mouse Model Based on the Ins2Akita Mutation for Analyses of Human Islet and Beta Stem and Progenitor Cell Function," *Diabetologia,* Vol. 51, 2008, pp. 1449–56.

[15] Fleischer, N., et al., "Functional Analysis of a Conditionally-Transformed Pancreatic β-Cell Line," *Diabetes,* Vol. 47, 1998, pp. 1419–25.

4

Generation of Insulin-Producing Cells from Adult Mouse Intrahepatic Biliary Epithelial Cells In Vitro

Masaki Nagaya and Gordon C. Weir*

Research Division, Section on Islet Transplantation and Cell Biology, Joslin Diabetes Center, One Joslin Place, Boston, Massachusetts 02215, and Department of Medicine, Harvard Medical School, Boston, Massachusetts, 02215

*Correspondence: Section on Islet Transplantation and Cell Biology, Joslin Diabetes Center, One Joslin Place, Boston, MA 02215
Phone: 617–732–2581; Fax: 617–732–2650
E-mail: gordon.weir@joslin.harvard.edu

Abstract

There is great hope that transplantation of insulin-producing cells will be a cure-equivalent for type 1 diabetes. However, two major challenges remain: the shortage of available tissue and the need for immunosuppressive drugs. We hypothesize that transdifferentiation of cells from a patient's own liver into pancreatic β-cells could resolve the first problem. The bile duct epithelium in the liver seems to have considerable promise as a new source of insulin-producing cells. Because intrahepatic biliary epithelial cells (IHBECs) can serve as progenitors in response to certain stimuli and the cells possess β-cell progenitor transcription genes. The concept is to obtain a small piece of tissue from a diabetic individual by endoscopic surgery, followed by expansion in vitro and then generation of insulin-producing cells for subsequent autologous transplantation. Herein, we present a strategy for the concept and protocols to expand the IHBECs and generate insulin-producing cells from IHBECs by the ectopic expression of critical β-cell transcription factors.

Key terms lliver, biliary epithelial cell, Adenovirus, insulin, and transdifferentiation

4.1 Introduction

The transplantation of insulin-producing cells for curing type 1 diabetes has several benefits [1, 2]. A minimally invasive procedure might be performed in the future, which would require only a brief hospitalization, lower the cost, and reduce complications of the intervention. However, the paucity of available islets for transplantation as well as the immunological barriers faced in allogeneic transplantation represents a tremendous barrier to the treatment. To resolve the difficulty of availability of tissue, efforts are underway to produce sufficient quantities of β-cells for transplantation. Promising strategies include conversion of either pancreatic adult stem/progenitor cells [3], expansion of existing β-cells [4], and differentiation of embryonic stem cells [5] or induced pluripotent stem (iPS) cells [6].

Some researchers have pursued transdifferentiation using liver tissue both in vivo and in vitro [7–10]. Ectopic expression of NeuroD and pancreatic and duodenal homeobox-1 (Pdx-1) has been found to promote the differentiation of liver towards pancreas [7, 8, 10]. The advantage of the liver is that both liver and pancreas develop embryologically from very similar pathways via the foregut endoderm [11–13].

Among the cell populations in the liver, we focused our attention on IHBECs as a new source of insulin-producing cells for the following reasons: (1) A considerable amount of liver tissue can be surgically removed without much difficulty [14], (2) the liver has considerable capacity for regeneration [15], (3) IHBECs can serve as progenitors in response to certain stimuli [16], and (4) IHBECs possess islet progenitor transcription factors such as HNF1β, HNF3β, HNF4β, and HNF6 [11]. Transdifferentiation of cells from a patient's own tissue into pancreatic β-cells could address one obstacle of islet transplantation. Liver tissue could be obtained from a diabetic individual by endoscopic surgery, followed by in vitro expansion and differentiation into insulin-producing cells for subsequent autologous transplantation. In the meantime, the liver can regenerate the cells lost during the removal of the lobe.

All of the above circumstances stimulated us to develop new methods for generating insulin producing cells from IHBECs.

Even though IHBECs are attractive targets for transdifferentiation, previous techniques for isolation of IHBECs have been cumbersome and associated with a low yield of cells [17–19] Furthermore, in vitro studies with freshly isolated cells or intrahepatic bile duct unit (IBDU) fragments are limited by lack of long-term (>24h) viability [19]. Therefore, expansion to obtain significant quantities of healthy IHBECs is an important possibility. Therefore, we developed a new culture method termed collagen embedded floating culture method (CEFCM). It has been found that three-dimensional culture conditions can allow polarized cells to keep their original character [20, 21]. We postulated that this circumstance would promote expansion and improve viability.

In this chapter, we describe progress on reprogramming of IHBECs to insulin-producing cells based on the overexpression of pancreatic transcription factors in IHBECs. The expression of the master switch gene can, under suitable circumstances, trigger an alternate program of gene expression that results in the new cell type [22]. The protocols below describe the overexpression of pancreatic transcription factors by adenoviral transduction. Adenoviral vectors are an excellent vehicle for gene delivery as they are capable of transducing nondividing cells such as hepatocytes. Furthermore, genes delivered by adenoviral vectors do not integrate into the host genome, thus their expression is short-lived [23]. This is advantageous from a safety point of view since there is little risk of insertional mutagenesis. Under suitable circumstances, transient overexpression of regulatory transcription factors can cause a transdifferentiation by activation of the cell's endogenous machinery to maintain the newly acquired phenotype. It should be noted, however, that the methods described do not lead to complete transdifferentiation to β-cells. At present, the numbers of cells that are trans-differentiated into insulin-producing cells and the amount of insulin production by these cultures are relatively low. Further refinement of the conditions will be necessary to achieve this goal.

4.2 Experimental Design

4.2.1 Generation of Insulin-Producing Cells from IHBECs

We have developed a five step-protocol for inducing the differentiation of IHBECs to insulin-producing cells. In step 1, the IHBEC-rich fraction is isolated from adult mouse liver, then in step 2 their number is expanded with CEFCM. At day 14 dispersed IHBECs are seeded onto collagen-coated dishes and cultured for 3 more days (step 3). For step 4, the media was changed to differentiation medium (DM) and the IHBECs were cultured for an additional 2 days (day17). We focused first on generating insulin-expressing cells by testing soluble factors. After 7 additional days (step 5) in DM control, IHBECs started to express some endocrine progenitor genes (Neurogenin 3(Ngn 3), NeuroD, Nkx6.1, and Pdx-1), but lacked insulin mRNA. Therefore, we decided to transduce the IHBECs with Ad-Pdx-1, -Ngn3, -Neuro D, and -Pdx-1/VP16 at day 19. At 7 days after transduction, the cells were evaluated (at day 26).

4.3 Materials

4.3.1 Reagents and Equipment

- C57/BL6 mice (Taconic Farms, Germantown, NY) adult 8-week male. *Note that all animal experiments must comply with national and institutional regulations.*
- L-Ascorbic acid 2-phosphate (Sigma-Aldrich Corporation, cat. no. A8960) (see Section 4.3.3).
- Collagenase (Roche, cat. no. 11 088 874 103).
- Dexamethasone (Sigma, cat. no. D2915) (see Section 4.3.3).
- DMEM/nutrient mixture Ham F-12 (DMEM/F12) (GIBCO Invitrogen Corpoation, cat. no. 11330-057).
- Nicotinamide (Sigma, cat. no. N0636).
- Epidermal growth factor (EGF, Fitzgerald, cat. no. RDI-115) (see Section 4.3.3).
- Hepatocyte growth factor (R&D systems, cat. no. 1025CE) (see Section 4.3.3).
- Forskolin (Calbiochem, cat no. 344270) (see Section 4.3.3).
- L-3,3,5-triiodo-L-thyromine (Calbiochem) (see Section 4.3.3).
- NuSerum IV (BD, cat. No.355104).
- MEM vitamin solution (GIBCO, cat no. 11120-052).
- MEM nonessential amino acids solution (GIBCO, cat no. 11140-050).

- GlutaMAX™-I supplement (GIBCO, cat no. 35050-061).
- Transferrin (GIBCO, cat no. 0030124SA) (already made) DAPT (1 μM).
- Betacellulin (R&D systems, cat no. 1025CE) (see Section 4.3.3).
- Insulin growth factor-1 IGF-I (R&D systems, cat no. 791-MG) (see Section 4.3.3).
- Excendin-4 (Sigma, cat no. E7144).
- EGTA (Sigma, cat. no. E-0396).
- Gentamicin solution (GIBCO, 15750-064).
- Collagen type 1 rat tail (Becton Dickenson (BD), cat. No. 354236).
- 1×HANKS Ca,Mg free (GIBCO, 14170-161).
- 1×HANKS with Ca, Mg (GIBCO, 24020-117).
- Wash solution (see Section 4.3.3).
- HEPES (Sigma, cat. no. H-3784).
- Insulin (Sigma, cat. no. I-5500) (see Section 4.3.3).
- Insulin-transferrin-selenium (ITS, GIBCO, cat no. 51500-056).
- AnaSed® INJECTION, Xylazine Sterile Solution 20 mg/mL (LLOYD Incorporated, NADA 139-236).
- Ketaset (ketamine hydrochloride) 100 mg/mL (Fort Dodge, NADA 45-290) (see Section 4.3.3).
- Nicotinamaide (Sigma, cat. no. N3376) (see Section 4.3.3).
- Penicillin-streptomycin solution (Cellgro, cat no. MT-30-002-CI).
- Gentamicin (GIBCO, cat no. 15750-064).
- Trypan blue (Fluka, cat no 93595).
- Preperfusion solution (see Section 4.3.3).

4.3.2 Equipment

4.3.2.1 Surgical Instruments

- Leica Microsystem (macroscope).
- One pair of sharp round-ended Mayo scissors.
- Two pairs of blunt curved forceps.
- Angiocath (Angiocath St 22GA X1IN BLUPK50) BD Falcon, cat. no. 381123).
- Surgical silk 3-0 (Ethicon Inc., cat. no. A54).
- Peristatic pump (Ranin Dynamax, Model RP-1).
- Silicon tube (NALGENE 50 Platinum-Cured Silicone Tubing, cat no. 8060-3016).
- T-shape stopcock.
- Tray.
- Rack.
- Water bath (Thermo Haake® immersion circulator, Aldrich, cat. no. Z320722).
- Autoclaved mesh 207um (Newark Wire Cloth company).
- Funnel (Hutzler, cat no. 04-0198).
- Cell strainer, 100, 40 um filter (BD Falcon, cat. no. REF352360, 352340).
- Dish 60 mm (Petri dish, Falcon, cat. no. 351007).

- Hemocytometer (Fisher, cat. No 026711).
- Plate Collagen I rat tail Multiwell plate 12 well (BD, cat. no. 354500).
- BD BioCoat Collagen I 35 mm Culture Dishes (BD, cat. no. 354456).

4.3.3 Reagent Setup

- *1,000 X insulin stock solution (500 ug ml^{-1}).* Add 100 mg insulin to 100 ml ddH_2O and then add 1.2 ml 1 N HCl. Adjust to 200 ml with ddH_2O and filter with a 0.2-um filter. *Critical: Insulin dissolves in acidic solution.*
- *Preperfusion solution.* To 500 ml Ca^{2+}, Mg^{2+}-free Hanks' solution, add 2 ml penicillin-streptomycin solution, 85 mg EGTA, 0.5 ml 1000×insulin stock solution, 0.5 ml dexamethasone stock solution. Distribute 50 ml into each 50 ml Falcon tube. Store at 4°C until use. *Critical: EGTA solution should be warmed to 37°C before use.*
- *Perfusion solution.* To 500 ml Hanks' with Ca, Mg solution, add 2 ml 0.5 ml insulin stock solution, 0.5 ml dexamethasone stock solution. Distribute 50 ml into each 50 ml Falcon tube. Store at 4°C until use. To 50 ml Hanks' solution add collagenase (50 mg). Shake gently. *Critical: Prepare Hanks' with insulin before the experiment. Add collagenase to prewarmed perfusion solution just before use and immediately dissolve by gentle shaking.*
- *Wash solution.* To 500 ml Hanks' with Ca, Mg solution, add 0.5 ml 1,000×insulin stock solution, 2 ml penicillin–streptomycin solution, and 0.5 ml gentamicin solution.
- *Dexamethasone stock solution.* Add 39.2 mg dexamethasone to an autoclaved brown bottle. Add 10 ml ethanol (10^{-2} M stock solution); dilute to 100×(10^{-4} M) with autoclaved ddH_2O. Store at 4°C until use. *Critical: 10^{-4} M stock solution must be used within 3 months.*
- *EGF stock solution (10 ug ml^{-1}).* Add 10 ml autoclaved ddH_2O to an EGF vial. Store 1-ml aliquots in microcentrifuge tubes at –20°C until use.
- *HGF stock solution (10 ug ml^{-1}).* Add PBS containing at least 0.1% bovine serum albumin to the vial to prepare a stock solution of no less than 10 μg/mL. Store 1-ml aliquots in microcentrifuge tubes at –20°C until use.
- *Betacellulin stock solution (10 ug ml^{-1}).* Same as HGF.
- *IGF-1 stock solution (10 ug ml^{-1}).* Same as HGF.
- *Exendin-4 solution (25 ug ml^{-1}).* Same as HGF.
- *Forskolin (4 μg ml^{-1}).* Dissolve 10 mg Forskolin to 1 ml dimethyl sulfoxide (DMSO).Up to 2.5 ml by DMSO. Store 0.5-ml aliquots in microcentrifuge tubes at –20°C until use.
- *L-3,3',5-triiodothyronine.* Storage: refrigerate (+4°C). *Protect from light.* This product is stable for 3 years as supplied; reconstitute just prior to use. 1.7 mg L-3,3',5-triiodothyronine is soluble in 100 mM NaOH.
- *Icotinamide stock solution (1M).* Add 12.21g nicotinamide to 100 ml PBS. Filter with a 0.2-um filter and store at 4 °C until use.
- *Asc2P stock solution (100 mM).* Add 2.90g Asc2P to 100 ml PBS, filter with a 0.2-um filter and store at 4°C in a 100-ml brown bottle until use.

- *Preparation of Anesthesia.* Add 3.1 ml of AnaSed (xylazine) to 10 ml of Ketaset (ketamine hydrochloride).
- *Preparation of growth medium.* 500 ml DMEM/F12, nicotinamide stock solution (5.50 ml), Asc2P stock solution (5 ml), ITS (5 ml), EGF stock solution (0.5 ml), dexamethasone stock solution (0.5 ml), and gentamicin (0.5 ml), FBS (25 ml), NuSerum IV (25 ml), forskolin (0.5 ml) 3,3',5-triiodo-L-thyromine (1.7mg). This medium is a modification of that described by Alpini et al [19, 25], which we term growth medium (GM); it was changed every 2 days. With culture the collagen gel containing the aggregates gradually contracted.
- *Preparation of differentiation medium (DM).* This medium is a serum- and insulin-free differentiation medium. 500 ml DMEM/F12, nicotinamide stock solution (5.50 ml), Asc2P stock solution (5 ml), and gentamicin (0.5 ml), MEM vitamin solution (5 ml), MEM nonessential amino acids solution (5 ml), GlutaMAX™-I supplement (5 ml), transferrin (0.65 ml), betacellulin stock solution (0.5 ml), IGF-I stock solution (0.5 ml), and exendin-4 stock solution (0.5 ml).
- *Equipment setup; perfusion system.* The system is composed of a water bath, peristatic pump, silicon tube ($\varphi 4.76 \times 7.94\ mm^2$) macroscope, angiocath, and surgical silk 3-0.

4.3.4 Adenovirus Transduction

1. Preparation of recombinant adenoviruses: Recombinant adenovirus expressing Pdx-1, -Ngn3, -NeuroD, and -Pdx-1/VP16 driven by a cytomegalovirus (CMV) promoter were prepared with the AdEasy system as previously described [25]. These AdVs all carried the reporter GFP. The control adenovirus (Ad-GFP) was prepared in the same manner. The integration of each gene into the adenovirus was done by transfection into the adenovirus packing cell line 293 cells according to the manufacturer's instructions.
2. Adenovirus titers were increased up to 1×10^{10} plaque forming units (PFU)/ml with Vivapure AdenoPACK 100 purification kits (Vivascience, cat no. VS-AVPQ101).
3. Adenoviruses of known titer, measured as the number of infectious units per ml (I.U./ml), are prepared. The viruses should always be stored at –80°C until required.
4. Serum-free DMEM.
5. 10% sodium dodecyl sulfate (SDS) and 70% ethanol (disinfection).

4.3.5 Reverse Transcription Polymerase Chain Reaction (RT-PCR)

1. For RNA extraction: RNeasy Mini Kit (Qiagen, cat. no. 74104).
2. Spectrophotometer.
3. For reverse transcription: Omniscript RT Kit (Qiagen, cat. no. 205111). Ribonuclease inhibitor, Cloned (RNase inhibitor, Invitrogen. cat. no. 15518-012 and Oligo $(dT)_{12\text{-}18}$ Primer (Invitrogen. cat. no. 18418-012).
4. For the PCR: PCR ReddyMix™ Master mix (Thermo Scientific, cat. no. AB-0575/LD/a). The forward and reverse primers for the genes of interest are diluted to 1 μg/μl in UltraPure™ DEPC-Treated Water (Invitrogen, cat. no. 750024).

4.3.6 Immunocytochemistry

1. For fixation: cold absolute ethanol.
2. For blocking: 10% goat serum in PBS.
3. For dilution of antibody: 1% bovine serum albumin (BSA).
4. Mounting media: propidium iodide (PI) or 4′, 6-diamino-2-phenylindole (DAPI) (VECTASHIELD Mounting Medium with PI, cat. no. H-1300, H-1200).
5. Primary antibodies (Table 4.1) and appropriate fluorescent secondary antibodies. All antibodies are diluted in the 1% BSA for use.

4.3.7 Measurement of Insulin Production

Ultrasensitive insulin ELISA (Crystal Chem, Inc., cat. no. 90080).

4.3.8 Analysis of Single Cells by RT-PCR

See our findings in [21].

4.4 Methods

4.4.1 Step 1

The protocol used for mouse IHBECs isolation is based on a modification of the technique previously described by Mitaka and our colleagues [16, 25, 26]. There are a number of excellent papers on the subject of adenovirus preparation [23, 24]; hence it will not be discussed here. The success of an adenoviral infection can be followed by the detection of various reporters incorporated into the transgene construct. The novel pancreatic genes expressed as a consequence of the introduction of the transgene can be detected by RT-PCR and single cell PCR. Detailed instructions for the fluorescent RT-PCR and immunocytochemistry protocols will be given below. ELISA and single-cell PCR information are available in our recent publication [21].

4.4.2 Mouse Biliary Epithelial Cell Isolation and Culture

1. Preperfusion solution and perfusion solution should be prewarmed in a water bath to 37°C before use.
2. Dissolve collagenase (Roche) in 50 ml perfusion solution just before use (final concentration; 0.04%). *Note: The weight of collagenase is only a guide. It is crucial to batch test collagenase lots from different suppliers.*

Table 4.1 Antibodies for Use in Detecting Pancreatic Marker Expression

Antigen	Supplier	Dilution	Species
NeuroD	Cell Signaling Technology	1:200	Rabbit
Pdx-1(IPF-1C)	Dr. J.M.W.Slack, University of Minnesota	1:1000	Rabbit
C-peptide	Cell Signaling	1:200	Rabbit
Insulin	Linco	1:200	Guinea pig
Cytokeratin 7	Dako	1:25	Mouse

3. Prepare surgical instruments and perfusion apparatus comprising a peristaltic pump and tubing.
4. Inject anesthesia. Note: for an 8-wk-old mouse, 0.05 ml of mixture would be enough. The operation should be performed in living mice. The yields of the cells from dead mice are extremely low.
5. Cannulation and perfusion. General two-step liver perfusion is performed [16, 21, 25, 26].
6. Settle the perfusion apparatus in the warmed water bath and pour the preperfusion solution into the apparatus.
7. After light anesthesia by mixed anesthesia solution, anesthetize a mouse with an i.p. injection of nembutal (0.5 ml).
8. Cut the abdominal wall using surgical scissors and open the abdominal cavity to look at the portal vein.
9. Insert an angiocath filled with preperfusion solution into the portal vein 1.5–2.0 cm from the bifurcation of the portal vein; stop the tip of the angiocath at a position close to the bifurcation of the portal vein and ligate the angiocath with surgical silk 3-0 tightened around fat tissue.
10. Start the perfusion at a flow rate of 15 ml min^{-1}.
11. Cut the inferior vena cava and heart as soon as the flow starts. *Critical step: Cut the inferior vena cava at the portion beneath the right kidney and the thoracic cavity, and then cut the heart to allow flow of the perfusate out of the cadaver. Washing out the blood completely from the liver and preventing an increase of intrahepatic pressure is important to the success of the preparation.*
12. Continue perfusion until 100 ml of the preperfusion solution has run through. When the amount of the preperfusion solution becomes small, add 50 mg of collagenase in 50 ml perfusion solution and then pour it into the perfusion apparatus. *Critical step: To avoid a decrease in collagenase activity, dissolve it after the preperfusion solution flows. Do not mix vigorously.*
13. Have the solution flow at a rate of 7 ml min^{-1}. *Critical step: The flow rate should be decided by the perfusion condition of the liver.*
14. Stop the flow before air bubbles move into the liver when the solution flows out from the reservoir.
15. Remove the liver from the abdominal cavity and transfer it to a sterilized Petri dish.
16. Prepare a 100-ml beaker with 50-ml wash solution and add a small amount of the wash solution to the Petri dish. *Critical step: From this step onward, all procedures should be done in sterilized conditions under cabinet.*
17. Trim the gall bladder and connective tissue carefully. Peel the hepatic capsule as carefully as possible; to drop the digested cells, shake the liver gently into the beaker. Transfer the remnant tissue in the flask with wash solution. *Critical step: This treatment serves to remove residual hepatic parenchymal cells from biliary tissue fragments.*
18. The remnant tissues are collected and minced. Transfer to new conical tubes.
19. Centrifuge at 150g for 1 min at 4°C and discard the supernatant.
20. Transfer the tissue into a flask with perfusion solution, and then treat with 0.04% collagenase solution for 7 min. *Critical step: This procedure is carried out to remove damaged hepatocytes contained in the suspension and make a small aggregate of IHBECs.*

Quantity of collagenase: there are differences between lots and suppliers, so adjust the quantity of collagenase by units, not percentage.

21. Centrifuge at 150g for 1.5 min at 4°C. Change the media to wash solution.
22. Filter the cell suspension through a 250-μm nylon filter net into a new 100-ml beaker using a 25-ml pipette and then adjust each tube to an equal volume with the wash solution (approximately 40 ml).
23. Centrifuge at 150g for 1.5 min at 4°C and discard the supernatant.
24. Resuspend the pellet in the wash solution, and filter sequentially through 100-μm and then 40-μm mesh.
25. Retrieve small cell aggregates on the 40-μm mesh.
26. Suspend the aggregates with GM.
27. Centrifuge at 150g for 1.5 min at 4°C and discard the supernatant. Gather the suspension into one tube.
28. Add 0.5 ml cell suspension to 1.5 ml trypan blue solution and pipette gently. Critical step: Tubes should be on ice. Count the viable cells as soon as trypan blue is added. For counting, use a hemocytometer. Count the number of cells with trypan blue–negative nuclei, which exist in nine masses. *Critical step: Cells with trypan blue–positive nuclei are dead.*
29. Typical recovery is 1×10^5 viable cells with essentially pure biliary tissue obtained.

4.4.3 Step 2

4.4.3.1 Collagen-Embedded Floating Culture Method (CEFCM)

To expand the number, the cells are cultured in our original culture method for 14 days.

1. Suspend the aggregates in GM on ice with *0.06 mg/mL collagen* (Collagen type I rat tail, BD). *The volume of cell suspension is 10 ml. There is no need to neutralize this mixture.* Then, plate 1.0×10^4 cells/ml/well in 12-well dishes (Figure 4.1(a), left). The collagen-containing aggregates are allowed to solidify at 37°C for 2h.
2. Place dishes in a 5% CO_2/95% air incubator at 37°C.
3. Pour 1 ml growth media gently on the collagen gel containing aggregates.
4. Renew medium every 2 days. *Critical step: Two times a week is enough when the number of expanded cells is small.*
5. With culture, the collagen gel containing the aggregates gradually contracted. After about 5 days with CEFCM, IHBECs formed three-dimensional ductal cysts (Figure 4.1(a), middle) and rapidly expanded their number about 15-fold within 2 wk. In cultures without CEFCM (CEFCM-), IHBECs initially expanded but then started to die at around 8 days.

4.4.4 Step 3

4.4.4.1 Monolayer Culture for IHBECs

1. After 14 days in CEFCM culture, the aggregates were washed in Ca^{2+}-free HEPES-buffered saline, then a 0.1% collagenase solution (Roche) for 10 min to dissolve the collagen gel and release the aggregated IHBEC fragments.

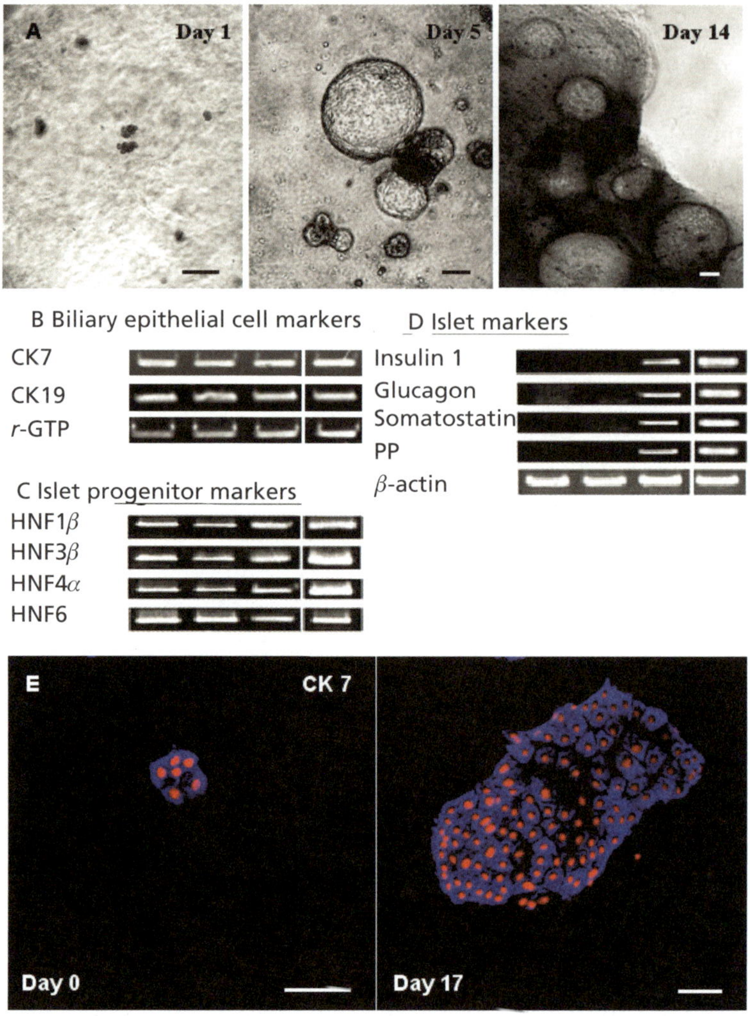

Figure 4.1 IHBECs cultured with CEFCM. (a) Nomarski differential interference contrast (DIC) image of cultured IHBECs. Isolated IHBECs were suspended in rat tail collagen plus GM. Left: Day 0 with CEFCM; very small aggregates of IHBECs. Middle: After 5 days with CEFCM, IHBECs formed three-dimensional ductal cysts. Right: IHBECs at day 14 with CEFCM. The collagen gel containing the IHBECs gradually contracted with time. Scale bar = 100 μm. (b) Cells in all groups expressed BEC markers. (c) IHBECs expressed a number of HNF family genes considered to be islet progenitor markers [12]. (d) Insulin gene expression induced in IHBECs as a consequence of adenovirus expressing the pancreatic transcription factor Pdx-1/VP16. Gene expression studies in all groups were repeated at least three times independently. (b–d) Gene expression profiles of IHBECs. RT-PCR was carried out to clarify the character of IHBECs at day 0, day 17, and at 7 days after transduction with Ad- Pdx-1/VP16 (day 26). BEC markers (cytokeratin [CK] 7 and 19, β-glutamyltranspeptidase [γ-GTP], hepatocyte nuclear factor (HNF) family genes (HNF1β, HNF3β, HNF4β, HNF6), and islet markers (insulin 1, glucagon, somatostatin, and pancreatic polypeptide [PP]). Lane 1: IHBECs day 0; lane 2: IHBECs day 17; lane 3: at 7 days after transduction with Ad- Pdx-1/VP16 (day26); and lane 4: control. β-Actin controls are also shown. Mouse liver for (b) and islets for (c) were used as controls.(e) Immunofluorescent analysis for CK 7 in IHBECs. IHBECs were positive for both for CK 7 (blue) at days 0 (left) and 17 (right). Propidium Iodide (PI, red) was used to show nuclei. The cells were positive for both for CK7 (Blue) and CK19 (data not shown). Scale bar = 50 μm.

2. The dispersed cells were counted on the hemocytometer with trypan blue, allowed to settle, then resuspended in GM, plated onto collagen-coated, 12 well dishes (BD) and maintained at 37°C in a humidified 5% CO_2 incubator. (For carrying out immunocytochemistry, we always use 35-mm collagen-coated dishes.)

4.4.5 Step 4

4.4.5.1 Change to DM

1. After 3 days in GM, the media was changed to DM. These IHBECs formed colonies and expanded about 2.4-fold during these 3 days. The purity of IHBECs averaged >90% at day 17 as assessed by immunostaining for CK7 (Figure 4.1(e)) and CK19 (data not shown).

4.4.6 Step 5

4.4.6.1 Adenovirus Transduction

1. Based on the virus titer and the number of cells per dish, calculate the volume of virus required to infect at a total multiplicity of infection (MOI) of 1–200.
2. In a class II biosafety cabinet, add the appropriate amount of virus made up in serum-free DMEM, to each dish and then transfer the dishes to a 37°C incubator for infection to take place.
3. After 1h, replace the virus-containing culture medium in each dish with fresh DM. Discard any virus-contaminated solutions into a bottle of surface active agent (surfactant) and dispose accordingly.
4. After 48h, the success of infection can be visualized based on GFP reporters (Figure 4.2(a)).
5. Transduced IHBECs with Ad- Pdx-1/VP16.MOI of 100 MOI for Pdx-1/VP16 with recombinant adenoviruses resulted in efficient expression of the transgenes with 30% to 50% transduction efficiency and high cell survival (Figure 4.2(b)). If we increased the MOI from 100 to 200, transduction efficiency is increased up to around 60%. However, the number of transduced Ad-Pdx-1/VP16 with IHBECs sharply dropped from 3 days after transduction. By about 9 days after transduction, almost no cells are seen in the dishes.

4.4.7 Evaluation

4.4.7.1 RT-PCR

1. Total RNA was isolated from cultured cells at day 26 using RNeasy Mini Kits according to manufacturer-suggested protocols, and RT-PCR is conducted.
2. The extracted total RNA was reverse-transcribed into cDNA with Omniscript RT Kit using standard procedures.
3. For the PCR: PCR ReddyMix™ Master mix (AbGene) is used. Run the PCR with 1 μl of the cDNA and 500 ng of the forward and reverse primers in 20 μl of the PCR mastermix. β-Actin can be used as a good control gene for hepatocytes (Figure 4.1(b)).

4.4.7.2 Immunocytochemistry

1. Remove the culture medium from the 35-mm dishes and wash the cells with PBS.
2. Block in 10% goat serum for 1 to 2h to minimize any unspecific binding of antibodies.

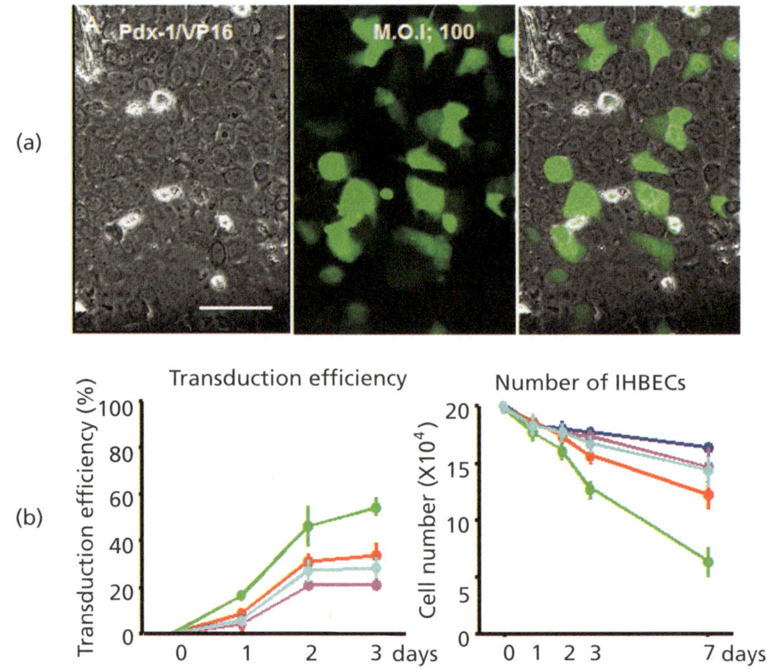

Figure 4.2 IHBECs transduced with Ad- Pdx-1/VP16. (a) Adenovirus transduction was carried out at day 19 for the monolayer cultured IHBECs. An MOI of 100 for Pdx-1/VP16 led to 30%–50% transduction efficiency and high cell survival rate. Pdx-1/VP16 coexpressed GFP as a reporter. Cells that appear white in Nomarski differential interference contrast (DIC) image are apoptotic. Scale bar = 100 μm. The success of an infection can be determined by following the expression of reporters such as GFP. (b) Transduction efficiency and number of IHBECs. High MOI resulted in toxicity and loss of cells. Blue line: MOI 0; pink line: MOI 25; light blue line: MOI 50; red line: MOI 100; green line: MOI 200.

3. Add the primary antibody diluted in PBS containing 1% BSA and incubate at 4°C overnight.
4. Wash off the primary antibody (3×15 min) with PBS. Add the secondary antibody diluted to appropriate dilution (see Table 4.1) in PBS containing 1% goat serum and incubate at room temperature in the dark for 1 to 2h.
5. In the dark, wash off the secondary antibody (3×15 min) with PBS.
6. Mount the coverslips onto dishes using a mounting media.
7. Observe under a fluorescent microscope (Figure 4.3). *Critical step: Apoptotic cells can take up exogenous insulin from the culture medium [27]. Evaluate the insulin positive cells carefully.*

4.4.8 Statistical Analysis

Experimental results were expressed as the mean ± standard error of the mean (mean ± SEM). Student's *t*-test, ANOVA, and Fisher's protected least significant difference test were used, and a p value <0.05 was considered significant. All experiments were repeated at least three times.

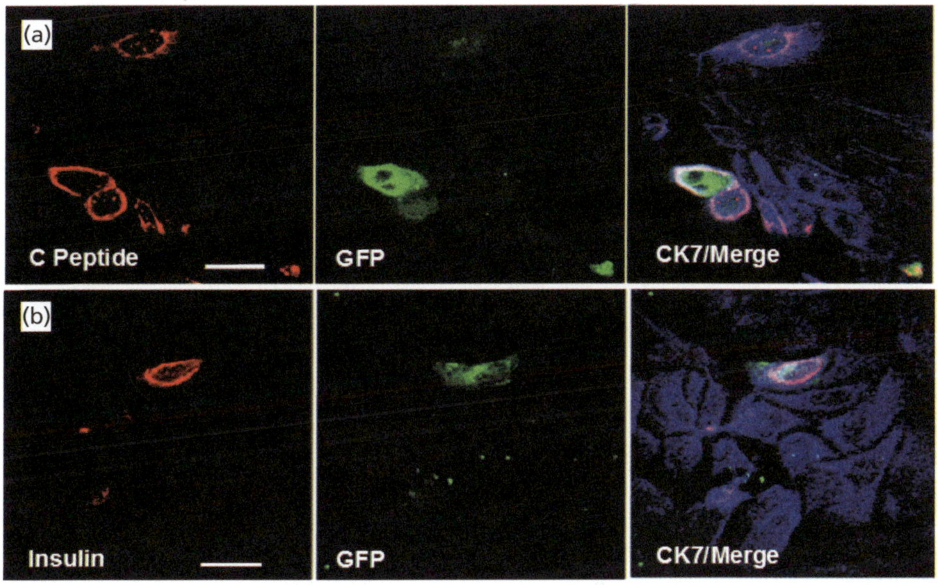

Figure 4.3 Immunofluorescence analysis of Ad- Pdx-1/VP16 transduced IHBECs at 7 days after transduction (day 26). (a) C-peptide staining for Ad- Pdx-1/VP16 transduced IHBECs 7 days after transduction. Triple staining was for C-peptide (red), GFP (green), and CK 7 (blue). Of 4,000 cells counted, 2.8% in Pdx-1/VP16 transduced cultures stained strongly for C-peptide. The 4,000 counted cells were from five independent experiments. (b) Insulin staining for Ad- Pdx-1/VP16 transduced IHBECs 7 days after transduction. Triple staining was for insulin (red), GFP (green), and CK 7 (blue). Scale bars = 50 μm.

4.5 Data Acquisition, Anticipated Results, and Interpretation

To avoid contamination of the preparation with hepatocytes, timing of the digestion is critical. If the digestion time is too short, there will be more hepatocyte contamination, and if the digestion is too long, the IHBECs can be killed. The concentration of collagenase we described is only a guide. Perform batch testing of different collagenase lots from various suppliers. It is necessary to carefully check each step: perfusion, duration of stirring, and time for digestion, to obtain the right balance between hepatocyte contamination and IHBEC damage. If the cystic structure has not appeared within 5 days with CEFCM, it is recommended to start over.

4.6 Discussion and Commentary

The first report of transdifferentiation of liver cells into insulin-producing cells was by the group of Ferber [7]. In vivo transduction of mice with an adenovirus containing Pdx-1 induced expression of endogenous Pdx-1, other β-cell markers and some insulin. With a similar approach using injections of helper-dependent adenoviruses, expression of both betacellulin (a mitogen) and NeuroD (a β cell transcription factor) led to reversal of streptozocin diabetes [8]. With these two studies, however, the origin of these cells in the liver was never determined and there is still uncertainty about how much insulin was actually being produced by liver and how much β-cell-like differentiation had occurred. The question of which liver cell adopts the insulin-producing phenotype remains unanswered. To clarify these questions, we chose IHBECs to determine whether these cells could differentiate into insulin-producing cells.

In this chapter, we describe our technique for isolating IHBECs, expanding the cells, and inducing insulin-producing cells. We characterized the differentiation process of transduced IHBECs at the RNA level using RT-PCR, and at protein levels with immunofluorescence and ELISA.

Initial experiments with IHBECs at steps 1 and 2 were performed to determine their behavior in tissue culture and their potential for transdifferentiation. RT-PCR and immunofluorescent staining were carried out with IHBECs (Figure 4.1). Cells at day 0 expressed a number of hepatocyte nuclear factor (HNF) family genes, (HNF1β, HNF3β, HNF4β, HNF6) considered to be endoderm progenitor markers [11] and but did not express islet markers (Figure 4.1(b) and (d)).

The cells were also positive for both for CK7 (blue, Figure 4.1(e), left) and CK19 (data not shown). Then, we developed a culture system which is able to expand the number of IHBECs. This new culture method of CEFCM makes it possible to obtain significant numbers of healthy cells (Figure 4.1(e), right). The advantage of this method is not only induction of proliferation but also prevention of cell death compared with flat culture [21]. In the end, we think it likely that the three-dimensional culture conditions are important to promote the cell expansion.

At day 14, dispersed IHBECs were seeded onto collagen-coated dishes and cultured for 3 more days (Step 3). These IHBECs formed colonies and expanded about 2.4-fold during these 3 days. The monolayers of homogenous appearing cells maintained their BEC character even at day 17 (Figure 4.1(d), right) and proliferated efficiently in culture. The purity of IHBECS averaged >90% at day 17 as assessed by immunostaining for CK7 and CK19.

To analyze the molecular events occurring in IHBECs during the series of culture steps, gene expression profiles of transcription factors and pancreas-related genes were determined by RT-PCR. IHBECs at Step 1 only expressed biliary epithelial cell markers (Figure 4.1(b)) and islet progenitor markers (Figure 4.1(c)). From step 4, our serum- and insulin-free DM was used in hopes of priming the cells for transdifferentiation.

We assume the key for priming the cells for transdifferentiation was the combined use of a two-dimensional culture environment and DM to accelerate differentiation pathways toward pancreatic fate. Indeed, these cells began to express the pancreatic progenitor markers NeuroD, Ngn3, Nkx6.1, Pdx-1, and Ptf1β at Step 4. These genes were not observed in cultured IHBECs at day 14; therefore, the IHBECs assumed these characteristics in DM culture. However, this failed to induce insulin gene expression. Therefore, we added the transduction by the adenoviral transgene process. Seven days after transduction with Ad-Pdx-1, NeuroD, or Pdx-1/VP16, insulin expression was found by RT-PCR, immunofluorescence and ELISA. Transduced expression of Pdx-1, NeuroD, or Pdx-1/VP16 led to expression of not only both insulins but also GLUT2 and prohormone convertases 1 and 2 (PC1 and 2) (day 26). Pdx-1 or Pdx-1/VP16 transduced cells faintly expressed amylase [21]. Neither Ad-Ngn3 nor Ad -Pax4 led to expression of insulin; therefore, their use was discontinued.

Pdx-1/VP16 transduction led to the appearance of endocrine precursor markers such as Neuro D and Pdx-1 and islet related markers by immunofluorescence analysis. C-peptide and insulin staining confirmed the synthesis of insulin by IHBECs (Figure 4.3). We found that 30.1% and 2.8% of 4,000 cells counted in Pdx-1/VP16 transduced cultures

Troubleshooting Table

Problem	Explanation	Potential Solutions
Reagent bacterial or viral contamination or lost activity.	Storage conditions.	Stock solution is kept at 4°C. To avoid bacterial contamination, a small amount of the stock solution should be prepared for several experiments. Avoid long-term storage (more than 1 month).
Quantity of collagenase.	Difference between lots and between suppliers.	Adjust the quantity of collagenase by units, not percentage.
		The weight of collagenase we described is only a guide. Perform the batch test of collagenase lots as well as those from different suppliers.
Poor yield or viability, and cell aggregation cavity causes the early rupture of hepatic capsules and inadequate digestion.	Optimize the anesthesia.	Light anesthesia is crucial. Avoid the death of the animal due to a high dose of anesthesia.
	Collagenase.	Collagenase activity is best at 37°C. Long-term storage (more than 1 yr) causes loss of activity.
	Length of perfusion time.	Although collagenase works efficiently at a slow flow rate, too slow a flow causes partial perfusion of the liver, and the yield and viability of the cells become worse. A rapid change of liver color from red-brown to ocher is a sign of good circulation.
	Mince.	Mince the cells collectively. Use dull shape scissors instead of blade.
	Stir.	The weight of collagenase for further digestion in beaker and time we described are only a guide. Optimize them.
Poor growth of IHBECs in CEFCM.	Few IHBEC aggregates.	Avoid plating too small a number of cells. We found that fewer than 1×104 cells/well/12 well plate resulted in a smaller number of IHBEC colonies than we expected.
	Disrupt the collagen.	If you cannot see cystic structure in CEFCM until 5 days after starting the culture, the possibility of cell growth is quite low. Perform another isolation.
		Carry out the media change very gently until the cells make cystic formation.
Poor yield, viability, and/or cell growth of IHBECs in monolayer culture.	Insufficient dissolution of collagenase.	Wash the collagen embedded aggregates in Ca2+-free HEPES-buffered saline several times, and then immerse it in 0.1% collagenase solution. Incubate the embedded aggregates for 10 min.
		Carry out pipetting several times gently with pipette.
	Few IHBEC colonies.	Avoid plating too small a number of cells. We found that fewer than 1×105 cells/ well/ 12 well plate resulted in a smaller number of IHBEC colonies than we expected.
Low titer, transduction efficiency of adenovirus.	Thawing and freezing.	Thawing and freezing the virus will reduce its titer. Make aliquots.
	No purification.	We recommend high purity and high titer for adenovirus transduction.
		For viral purification, we recommend using an appropriate purification kit, which yields higher titers.
	Transduction.	Viruses of lower titer will result in little induction of insulin+ cells in our hands. The MOI for an infection is the theoretical mean number of infectious particles per cell. A low MOI will result in inefficient gene delivery. We found that MOIs of 100–200 resulted in toxicity and loss of cells.

were GFP⁺ and C-peptide⁺, respectively. Although the amount of insulin was low, the cells also released insulin into the culture medium by 48h incubation at day 7 after transduction [21].

To further characterize these cells, we examined islet hormone gene expression in single cells that had been transduced with Pdx-1/VP16 in our recent publication. About 23% of GFP⁺ of Pdx-1/VP16 transduced cells express the insulin gene and 91% of insulin-positive single cells expressed all four islet hormone genes [21].

Data is only presented from the experimental conditions described in Section 4.4.. We have also used various doses and combinations of other agents including nicotinamide, betacellulin, activins, hepatocyte growth factor, and exendin-4, but these efforts have not provided better results. We have not yet defined the optimal conditions for IHBEC-derived insulin-producing cells; therefore, we have only included the approach in the chapter that produced the results we reported [21].

4.7 Summary Points

We have generated insulin-expressing cells, but the cells continued to express biliary epithelial cell markers. In addition, it is known that fetal islets also express all four islet hormone genes and respond to multiple secretory stimuli but not to glucose [28, 29]. We suspect that the insulin-expressing cells we have generated are similar to immature fetal â-cells. Future research will focus on further definition of the conditions that promote differentiation of insulin-producing cells to mature pancreatic β cells.

Acknowledgments

We thank Drs. Susan Bonner-Weir, Hitoshi Katsuta, and Hideaki Kaneto for their assistance. We are grateful to Dr. Atsushi Miyajima (Tokyo University, Tokyo, Japan) and Dr. J. M. W. Slack, (University of Minnesota, MN) for the generous gifts of the rabbit anti-CK19 antibody and rabbit anti-Pdx-1 (IPF-1C antibody), respectively. This research was supported by grants from the Juvenile Diabetes Research Foundation, the Diabetes Research and Wellness Foundation, and an important group of private donors. Help was also provided by the Advanced Microscopy, Flow Cytometry and Specialized Assay Cores of the National Institutes of Health supported Diabetes and Endocrinology Research Center (DERC, DK 36836) of the Joslin Diabetes Center. The authors are filing a patent based on the results reported in our recent publication [21].

References

[1] Shapiro, A. M., et al., "Islet Transplantation in Seven Patients with Type 1 Diabetes Mellitus Using a Glucocorticoid-Free Immunosuppressive Regimen," *N. Engl. J. Med.,* Vol. 343, No. 4, 2000, pp. 230–8.

[2] Ricordi, C., and T. B. Strom, "Clinical Islet Transplantation: Advances and Immunological Challenges," *Nat. Rev. Immunol.,* Vol. 4, No. 4, 2004, pp. 259–68.

[3] Bonner-Weir, S., et al., "In Vitro Cultivation of Human Islets from Expanded Ductal Tissue," *Proc. Natl. Acad. Sci. USA,* Vol. 973, No.14, 2000, pp. 7999–8004.

[4] Gershengorn, M. C., et al., "Epithelial-to-Mesenchymal Transition Generates Proliferative Human Islet Precursor Cells," *Science,* Vol. 306, No. 5705, 2004, pp. 2261–64.

[5] Kroon, E., et al. "Pancreatic Endoderm Derived from Human Embryonic Stem Cells Generates Glucose-Responsive Insulin-Secreting Cells In Vivo," *Nat. Biotechnol.,* Vol. 26, No. 4, 2008, pp. 443–52.

[6] Tateishi, K., et al., "Generation of Insulin-Secreting Islet-Like Clusters from Human Skin Fibroblasts," *J. Biol. Chem.,* Vol. 283, No. 46, 2008, pp. 31601–07.

[7] Ferber, S., et al., "Pancreatic and Duodenal Homeobox Gene 1 Induces Expression of Insulin Genes in Liver And Ameliorates Streptozotocin-Induced Hyperglycemia," *Nat. Med.,* Vol. 63, No. 5, 2000, pp. 568–72.

[8] Kojima, H., et al., "NeuroD-Betacellulin Gene Therapy Induces Islet Neogenesis in the Liver and Reverses Diabetes in Mice," *Nat. Med.,* Vol. 9, No. 5, 2003, pp. 596–603.

[9] Sapir, T., et al., "Cell-Replacement Therapy for Diabetes: Generating Functional Insulin-Producing Tissue from Adult Human Liver Cells," *Proc. Natl. Acad. Sci. USA,* Vol. 102, No. 22, 2005, pp. 7964–69.

[10] Tang, D. Q., et al., "Reprogramming Liver-Stem WB Cells into Functional Insulin-Producing Cells by Persistent Expression of Pdx1- and Pdx1-VP16 Mediated by Lentiviral Vectors," *Lab. Invest.,* Vol. 86, No. 1, 2006, pp. 83–93.

[11] Wilson, M. E., Schell, D. and German, M. S., "Gene expression cascades in pancreatic development," *Mech. Dev.,* Vol 120, No. 1, 2003, pp. 68–80.

[12] Lemaigre, F., and K. S. Zaret, "Liver Development Update: New Embryo Models, Cell Lineage Control, and Morphogenesis," *Curr. Opin. Genet. Dev.,* Vol. 14. No. 5, 2004, pp. 582–90.

[13] Zaret, K. S., "Genetic Programming of Liver snd Pancreas Progenitors: Lessons for Stem-Cell Differentiation," *Nat. Rev. Genet.,* Vol. 9. No. 5. 2008, pp. 329–40.

[14] Antolovic, D., et al., "Hepaticojejunostomy—Analysis of Risk Factors for Postoperative Bile Leaks and Surgical Complications," *J. Gastrointest. Surg.,* Vol. 11, No. 5, 2007, pp. 555–61.

[15] Michalopoulos, G. K., and M. C. DeFrances, "Liver Regeneration," *Science,* Vol. 276, No. 5309, 1997, pp. 60–6.

[16] Nagaya, M., et al., "Thermoreversible Gelation Polymer Induces the Emergence of Hepatic Stem Cells in The Partially Injured Rat Liver," *Hepatology,* Vol. 43, No. 5, 2006, pp. 1053–62.

[17] Alvaro, D., et al., "Effect of Secretion on Intracellular pH Regulation in Isolated Rat Bile Duct Epithelial Cells," *J. Clin. Invest.,* Vol. 92, No. 3, 1993, pp. 1314–25.

[18] Alpini, G., et al., "Functional Expression of the Apical Na+-Dependent Bile Acid Transporter in Large but Not Small Rat Cholangiocytes," *Gastroenterology,* Vol. 113, 1997, pp. 1734–40.

[19] Alpini, G., et al., "Development and Characterization of Secretin-Stimulated Secretion of Cultured Rat Cholangiocytes," *Am. J. Physiol. Gastrointest. Liver Physiol.,* Vol. 284, No. 6, 2003, pp. G1066–73.

[20] Grünert, S., M. Jechlinger, and H. Beug, "Diverse Cellular and Molecular Mechanisms Contribute to Epithelial Plasticity and Metastasis," *Nat. Rev. Mol. Cell. Biol.,* Vol. 4, No. 8, 2003, pp. 657–65.

[21] Nagaya, M., et al., "Adult Mouse Intrahepatic Biliary Epithelial Cells Induced In Vitro to Become Insulin-Producing Cells," *J. Endocrinol.,* Vol. 201, No. 1, 2009, pp. 37–47.

[22] Zhou, Q., et al., "In Vivo Reprogramming of Adult Pancreatic Exocrine Cells to Beta-Cells," *Nature,* Vol. 455, No. 7213, 2008, pp. 627–32.

[23] Stadtfeld, M., et al., "Induced Pluripotent Stem Cells Generated without Viral Integration," *Science,* Vol. 322, No. 5903, 2008, pp. 945–9.

[24] Kaneto, H., et al., "PDX-1/VP16 Fusion Protein, Together with NeuroD or Ngn3, markedly Induces Insulin Gene Transcription and Ameliorates Glucose Tolerance," *Diabetes,* Vol. 54, No. 4, 2005, pp. 1009–22.

[25] Chen. Q., et al., "Selective Proliferation of Rat Hepatocyte Progenitor Cells in Serum-Free Culture," *Nat. Protoc.,* Vol. 2, No. 5, 2007, pp. 1197–205.

[26] Hashimoto, W., et al., "Ductular Network Formation by Rat Biliary Epithelial Cells in the Dynamical Culture with Collagen Gel and Dimethylsulfoxide Stimulation," *Am. J. Pathol.,* Vol. 173, No. 2, 2008, pp. 494–506.

[27] Rajagopal, J., et al., "Insulin Staining of ES Cell Progeny from Insulin Uptake," *Science,* Vol. 299, No. 5605, 2003, pp. 363.

[28] Teitelman, G., et al., "Precursor Cells of Mouse Endocrine Pancreas Coexpress Insulin, Glucagon and the Neuronal Proteins Tyrosine Hydroxylase and Neuropeptide Y, but Not Pancreatic Polypeptide," *Development,* Vol. 118, No. 4, 1993, pp. 1031–39.

[29] Alpert, S., D. Hanahan, and G. Teitelman, "Hybrid Insulin Genes Reveal a Developmental Lineage for Pancreatic Endocrine Cells and Imply a Relationship with Neurons," *Cell,* Vol. 53, No. 2, 1988, pp. 295–308.

5

Technologies to Improve Human Islet Transplantation

Hirofumi Noguchi,[1,2*] Naoya Kobayashi,[3] Shinichi Matsumoto[1*]

[1]Baylor All Saints Medical Center, Baylor Research Institute, Fort Worth, TX
[2]Institute of Biomedical Studies, Baylor University, Waco, TX
[3]Department of Surgery, Okayama University Graduate School of Medicine and Dentistry, Okayama, Japan

*Correspondence: Shinichi Matsumoto, Baylor Research Institute, 1400 8th Avenue, Fort Worth, TX 76104; Phone: 817–922–2570; Fax: 817–922–4645; E-mail: shinichm@baylorhealth.edu
or
Hirofumi Noguchi, Baylor Research Institute, 1400 8th Avenue, Fort Worth, TX 76104
Phone: 214–820–9016; Fax: 214–820–4952; E-mail: hirofumn@baylorhealth.edu.

Abstract

Pancreatic islet transplantation is a treatment alternative for patients with type 1 diabetes who experience hypoglycemic unawareness despite maximal care. We recently established a new islet isolation protocol based on the Ricordi method. It includes pancreatic ductal protection, a modified two-layer (modified Kyoto solution/perfluorochemical [PFC]) method for pancreas preservation, and a new islet purification solution (iodixanol-based solution). These techniques are now used for clinical islet transplantation from brain-dead donor (BDD) pancreata, non-heart-beating donor (NHBD) pancreata, or a living donor pancreas. We have a greater than 80% success rate for islet transplantation from BDD and NHBD pancreata. The method enables us to make diabetic patients insulin-independent using islets from half a pancreas from a living donor. Our method makes it feasible to use these pancreata efficiently for islet transplantation in patients with type 1 diabetes.

Key terms islet transplantation, islet isolation, ductal injection, modified two-layer method, modified Kyoto solution, iodixanol-controlled purification

5.1 Introduction

Pancreatic islet transplantation is one of the treatment strategies for a select group of individuals with type 1 diabetes mellitus who experience difficulty with blood glucose control problems despite maximal diabetes care. These patients have complied with various regimens of insulin preparations and have frequent dangerously low blood glucose levels and hypoglycemic unawareness, a situation that can lead to death or neurologic damage in the patient. In 2000, Shapiro et al. described that each of the first seven recipients who underwent isolated islet transplantation achieved insulin independence. Their treatment protocol, known as the Edmonton protocol, stressed the need to transplant a critical islet cell mass to achieve insulin independence (≥10,000 islet equivalents (IEs) per kg of recipient body weight), using a corticosteroid-free immunosuppression regimen. This islet mass could be achieved in most cases by infusion of islets obtained from two pancreata from two separate deceased donors.

The Edmonton protocol has been replicated by advanced islet transplantation centers [1]; more than 600 patients with type 1 diabetes at more than 50 institutions have received islet transplantation to treat their disease. Moreover, we have performed islet transplantations from living [2] and non-heart-beating donors (NHBDs) [1] using the modified Edmonton protocol (modified Ricordi method islet isolation method). These advances were based on advanced pancreas preservation [3,4] and transport systems [5], improved islet isolation methods [6], enhanced islet engraftment, and revised immunosuppressant protocols. Although experiments on β cell regeneration from stem cells have proceeded, there is still no reliable method for producing β cells. Until a new method to generate β cells is developed, improving the efficacy of islet transplantation seems the most realistic and prudent method to cure diabetes.

In this chapter, we show several advanced technologies to improve human islet isolation/transplantation for the treatment of diabetes.

5.2 Materials

5.2.1 Pancreas Procurement and Preservation

1. Viaspan (UW solution) (1000-46-06 Duramed Pharmaceuticals).
2. Kyoto solution (Otsuka Pharmaceutical Factory, Inc.).
3. Perfluorodecalin (Fluoromed, LP, Round Rock TX).
4. Two-layer method plate in 1L Nalgene jar (Umihira, Kyoto, Japan) (see Figure 5.1).
5. Bulb-tip cannula (18G, 20G).

5.2.2 Islet Isolation

1. Hanks' balanced salt solution (HBSS), 1×.
2. Cefazolin sodium USP.
3. Povidone iodine USP, 10%.

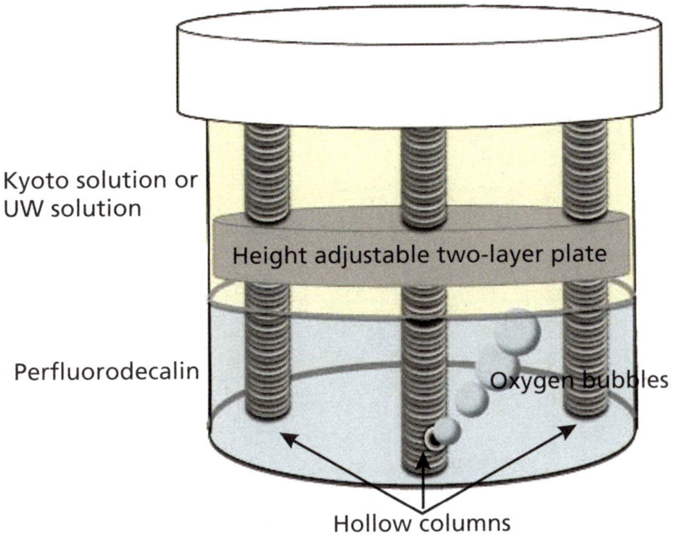

Figure 5.1 Oxygen-charged static two-layer method consists of perfluorodecalin and Kyoto solution or UW solution, the two-layer plate, and hollow columns in 1L nalgene jar. Oxygenation is performed via hollow columns at organ procurement site.

4. Collagenase NB 1 (GMP Grade 17452 / Premium Grade 17455 SERVA/Nordmark).
5. Neutral protease NB (GMP Grade 30303/NB Premium Grade 30301 SERVA/Nordmark).
6. Perfusion solution (99-781 Mediatech).
7. Priming solution (99-782 Mediatech).
8. Dilution solution (99-783 Mediatech).
9. Wash solution (99-784 Mediatech).
10. Albumin human USP, 25% Solution.
11. Heparin sodium injection USP, preservative-free.
12. Viaspan (UW solution) (1000-46-06 Duramed Pharmaceuticals).
13. Kyoto solution (Otsuka Pharmaceutical Factory, Inc.).
14. OptiPrep (1114542 Nycomed).
15. Hydrochloric acid NF, 1N.
16. NaOH solution 1N.
17. Perfusion system (Umihira, Kyoto Japan) (see Figure 5.2).
18. Ricordi Digestion System (see Figure 5.3).
19. COBE 2991 Cell Processor (Gambro Laboratories).

5.2.3 Islet Culture/Preservation (Optional)

1. CMRL-based Miami-defined media #1 (MM1; Mediatech-Cellgro, Herndon, VA).
2. Albumin human USP, 25% solution.
3. Viaspan (UW solution) (1000-46-06 Duramed Pharmaceuticals).

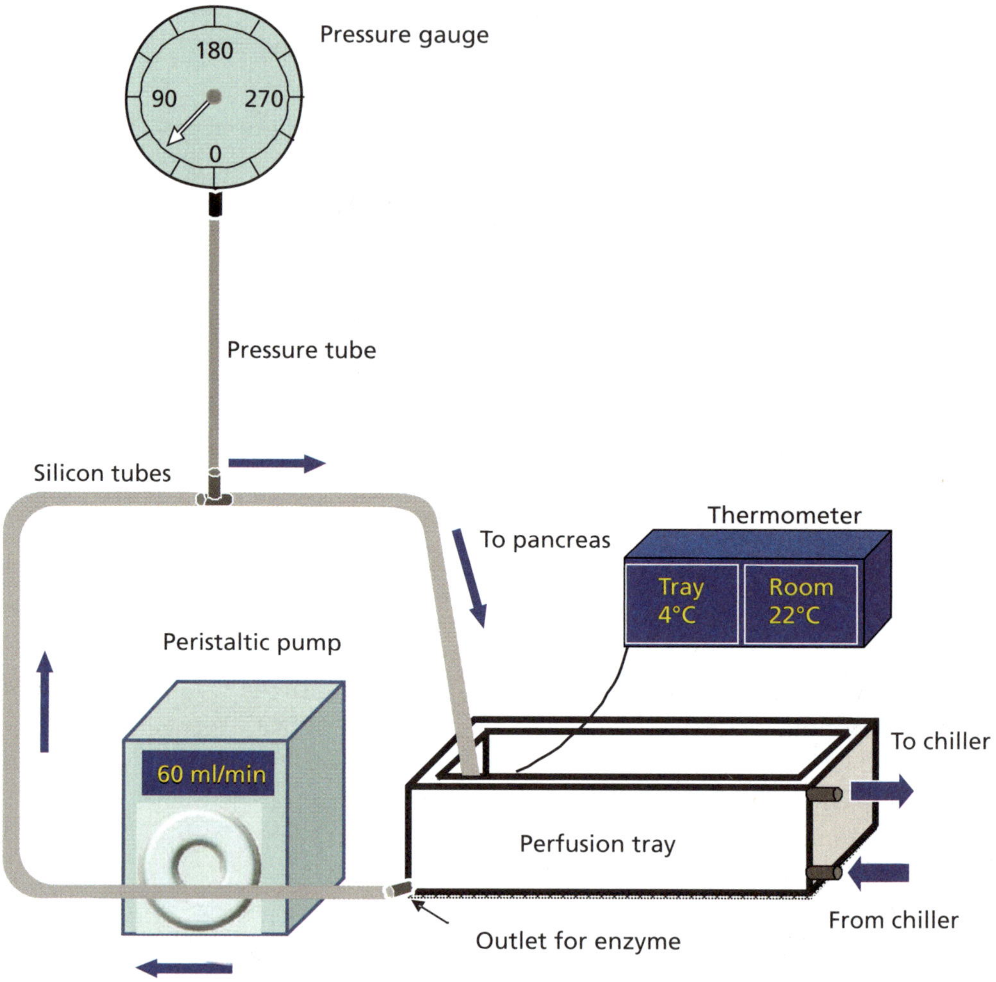

Figure 5.2 Collagenase perfusion system consists of a perfusion tray, silicon tubes, peristaltic pump, thermometer, and pressure gauge. Collagenase is perfused in a pressure-controlled manner. Perfusion tray keeps cold by circulating chilled water in the perfusion tray with a chiller.

5.2.4 Islet Transplantation

1. CMRL 1066 transplant media, containing HEPES without sodium bicarbonate (98-291 Mediatech).
2. Heparin sodium injection USP, preservative-free.
3. Albumin human USP, 25% solution.
4. Ricordi Infusion Bag (IB-01 Biorep Technologies, Inc.).

5.3 Methods

5.3.1 Pancreas Procurement for Islet Isolation

Pancreas procurement is a critical step for islet isolation. It has been shown that pancreas procurement by the islet team significantly improved the outcome of islet isolation [7]. A majority of pancreata are procured from brain-dead cadaveric donors but

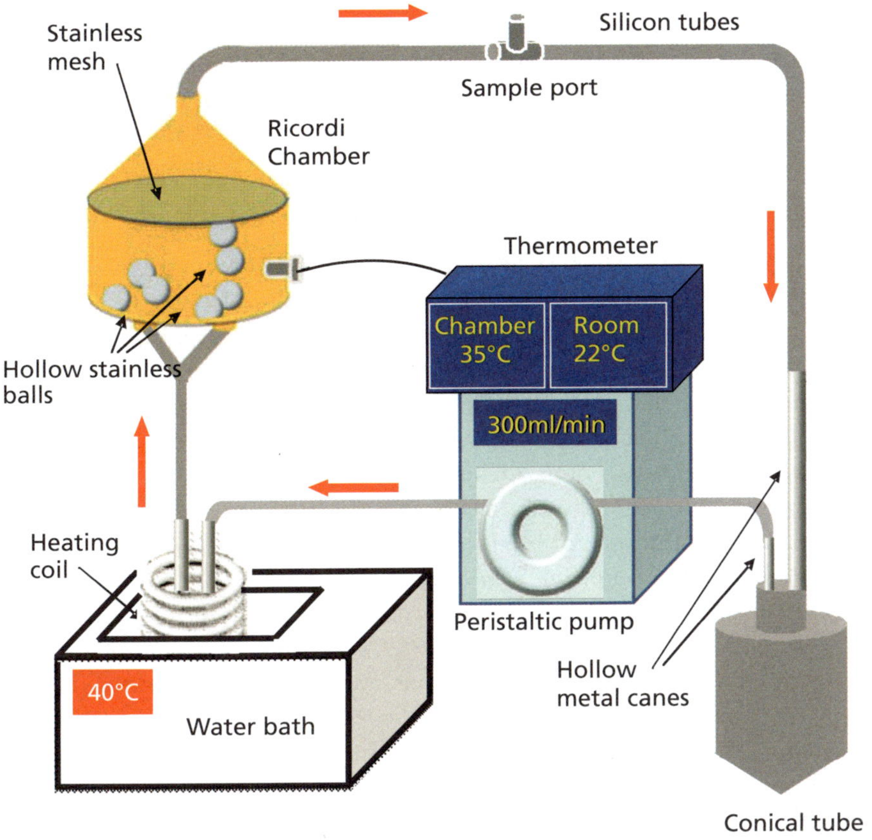

Figure 5.3 Ricordi system consists of Ricordi chamber, silicon tubes, peristaltic pump, water bath, heating coil, conical tube, and sample port. During the digestion period, collagenase containing solution is recirculating in this system. Once the pancreas is digested, digested tissue is collected using dilution solution without recirculation.

non-heart-beating donors (NHBDs) were also used. The methods for pancreas procurements are different between them [8, 9].

5.3.2 Pancreas Procurement from Heart-Beating Brain-Dead Donors

A midline incision is made from the suprasternal notch to the pubic symphysis. After a thorough laparotomy, the aorta and inferior mesenteric vessels are identified and prepared for later cannulation. The lesser sac is opened by dividing the gastrohepatic and the gastrocolic ligaments, and the pancreas was visually inspected. The short gastric arteries and veins are then ligated and divided, and the stomach is completely separated from the spleen. With division of the gastrocolic ligament, the splenic flexure of the colon is taken down. The middle colic vessels are ligated and divided and the first and fourth portions of the duodenum are divided with a GIA linear cutting stapler (Endo GIA, U.S. Surgical, Norwalk, CN). After division of the duodenum, the spleen and body and tail of the pancreas are mobilized toward the midline by dissection from their retroperitoneal attachments. The aorta is cannulated just proximal to its infrarenal bifurcation in the usual fashion and the inferior mesenteric vein (IMV) is also cannulated. After cannulation of the aorta and IMV, University of Wisconsin (UW) solution (DuPont Pharmaceutical, Wilmington, DE) is perfused through the aortic and venous cannulas. After the UW perfusate is started, the entire pancreas is packed both anteriorly and posteriorly with copious amounts of saline slush. Upon completion of the flush and before or after liver procurement, the pancreas is rapidly excised en bloc with the spleen.

5.3.3 Pancreas Procurement from Non-Heart-Beating Donors

After confirmation of brain death status, a double balloon catheter is inserted for in-situ regional cooling. Before cardiac arrest, a tip of the double balloon catheter is placed above the celiac axis in the aorta via the femoral artery. Ultrasound is used to confirm whether the celiac axis and renal arteries are placed between the two balloons. A venous catheter is also placed in the inferior vena cava via the femoral vein for drainage of the perfusate and blood.

In situ regional cooling for the pancreas and kidney is achieved by pump or drip infusion of a hypothermic lactated ringer solution beginning immediately after cardiac arrest and continuing until the end of the nephrectomy and pancreatectomy.

After declaration of cardiac death, the cadaveric donor is brought to the operating room for organ procurement. After laparotomy, the lesser sac is opened to determine whether the pancreas had uniform perfusion efficacy by means of in-situ regional cooling. After a visual check of the pancreas, 500 mL of sterile crushed ice is placed on it.

To expose the left kidney, the splenic flexure and the attachments of the descending colon to the lateral peritoneum are divided. When it is totally exposed, the splenic flexure of the colon is mobilized toward the middle until the renal artery can be entirely separated from the aorta.

Then, the right kidney is also exposed by dissection in the avascular plane between Gerota's fascia and mesocolon. Mobilization of the duodenum is started to separate the right renal artery and the right renal veins. While the duodenum is mobilized, the proper hepatic artery (PHA) is ligated and divided to avoid inflow of perfusate into the liver. The gastroduodenal artery (GDA) and the superior mesenteric artery (SMA) are left as intact as possible to maximize perfusion of the pancreas. The aorta and inferior vena cava are dissected and both kidneys are harvested en bloc.

After ligation of the major vessels, perfusion is terminated. Then the first and fourth portions of the duodenum are divided with a proximate 55-mm linear cutter (Ethicon; Johnson and Johnson K.K., Suture Div, Tokyo, Japan.). After division of the duodenum, the attachment between retroperitoneal and body of the pancreas is dissected toward the spleen. The superior surface of the pancreas is divided toward the spleen and the short gastric arteries and vein are dissected until the stomach is separated from the spleen. The pancreas is then rapidly excised en bloc together with the spleen. The spleen is subsequently removed on the back table.

5.3.4 Pancreas Preservation

The procured pancreas is trimmed in the chilled saline or lactate ringer solution at the back table. After the removal of the spleen and duodenum, an 18G bulb-tip cannula is inserted into the main pancreatic duct. The pancreas is weighed and 1 mL/g pancreas weight of chilled Kyoto solution is infused through the intraductal cannula. Then, the pancreas graft is preserved by the oxygen charged static two-layer method [10]. The two-layer method consists of Kyoto solution or UW solution, oxygenated perfluorodecalin, a two-layer plate (Umihira, Kyoto, Japan), three hollow columns (Umihira) in the Nalgene® polycarbonate jar (Figure 5.1). Perfluorodecalin is heavy (density is approximately 1.9 g/ml) and hydrophobic, therefore this solution clearly separates from the other solution. After pouring chilled perfluorodecalin and Kyoto solution (or UW solution) into the Nalgene jar, the two-layer plate is placed into the solution. Then oxygen is provided through one of the hollow columns at the procurement site. For full saturation of perfluorodecalin with oxygen, 100% oxygen is perfused at least 30 minutes at the flow rate 2 to 3 L/min. Since oxygen may escape from the perfluorodecalin, oxygenation is usually continued until pancreas is ready for preservation. Height of the two-layer plate may need to be adjusted to insure that at least two-thirds of the pancreas is immersed in the oxygenated perfluorodecalin. Then the pancreas is put under the two-layer plate. The Nalgene jar is put in two sterile plastic bags and placed into an ice-packed Styrofoam box.

5.3.5 Islet Isolation

Human islet isolation is conducted using the standard Ricordi technique with modifications introduced in the Edmonton protocol. The procured pancreas is dipped into chilled povidone iodine solution, then placed into a chilled antibiotic solution for 5 min and rinsed with chilled HBSS for the decontamination step. After decontamination of the pancreas, the duct is perfused in a pressure controlled fashion with a cold enzyme blend of Serva collagenase NB1 (20 PZ U/ml: 1 PZ unit according to Wünsch catalyzes the hydrolysis of 1 μmol 4-phenylazobenzyloxycarbonyl-L-prolyl-L-leucylglycyl-L-prolyl-D-arginine per minute at 25°C, pH 7.1) with neutral protease (1 DMC U/ml: 1 DMC unit catalyzes the cleavage of 1 μmol peptide bond from dimethylcasein per minute at 25°C, pH 7.0) or 1 vial of Liberase MTF (Roche Diagnostics Corp., IN) with 1 vial of Thermolysin (Roche Diagnostic Corp.) (Figure 5.2). Pressure of perfusion is adjusted between 60 and 80 mmHg for the initial 5 min followed by 160 to 200 mmHg for the next 5 min by changing the pump speed. Temperature of the enzyme solution is also monitored. We use a special perfusion tray (Umihira) that has an outlet for perfusion of enzymes. The entire tray is made of stainless steel and there is a space between the inner tray and outer tray. Chilled water circulates in this space using a water chiller.

The distended pancreas is then cut into 7 to 9 pieces, placed in a Ricordi chamber (Biorep Technologies, Inc., Miami, FL) with seven hollow stainless balls, and shaken gently (Figure 5.3). While the pancreas is being digested by recirculation of the enzyme solution through the Ricordi chamber at 37°C, we monitor the extent of digestion with dithizone staining by taking small samples via a sample port from the system. Once digestion is confirmed to be complete, dilution solution is introduced into the system and the digested tissue is collected in 250-mL conical tubes containing 15 mL of 25% HSA. Then, the system is cooled to stop further digestive activity. The collected digested tissue is washed with fresh medium to remove the enzyme.

Islets are purified with a continuous density gradient of iodixanol (Optiprep®, Sigma-Aldrich, St. Louis, MO)-Kyoto solution (IK solution) in an apheresis system (COBE 2991 cell processor, Gambro Laboratories, Denver, CO). Low-density (1.075 g/cm^3) and high-density (1.085 g/cm^3) solutions are produced by changing the volumetric ratio of iodixanol and Kyoto solution (IK), as shown in Table 5.1. Before purification by IK solution, we calculate the density of digested tissue (called density determination). Digested

Table 5.1 Characteristics of Purification Solution and Volumetric Ratio of Iodixanol and High-Density IK Solution

IK Solution	Kyoto* (ml)	Iodixanol †(ml)	Density (g/cm^3)	Osmolality (mOsm)
High Density	500	100	1.0850	411
Low Density	500	75.5	1.0750	401

* Density of Kyoto solution is 1.038 g/cm^3.
† Density of iodixanol solution is 1.320 g/cm^3.

tissue (0.2 mL) (after incubation in UW solution and prior to purification) is added to six 5-mL test tubes of different densities (1.085, 1.090, 1.095, 1.100, 1.105, 1.110 g/cm^3), and these tubes are centrifuged at 1,000 rpm for 5 min. The density at which most of the digested tissue was floating is defined as the density of the digested tissue. According to the outcome of the density determination step, we determine the necessary density of the high-density IK solution and add an appropriate amount of iodixanol into 130 mL of high-density IK (1.085 g/cm^3) solution as shown in Table 5.2. The purification procedure by IK solution is performed by top-loading digested tissue (=20 mL of tissue/run) in UW solution on the high-density and low-density gradients in doughnut-shaped COBE purification bags. After a 5-min centrifugation at 1,000 rpm, approximately 10 fractions (50 ml each) are collected and examined for purity.

5.3.6 Islet Culture/Preservation (Optional)

Isolated islets from human pancreata are cultured at 22°C to 37°C and 5% CO_2 in culture medium or preserved at 4°C in UW solution for overnight to 48h. Culture medium is CMRL-based Miami-defined media #1 (MM1; Mediatech-Cellgro, Herndon, VA) containing 0.5% human serum albumin. Islets are seeded as 100 to 150 IE/cm^2 dish area and the height of the media was 1.5 to 2.0 mm (20,000 to 30,000 IE per a T-175 flask, 30 to 50 ml per a T-175 flask).

5.3.7 Islet Transplantation

Isolated islets in 100 mL of transplant media with heparin are transferred to an infusion bag through a 50 mL syringe. Transplant media (50 mL) is added to the conical tube to rinse and then the contents are transferred into an infusion bag. Another 50 mL of transplant media is used to rinse the conical tube again and then transferred into an infusion bag. After transferring the entire final product to the infusion bag, the air is removed using a "burping" technique and then the port is clamped with a hemostat so that no air enters the bag. The islet cell infusion is an interventional radiology procedure. The med-

Table 5.2 High Density

IK Solution (ml)	Iodixanol (ml)*	Total Volume (ml)	Final Density (g/cm^3)
130.0	0.0	130.0	1.085
130.0	2.8	132.8	1.090
130.0	5.8	135.8	1.095
130.0	8.9	138.9	1.100
130.0	12.1	142.1	1.105
130.0	15.5	145.5	1.110

* Density of iodixanol solution is 1.320 g/cm^3.

ical personnel involved in the procedure aim to minimize morbidity and discomfort of the patient related to the procedure. Monitoring the portal pressure during islet cell infusion will aid in prevention of complications related to this procedure.

5.3.8 Potential Pitfalls and Troubleshooting

There are several difficult procedures among our islet isolation method. At the time of pancreas procurement, finding the orifice of main pancreatic duct might be difficult. In such a case, find the bile duct first and cut the pancreas around the bile duct. You can find the duct in the middle of the pancreas, because sometimes pancreatic ducts contract and hide in the pancreatic parenchyma. If this does not work, the pancreas needs to be cut between the head and tail portion. In this case, two cannulations are necessary and perfusion of collagenase is less effective. During collagenase perfusion, perfusion pressure might become out of control. In such a case, check for any torsion of pancreas and correct. In some cases, the main pancreatic duct is very narrow. In such a case, the accessory duct might be suitable for perfusion. During pancreas digestion, islets may be overdigested or underdigested. When overdigested, reduce the temperature and when underdigested, raise the temperature. Shorter digestion time is associated with successful islet isolation [7], therefore take care to avoid prolonged digestion more than 30 min. Islet purification is the most difficult procedure. Even using density adjustments, purification may fail due to increasing density of islets after density measurements. In such a case, islets need to be repurified using higher-density gradients. The summary is shown in the Troubleshooting Table.

5.4 Data Acquisition, Anticipated Results, and Interpretation

5.4.1 Islet Isolations from Brain-Dead Donors

Currently we analyzed our 100 human islet isolations including both islet isolations from brain-dead donors and from chronic pancreatitis patients [7]. Among them, 79 islet isolates using brain-dead donors were analyzed. Multivariate analysis revealed that Baylor islet isolation method (BIIM) including pancreas procurement by islet team, pancreatic ductal injection, oxygen-static charged two-layer method, and density adjusted iodixanol-based purification significantly increased islet yields. When BIIM was applied we successfully obtained more than 300,000 islet equivalent (IEQ) islets in 27 out of 30 cases (90%). In addition, we applied pancreatic ductal injection by our islet isolate team

Troubleshooting Table

Problem	Solutions
Unable to find main pancreatic duct.	1. Cut pancreas around the bile duct. 2. Cut pancreas between head and tail parts.
Unable to control perfusion pressure.	1. Check the torsion of pancreas. 2. Insert cannula into accessory pancreatic duct.
Overdigestion during digestion.	Decrease the temperature of the system.
Underdigestion during digestion.	Increase the temperature of the system.
Failed islet purification.	Increase density of heavy solution and repurify.

in clinical islet isolations. Using this technique, we had seven consecutive successful clinical islet isolations. Among seven successful islet isolations, six islet preparations were transplanted into three type 1 diabetic patients and all three patients became insulin-free. Currently, success rate of islet isolation is approximately 50% and success rate for achieving insulin independence is approximately 70%. Therefore, introduction of BIIM for allogeneic islet transplantation should have significant impact on improving clinical outcomes.

5.4.2 Islet Isolations from Non-Heart-Beating Donors

When we used in situ regional cooling, oxygen-static charged two-layer method, ductal injection and density adjusted iodixanol-based islet isolations, our success rate of clinical islet isolation was 81% (17/21) [11]. We transplanted those islets into eight patients and all patients attained improved glycemic control. Among the eight patients, five received multiple islet transplantations and three of the five achieved insulin independence. However, none of patients could sustain insulin-free status even though they maintained excellent glycemic control long-term. Using this islet isolation technology islet transplantation from non-heart-beating donors is possible to achieve excellent glycemic control. It is obvious that the clinical outcomes using non-heart-beating donors seem inferior compared to using brain-dead donors. However, brittle type 1 diabetic patients whose blood glucose can not be controlled with the intensive insulin therapy should benefit from the islet transplantation using non-heart-beating donors.

5.4.3 Islet Transplantation from Living Donors

Our reliable islet isolation method allowed us to perform living donor islet transplantation [1]. Reliable islet isolation is critically important for living donor islet transplantation because failure of islet isolation is not acceptable.

We used the tail part of the pancreas from a living donor and obtained more than 400,000 IEQ islets. After islet transplantation, the patient achieved an insulin-free status. The patient could maintain excellent glycemic control for a relatively long period but insulin independence was not sustainable [12]. The donor has maintained excellent glycemic control even though she only has half of a pancreas. This case demonstrated that islet transplantation using half of a living donor pancreas is possible. However, considering the risk to the donor, this procedure should be applied in a prudent manner.

5.5 Discussion and Commentary

We introduced pancreatic ductal injection, oxygen-static charged two-layer method, and density adjusted iodixanol-based islet purification for islet isolation from brain-dead donors. In addition, in situ regional cooling is introduced for islet isolation from non-heart-beating donors.

Among these elements, the two-layer pancreas preservation method has been widely used. Recent meta-analysis of the two-layer method demonstrated that the two-layer method before islet isolation is beneficial when pancreata need to be preserved for a relatively long period [13]. Therefore it is recommended that pancreata for islet isolation should be preserved by the two-layer method when the preservation period exceeds 5h.

The other components (ductal injection, density adjusted iodixanol-based islet purification, in situ regional cooling for non-heart-beating donors) have limited experiences therefore further data are needed to confirm the beneficial effects. However, ductal injection [3] and density adjusted iodixanol-based islet purification [6] had significant improvements compared to the controls using human pancreata. Therefore, those two techniques also should have beneficial effects. In situ regional cooling for non-heart-beating donors before islet isolation has provided no significant data. However, there were many previous instances of in situ regional cooling for kidney transplantation using non-heart-beating donors. Thus, theoretically minimizing warm ischemic period by this technique should also be beneficial.

We are confident that introducing the islet isolation technology described in this chapter will improve the clinical outcome of islet transplantation.

5.6 Summary Points

1. Ductal injection of preservation solution. After procurement of the pancreas and removal of the spleen and duodenum, we immediately inserted a cannula into the main pancreatic duct. The pancreas is weighed and 1 mL/g pancreas weight of modified Kyoto (MK) solution is infused through the intraductal cannula.
2. Modified two-layer preservation method (MK/PFC). Pancreata are placed into an MK solution/oxygenated PFC two-layer preservation container at 4°C for less than 8h until the islet isolation procedure.
3. Iodixanol-controlled density gradient during islet purification. According to the density of digested tissue before purification, the density of the purification solutions was controlled by changing the volumetric ratio of iodixanol and the purification solutions.
4. Islet preservation (optional). Isolated islets from human pancreata are preserved at 4°C in UW solution for overnight to 48h. Islets are seeded as 100 to 150 IE/cm^2 dish area and the height of the media was 1.5 to 2.0 mm.

Acknowledgments

The author wishes to thank Ms. Sandy McCarthy and Mr. Jeffery A. SoRelle for their careful reading and editing of this manuscript. This work was supported in part by the All Saints Health Foundation.

References

[1] Matsumoto. S., et al., "Successful Islet Transplantation from Nonheartbeating Donor Pancreata Using Modified Ricordi Islet Isolation Method," *Transplantation,* Vol. 82, No. 4 2006, pp. 460–65.

[2] Matsumoto, S., et al., "Insulin Independence after Living-Donor Distal Pancreatectomy and Islet Allotransplantation," *Lancet,* Vol. 365, No. 9471, 2005, pp. 1642–44.

[3] Matsumoto, S., et al., "Seven consecutive Successful Clinical Islet Isolation with Pancreatic Ductal Injection," *Cell Transplant.,* Vol. 19, No. 3, 2010, pp. 291–97.

[4] Noguchi, H., et al., "Ductal Injection of Preservation Solution Increases Islet Yields in Islet Isolation and Improves Islet Graft Function," *Cell Transplant.,* Vol. 17, No. 1–2, 2008, pp. 69–81.

[5] Noguchi, H., et al., "Modified Two-Layer Preservation Method (M-Kyoto/PFC) Improves Islet Yields in Islet Isolation," *Am. J. Transplant.,* Vol. 6, No. 3, 2006, pp. 496–504.

[6] Noguchi, H., et al., "Iodixanol-Controlled Density Gradient during Islet Purification Improves Recovery Rate in Human Islet Isolation," *Transplantation,* Vol. 87, No. 11, 2009, pp. 1629–35.

[7] Takita, M., et al., "One Hundred Human Pancreatic Islet Isolations at Baylor Research Institute," *Proc. (Bayl. Univ. Med. Cent.),* Vol. 23, No. 4, 2010, pp. 314–48.

[8] Lee, T. C., et al., "Procurement of the Human Pancreas for Pancreatic Islet Transplantation," *Transplantation,* Vol. 78, No. 3. 2004, pp. 481–83.

[9] Nagata, H., et al., "Procurement of the Human Pancreas for Pancreatic Islet Transplantation from Marginal Cadaver Donors," *Transplantation,* Vol. 82, No. 3, 2006, pp. 327–31.

[10] Matsumoto, S., et al., "Efficacy of the Oxygen-Charged Static Two-Layer Method for Short-Term Pancreas Preservation and Islet Isolation from Nonhuman Primate and Human Pancreata," *Cell Transplant.,* Vol. 11, No. 8, 2002, pp. 769–77.

[11] Liu, X., et al., "Analysis of Donor- and Isolation-Related Variables from Non-Heart-Beating Donors (NHBDs) Using the Kyoto Islet Isolation Method," *Cell Transplant.,* Vol. 17, No. 6, 2008, pp. 649–56.

[12] Matsumoto, S., et al., "Follow-Up Study of the First Successful Living Donor Islet Transplantation," *Transplantation,* Vol. 82, No. 12, 2006, pp. 1629–33.

[13] Qin, H., et al., "A Meta-Analysis of Comparison of the Two-Layer and University of Wisconsin Pancreata Preservation Method in Islet Transplantation," *Cell Transplant.,* Nov. 19, 2010 (Epub ahead of print).

6

Pancreatic Islet Isolation from Swine

Suzanne Bertera*, Marco Marigliano, Rita Bottino, and Massimo Trucco

Children's Hospital of Pittsburgh of UPMC/University of Pittsburgh School of Medicine

*Correspondence: Rangos Research Center, Children's Hospital of Pittsburgh of UPMC, 4401 Penn Avenue, Pittsburgh, PA 15224; E-mail: subst5@pitt.edu

Abstract

Type 1 diabetes is an autoimmune disease that selectively destroys the islet beta cells of the pancreas. Without beta cells, insulin cannot be supplied in sufficient amounts to prevent high blood glucose levels and ketoacidosis, compelling patients to control their blood glucose with daily insulin injections for life.

Islet cell transplantation has long been considered a potential curative treatment for type 1 diabetes and recent clinical trials have further proven that donor replacement of human islets can successfully establish short-to-medium term insulin independence in immunosuppressed diabetic patients. Multiple islet cell infusions are usually required to achieve and maintain normal blood glucose levels, posing a problem of islet cell availability.

With the challenge of obtaining transplantable pancreatic endocrine tissues, which cannot be met by the limited availability of organ donors, an unlimited source of islets may be found in animals such as the pig. Efficient methods to isolate pig islets for clinical use are therefore required.

Key terms islets, pigs, pancreas,
islet isolation,
beta cell, diabetes

6.1 Introduction

Islets of Langerhans are composed mostly of beta cells. These cells have the capacity to synthesize and release insulin in response to ambient glucose changes. In type 1 diabetes, beta cells are the target of selective immune destruction, leading to insulin deficiency and consequent hyperglycemia. Beta cell replacement has been envisioned as a valid approach to supply insulin-producing tissue in diabetic patients. Pancreas transplantation allows for replacement of beta cell function in insulin dependent diabetes. Whole pancreas transplantation is a well-established surgical procedure that entails, however, major surgery and frequent secondary, often life-threatening, complications. In contrast, islet cell transplantation consists only of an infusion into the recipient's liver of isolated islets of Langerhans obtained from a deceased pancreas donor. The pancreas undergoes a process of mechanical and enzymatic digestion that allows release of islets that are further separated from the exocrine component of the pancreas by density gradients. In principle, islet cell transplantation via injection of the islets in the portal vein and deposition of the islets in the liver is a much simpler and safer procedure than whole pancreas transplantation, and it offers the advantage that it can be carried out under local sedation. Nonetheless, isolated islets suffer from the effects of early inflammatory events that occur as soon as the islets are injected in the bloodstream, and those that do engraft do not guarantee a long-term function. Islet infusion, therefore, must be repeated multiple times. Unfortunately, islet transplantation, as well as organ transplantation, is strongly limited by a paucity of donor organs [1]. The use of animal islets has therefore been a focus of investigation. For clinical applications, the animal of choice should be able to produce a kind of insulin that is very similar to human insulin and should offer islets in quantities that satisfy the metabolic requirements of diabetic patients. Based on these two parameters, the pig is the animal of choice. There is only one amino acid difference between human and pig insulin and this substitution is not particularly immunogenic. Also, the pig is large enough to supply sufficient amounts of transplantable islets to satisfy the needs of a human recipient of average weight. Suitable animals would also offer the possibility to select the pancreas harvesting time, eliminating all factors related to donor medical status and cause of death prior to organ donation, including brain death and life-supporting drugs and procedures. It is therefore predictable that pigs would offer qualitative superior islets than those obtained from deceased human donors.

Unfortunately, morphological conditions of the pig pancreas make it more challenging to isolate and purify the islets than from a human pancreas. For a long time, porcine islets were defined as the "elusive" islets, and for a long time, it was impossible to achieve reproducible islet isolation standards.

Morphologically, pig islets appear very different from mouse or human islets. Generally, pig islets have a less compact shape, the cells dissociate more easily than in human and rodent islets, and long culture times result in a reduction in islet size because of the spontaneous dissociation of the cells from the outer layers of the islet. "Retired breeders" represent an exception to this condition. These are defined as sows that have carried multiple pregnancies (more than two) to term and are usually 2 or more years of age. For reasons that are not fully understood, the islets obtained from these donors are more compact and exhibit morphological features that are more similar to rodents and humans. For experimental purposes, islets from retired breeders are favored over younger donors. However, from a practical point of view, using only retired breeders would entail higher costs in terms of maintenance and husbandry, and have a limited availability. Efforts have been therefore dedicated to obtaining good quality islets from pigs of both genders and for ages ranging from 5 to 8 months [2, 3].

Here we will present an overview of the methods currently utilized to isolate pig islets from pigs of all ages.

6.2 Materials

6.2.1 Animals

The main consideration when discussing islet yield is what type of pig to select. In terms of number and size of islets, by far, the highest yield is obtained from a retired breeder. A retired breeder is defined as a sow that has had two or more litters. Market-age pigs 5 to 8 months old are commonly utilized as well as market weight pigs of 100 to 125 kg. Several breeds including Landrace, Yorkshire, and Large White, have been successfully used as islet donors. Studies have shown differences in islet morphology, allowing for selection of more suitable breeds for good islet isolation outcome [4, 5]. Slaughterhouse donor pancreata have been utilized but elective organ donor harvesting in an authorized animal facility produces better results and higher qualitative properties due to a reduction of warm ischemia time and more suitable sterile conditions.

6.2.2 Younger Versus Older Pigs

There are distinctive differences in the morphology and function of pig islets that are age-dependent. Fetal and neonatal pig pancreata are organized in cell clusters that lack a discrete shape [6]. Endocrine tissue can be identified as single cells amid the other pancreatic tissue using dithizone staining. These single cells are very difficult to separate from the exocrine tissue and other nonislet cells of the pancreas. However, fetal and neonatal pig islet-like cell clusters (ICCs) have the ability to proliferate and for this reason, are potentially useful but insulin production and secretion requires several weeks in culture before the ICC become fully developed [7–9]. Another drawback to using pigs of this age is the large number of animals required to provide enough cells for normalization of a recipient after cell transplant.

Young pigs, 3 to 6 months of age and less than 50 kg, and young-adult pigs, 6 to 12 months of age and between 70 to 80 kg, also have single cells and small ICCs rather than consolidated islets [10, 11].

Market-weight pigs are usually between 100 to 125 kg and are not categorized by age. They have more defined islets than smaller pigs but the islets are still fragile and morphologically atypical. These pigs are the most economical choice for islet isolation and can produce a good yield with a little care [2, 12].

Adult pigs over 2 years of age and greater than 200 kg and especially retired breeders have large, well-formed cohesive islets with a uniform shape. In comparison, islets from younger pigs are smaller size islet cell aggregates with an irregular shape [11].

6.3 Equipment

6.3.1 Major Equipment

- Two laminar flow biological cabinets (NuAire, Plymouth, MN).
- Water bath with thermometer (Fisher Scientific, Pittsburgh, PA).
- Centrifuge with rotors for 250-ml conical tubes (Thermo Scientific).
- Peristaltic pumps (Masterflex, Cole-Palmer, Vernon Hills, IL).
- COBE 2991 cell separator (GAMBRO, Lakewood, CO).
- 750-ml Ricordi digestion chamber (Biorep Technologies, Miami, FL).

6.3.2 Small Equipment and Nondisposable Items

- One large and one small stainless steel surgical pan (Fisher Scientific).
- One stand, base, and clamp (Fisher Scientific).
- One stainless steel tubing coil (Biorep Technologies, Miami, FL).
- One stainless steel screen (Biorep Technologies, Miami, FL).
- Three to four siliconized glass marbles (1-cm diameter) (Biorep Technologies, Miami, FL).
- Surgical pack consisting of large curved scissors, small scissors, large forceps, four to six hemostats (Alpin Surgical Specialties, Inc., Butler, PA).
- Digital thermometer with dual readouts to report the temperature of the water bath and inside the digestion chamber (Fluke Electronics, Everett, WA).
- Two to three each 600- and 1,000-ml beakers (Fisher Scientific).
- Two 1,000-ml collecting bottles with caps (Corning, Fisher Scientific).
- Vacuum collection flask (Fisher Scientific).
- Three ice buckets and one large plastic bin for ice (Fisher Scientific).
- Racks for the 250 conical tubes (Fisher Scientific).

6.3.3 Disposable Items

- 15- and 250-ml conical centrifuge tubes (Fisher Scientific).
- 5- and 10-ml pipettes (Fisher Scientific).
- 10-ml aspirating pipettes (Fisher Scientific).
- 10- and 60-ml syringes (VWR, Bridgeport, NJ).
- 14G-24G catheters (Cardinal Health, Dublin, OH).
- 2.0 silk suture (Owens and Minor, Richmond, VA).
- No. 10 disposable scalpels (Fisher Scientific).
- 10×35 mm Petri dishes (Fisher Scientific).
- COBE blood cell processing sets (GAMBRO, Lakewood, CO).
- Nontissue culture T-175 flasks (Sarstedt, Newton, NC).
- 500-ml filter flasks (0.2 μm) (Fisher Scientific).
- Temperature probe (Tyco).

- Flexible silicone tubing, 0.125 inch ID/ 0.250 inch OD (Helix Medical, Carpinteria, CA).
- Sterile drapes (Fisher Scientific).
- Personal protective equipment (gowns, gloves, sterile gloves, masks, bonnets, shoe covers, eye protection) (Fisher Scientific).
- Dithizone or diphenylthiocarbazone (Sigma, St. Louis, MO).
- Betadine solution (Cardinal Health, Dublin, OH).
- Polysucrose stock solution (Euro-Collins base) (Cellgro, Mediatech, Manassas, VA).
- 1.108, 1.096, 1.037 polysucrose islet density gradients (Cellgro, Mediatech).
- Serva NB1 or Roche Liberase DL, or Vitacyte CIzyme with Neutral Proteases
- HBSS 1X (Mediatech, Manassas, VA).
- PBS 1X (Gibco, Invitrogen, Carlsbad, CA).
- HEPES 1M solution. (Fisher Scientific).
- CMRL 1066 (Life Technologies, Invitrogen, Carlsbad, CA).
- RPMI 1640 (Mediatech).
- Porcine serum (heat inactivated) (Gemini Bio-Products, West Sacramento, CA).

6.4 Media

- *Culture media pH 7.4, store at 4°C.* CMRL 1066, heat-inactivated porcine serum (10%), nicotinamide (10 mmol/l), 100 units/ml penicillin, 0.1 mg/ml streptomycin, L-glutamine (2 mmol/l).
- *Dithizone, 0.2-μm filtered, store at –20°C.* Dithizone (30 mg), DMSO (9 ml), 30% ammonium hydroxide (300 μl).
- *2% serum media, pH 7.4, store at 4°C.* HBSS (300ml), porcine serum (2%), 100 units/ml penicillin, 0.1 mg/ml streptomycin, HEPES, 20 mM.
- *5% serum collecting media, pH 7.4, store at 4°C.* RPMI 1640, porcine serum (5%), 100 units/ml penicillin, 0.1 mg/ml streptomycin.

6.5 Methods

6.5.1 Harvesting the Pancreas

Pigs are routinely used as food and porcine pancreata can be easily obtained from local abattoirs. The drawbacks to this are the sometimes compromising warm ischemia time as well as a greater likelihood for bacterial and fungal contamination. A surgical method of pancreas removal under sterile conditions with experienced surgeons results in a more desirable outcome especially considering the effort and expense that go into an islet isolation.

The pig pancreas is a retroperitoneal organ similar in appearance, location, and anatomical orientation to humans. The "tail" is attached to the spleen and the distal end is covered with a cap of fibrous tissue. The "body" of the pancreas encircles the portal vein with a large anterior and a thin posterior ring and the "head" is attached to the duodenum. At the level of the body of the pancreas, the 2 mm diameter pancreatic duct is identified near the inferior edge of the pancreas [13].

1. Donor pigs are anesthetized using ketamine (10 mg/kg), zylazine (2 mg/kg), and atropine (0.05 mg/kg) given intramuscularly. Anesthesia is maintained with 2.5% halothane and 95% of oxygen.
2. The abdomen is shaved and cleaned with betadine and surgical drapes appropriately placed. A midline laparotomy incision is made from the xiphoid to slightly caudal to the umbilicus.
3. The diaphragm is then incised and the suprahepatic aorta and suprahepatic inferior vena cava (IVC) are isolated. Infrarenal aorta infrahepatic IVC are also isolated and the pig is anticoagulated with IV heparin.
4. After entering the peritoneal cavity, the small and large bowel are exteriorized and the stomach is retracted. The spleen is large and sufficiently mobile so it can be exteriorized with its vasculature intact. The pancreas is then completely exposed after dissecting through the posterior parietal peritoneum.
5. The aorta is then clamped and cannulated superior to the bifurcation. The aorta is also clamped above the diaphragm.
6. The animal is then perfused with several liters of either University of Wisconsin (UW) preservation solution (if longer than 1h cold ischemia time is expected) or Hanks' balanced salt solution (HBSS) if islet isolation will begin immediately after harvest. The solution is infused via the aortic cannula while the suprahepatic IVC is transected to release the perfusate. The infusion continues until the IVC effluent runs clear.
7. Proceeding to the level of the portal vein, blunt dissection is used to free the pancreas from surrounding tissues. The pancreatic tail is mobilized away from the splenic hilum. Caudal to the head of the pancreas, the superior mesenteric vessels are divided to prevent extensive dissection within the pancreas.
8. When the body and tail of the pancreas have been separated from all tissue, the head can be dissected out including the portion of duodenum to which it is attached. The vasculature of the pancreas is carefully maintained at all times.
9. The main pancreatic duct is located on the descending portion of the duodenum. The bile duct, located near the pancreatic duct, is ligated and transected.
10. Once the pancreatic duct is identified, a length of duodenum containing the ampulla of Vater is clamped on either side of the pancreatic head and resected.
11. The entire pancreas is then removed and placed into cold (4°C) UW solution or HBSS and transported immediately to the islet isolation facility [11].

6.5.2 Laboratory Setup

The biological safety cabinets (flow hoods) should be set up shortly before the pancreas arrives, typically after the surgeons have begun the pancreatectomy. The setup requires a sterile person and a nonsterile person. The nonsterile person opens the autoclaved packs containing instruments, equipment and glassware, and so forth, while the sterile person takes them out of the packs and places them in the hoods.

1. Since the hoods are not inherently sterile to begin with, sterile drapes are laid down to create a sterile field on which to put all the sterile equipment.

2. Hood #1 is for cleaning the pancreas initially and after this phase is completed, it is then set up for receiving the digested tissue as it is collected from the Ricordi chamber (Figure 6.1(a)). The COBE tissue separations are also performed in this hood after the tissue is collected.
3. Hood #2 is set up for digestion and collection of tissue utilizing the peristaltic pump, water bath, Ricordi chamber with temperature probe, stainless steel coil (in the water bath), recirculating beaker, collecting bottle, and digital thermometer. (Figure 6.1(b)). The coil is attached inline to the recirculating tubing and placed in the water bath to maintain the proper temperature during the digestive phase of the isolation.

6.5.3 Cleaning and Cannulating Pancreas

The pancreas of an adult pig is larger than a human organ and the shape is also different. The body of the organ forms a circle around the portal vein, while the tail is covered with a thick cap of tissue (Figure 6.2(a)). The pancreas is generally described as having two lobes. The lobe attached to the duodenum is comparable to what is described as the head of the pancreas in humans, and in the pig, this is sometimes considered a third lobe. The head is connected to the body and tail (together considered one lobe) by a thick bridge of tissue anteriorly, while the second lobe appears to be more separated from the head and body with a slender bridge of tissue posteriorly (Figure 6.2(b)). This second lobe usually needs to be cannulated independently to achieve good infusion of digestive enzymes [14].

1. Upon receipt of the organ, maintaining sterility, take a sample of the procurement fluid with a 10-cc syringe and place fluid in a 15-ml centrifuge tube to check for contamination if desired. Add about 1 ounce of betadine to the pancreas in the remainder of the procurement fluid and mix the betadine with the fluid for about 3 min.
2. Remove the pancreas from the fluid and rinse it in ~300 ml cold HBSS.
3. In a large stainless steel pan, add crushed ice to cover the bottom. Place a sterile drape over the ice and place the smaller cleaning pan containing 300 ml of the 2% serum

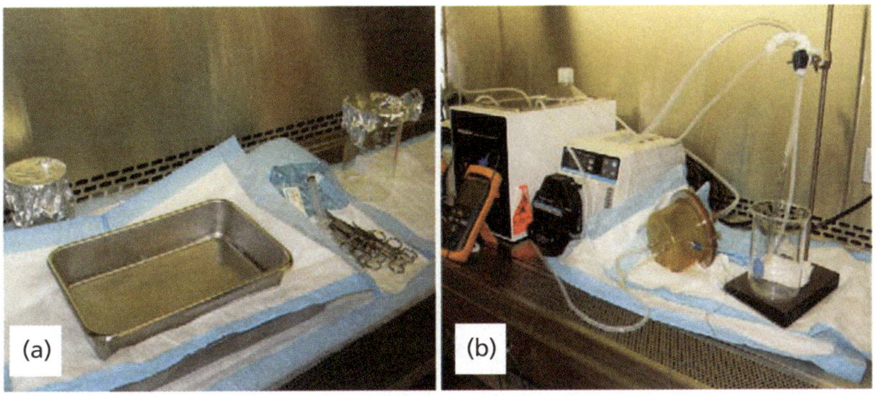

Figure 6.1 (a) Hood #1 with receiving pan over draped pan with ice, sterile instruments, and rinsing beaker. (b) Hood #2 setup for digestion and collection of tissue with Ricordi chamber, recirculating beaker, pump, and water bath. Stainless steel inline coil has been placed in the water bath to keep temperature constant as the fluid circulates through it.

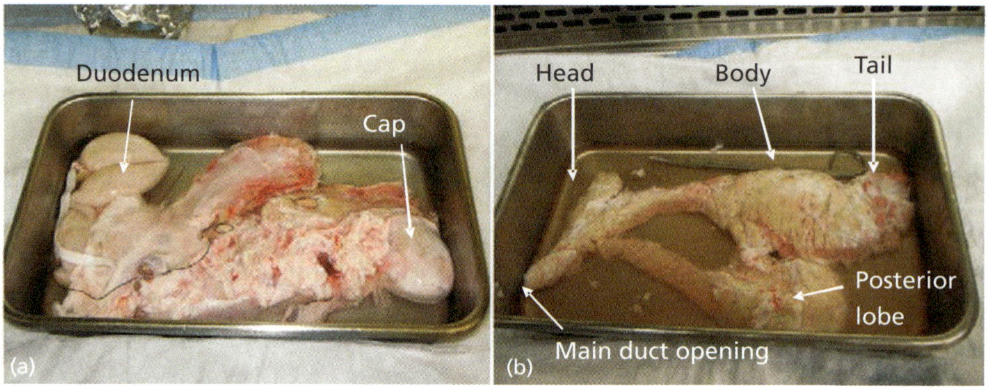

Figure 6.2 (a) Pig pancreas before cleaning with duodenal section attached to the head and tissue cap covering the tail. (b) Cleaned pancreas with head on the left, main duct facing down, body and tail at the top of the picture and the posterior lobe at the bottom.

medium on top of the pad (Figure 6.1(a)). This will keep the organ cold while maintaining sterility.

4. Move the pancreas to the cleaning pan. The first thing is to inspect the organ for cuts and to make sure the intestine is intact.
5. The intestine is removed first and discarded. Care should be taken not to cut into the intestine and potentially contaminate the islet tissue. The pancreatic duct drains into the intestine and the duct can be easily identified when the intestine is being separated from the rest of the organ. As the intestine is cut away, a good practice is to clamp the duct, close to the intestine, before severing it from the pancreas, to minimize contamination.
6. The pancreas is cleaned of fat that is easily identified and separated. Care must be taken not to cut the pancreas as this will create the possibility of leaking when the enzyme is injected.
7. Catheters are used to cannulate the ducts and are sutured into place with 2-0 braided silk suture (Figure 6.3(a)) Cannulate the main duct first and suture it into place.
8. Most often two catheters are used with the second one in the posterior lobe. To achieve this, the tissue bridge that leads to the second posterior lobe may be severed near the head, straight across the bridge (Figure 6.3(b)) The duct should be easily located and cannulated with a catheter of appropriate size in the direction of the sec-

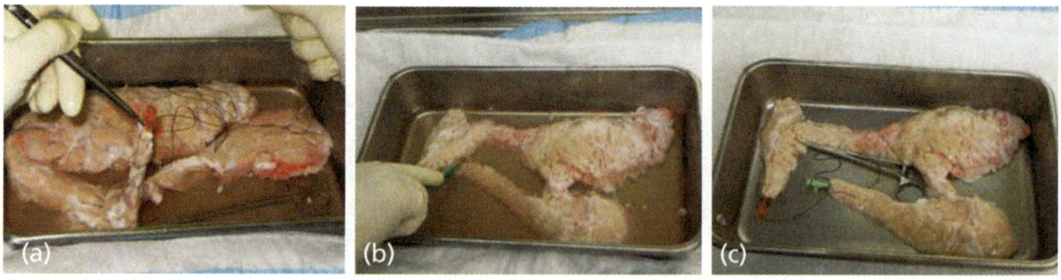

Figure 6.3 (a) Catheter in the main duct being sewn into place. (b) Proper place to cut the bridge, close to the head, to place the second catheter in the posterior lobe. (c) Both catheters placed and sewn in the ducts. Duct on the opposite side of the incision is clamped with a large hemostat to prevent enzyme leakage.

ond lobe. The duct on the other side of the incision, near the head, is then clamped to minimize enzyme leakage (Figure 6.3(c)) At this point, (before enzyme infusion) the pancreas is weighed and a piece of pancreas may be taken for histology if desired.

6.5.4 Enzyme Preparation

Several enzyme blends have been successfully utilized for the digestion of pig pancreatic organs. Most companies that produce collagenases usually have developed a blend that better suits the specific characteristics of the porcine extracellular matrix.

In our experience we have obtained excellent results using Liberase PI (Roche) at the concentration of 1 mg/ml of HBSS. After discontinuation of this blend of enzymes, good results have been obtained using CIzyme and Neutral Proteases (Vitacyte, Indianapolis, IN). Possible other options include Liberase DL low dispase (Roche, Indianapolis IN), and Collagenase NB1 with Neutral Proteases (Serva, Heidelberg, Germany). Lot-to-lot variations usually exist, and a good practice is to optimize enzymatic conditions prior to planning large-scale isolations. Each product may require a different preparation based on the company specifications. It is expected that the more concentrated the enzymes, and the higher the temperature is at injection, the faster the digestion occurs. Different approaches therefore entail flexibility to optimize the isolation conditions [15].

1. Based on the weight of the pancreas, mix together 7.5 Wunsch Units collagenase per gram tissue weight and 0.13 mg of proteases per gram tissue weight and keep on ice without agitation until collagenase is dissolved.
2. This mixture is resuspended in 500 ml of HBSS and the enzymes can be injected at either 4°C or prewarmed to 24°C. Our protocol uses the prewarmed collagenase.
3. Once the enzyme is at the desired temperature, remove the 2% serum media from the cleaning pan and, if infusing the enzyme at a warm temperature (e.g., 24°C), remove the pan from the ice.
4. Fill a 60-ml syringe with the enzyme and connect it to the first catheter. Begin hand-infusing with a slow and steady rate of about 20 to 30 ml/min. The pancreas should visibly distend without leaking. If there are leaks, clamp them with hemostats.
5. Refill the syringe and repeat with the second catheter. Alternate ducts until most of the enzyme has been infused and the pancreas is well distended (Figure 6.4, left panel).
6. After the enzyme has perfused throughout and the pancreas is well distended, the catheters are removed (Figure 6.4, right panel).
7. The pancreas usually has some clear gelatinous tissue that swells with the enzyme infusion and should be cut away before placing the organ in the Ricordi chamber. This is important to do because this "jelly" can clog the filter in the chamber and make it difficult to collect the digested tissue (Figure 6.5).
8. The pancreas is then cut into three to four pieces and placed in the Ricordi chamber with three to four siliconized glass marbles. The marbles are 1 cm in diameter and are placed in the chamber to gently break up the tissue when the chamber is rocked back and forth. The chamber is filled almost to the top with the enzyme, the stainless steel screen is placed over the opening of the chamber and the top is screwed into place. The screen is used to allow the free islets and all small tissue pieces to pass through

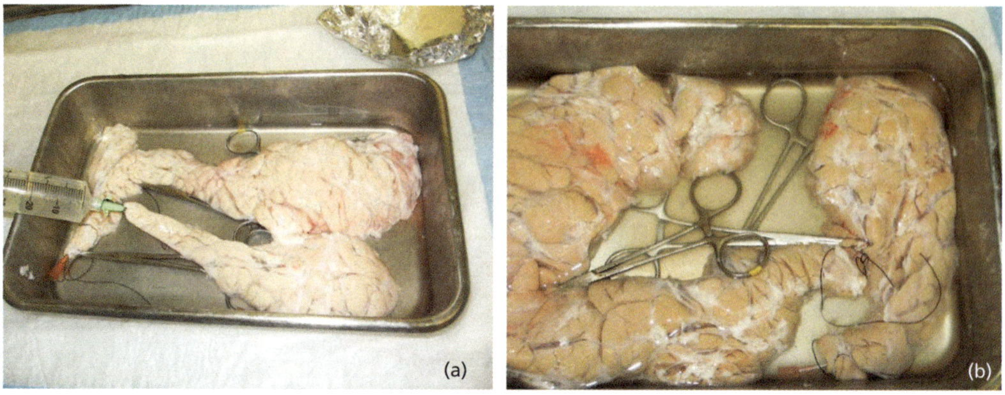

Figure 6.4 (a) Infusing enzyme through catheters using a large syringe. (b) Fully distended pancreas. Hemostats were placed to clamp leaks.

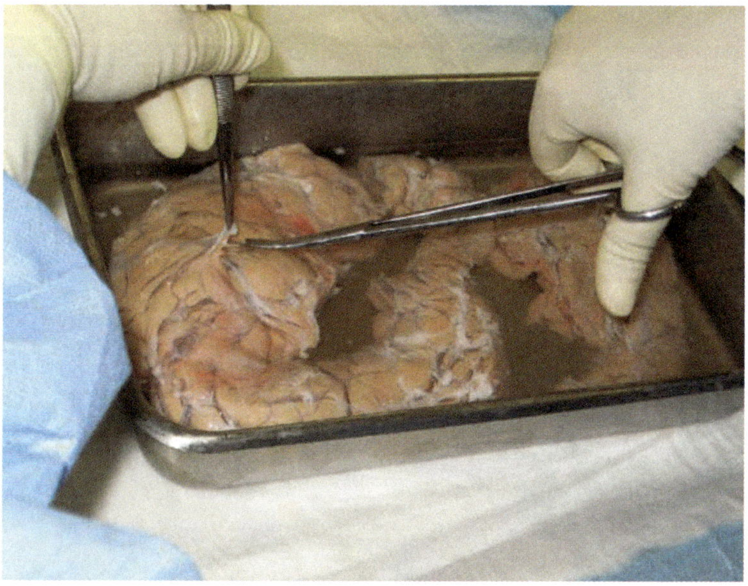

Figure 6.5 Cutting away gelatinous tissue from the surface of the pancreas after enzyme infusion.

the chamber and be removed from further enzymatic digestion while the undigested tissue remains in the chamber.

9. Any remaining enzyme is placed in the recirculating beaker. Turn on the peristaltic pump with a speed of 80 to 90 ml/min. The direction should flow into the bottom of the chamber and out the top. The temperature probe from the chamber is connected to a digital thermometer.
10. A bottle of HBSS (1L) is placed in the water bath to warm to 36°C. This is used to add more fluid to the recirculating beaker to fill the chamber. The more HBSS is added, the more the enzyme is diluted, so add only enough to finish filling the chamber (to expel the air) and complete the fluid circuit.
11. The stainless steel coil is used to maintain the desired temperature (36°C) as the fluid travels through the tubing circuit. It is placed in and out of the water bath as

necessary according to the digital thermometer. Once the fluid starts to drain back into the recirculating beaker, the circuit is complete.

12. Begin to gently shake the chamber, turning it over and back so the marbles hit the tissue (Figure 6.6). The shaking speed should be even and slow because pig islets are delicate (more so than human tissue) and vigorous shaking will damage them. The tissue should start to fall apart and begin exiting the chamber.
13. After about 1 min from the time the shaking is started, and once every 1 to 2 min after that, a sample (about 1 ml) of the fluid is taken to check for free islets. The degree of islet dissociation is monitored by dithizone staining and microscopic examination. The sample is placed in a 35-mm Petri dish with 80 μl dithizone. After about 30 seconds, 2 ml of PBS is then added and the sample is examined under 40X magnification.
14. The enzymatic digestion goes on until most of the islets seen in the samples are free from acinar tissue (Figure 6.7). This phase of the isolation usually takes less than 15 min.
15. When the islets are sufficiently digested, begin collecting the tissue into 1 L bottles instead of letting it return to the recirculating beaker. As the digested tissue is collected, it is cooled so the enzyme is inactivated and digestion stops. The remainder of the warm HBSS is added to the recirculation beaker and collected with the digested tissue as it exits the chamber.
16. The next fluid to be added to the chamber (via the recirculating beaker) is cold collecting medium (5% serum). More collecting medium is added to the beaker as it is removed with the tissue into the collecting bottles.
17. From the collecting bottles, the tissue is transferred to 250-ml centrifuge tubes and centrifuged for 3 min at 1,100 RPM.

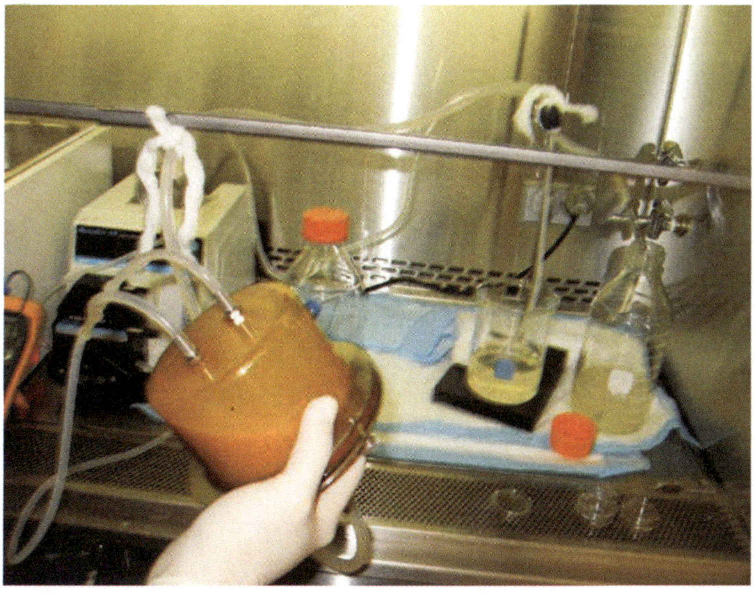

Figure 6.6 Ricordi chamber with pancreas being gently shaken by hand. In the background is recirculating beaker and collection bottle.

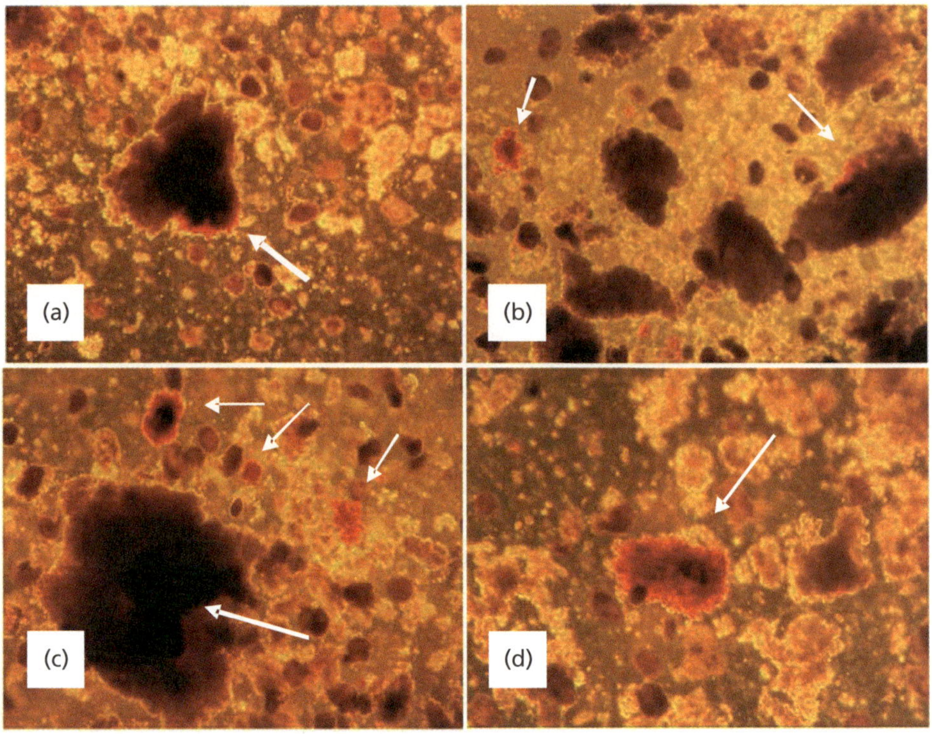

Figure 6.7 Stages of digestion. Each sample was taken from the recirculating beaker at one minute intervals. Endocrine tissue (islets) are stained red with dithizone. Arrows point to islets. (a) Islet trapped in acinar tissue. (b) One islet is trapped and one islet is free of acinar tissue. (c) More islet fragments appear free but some are still trapped. (d) No, or very few, trapped islets are seen. Free islets appear whole instead of fragmented. This is the point at which digestion is stopped and collection begins. Magnification 40X.

18. A 1L tissue collecting beaker is prepared to receive the digested tissue with 500 ml collecting medium that is supplemented with an additional 50-ml porcine serum. The beaker is placed on ice and kept in hood # 2.
19. The supernatant is poured off and discarded from the 250-ml tubes and the tissue pellet is added to the collecting beaker. This is repeated until all the tissue from the pancreas is combined in the beaker. It is helpful to record the amount of tissue in the packed pellet as it is added to the beaker to estimate the number of COBE purification runs needed (Troubleshooting Table).
20. The collection is stopped when most of the digested tissue has been pumped out of the Ricordi chamber. An indication of when to stop the collection is when the tissue pellets get smaller. The collection phase could take up to 90 min and 8 to 10 liters of collecting media.
21. When the decision is made to stop, collecting medium is no longer added to the recirculating beaker and all the fluid is pumped out of the Ricordi chamber and collected.
22. The remaining undigested tissue in the chamber is weighed and subtracted from the initial weight of the pancreas (Figure 6.8(a)). This is done to obtain a final weight of digested tissue and to determine how much of the pancreas was not digested (Troubleshooting Table). If the pancreas was not perfused completely during surgical

harvesting, the tissue that still has blood remaining will not digest. Blood remnants interfere with the enzyme activity (Figure 6.8(b)).

23. The tissue accumulated in the collecting beaker is centrifuged and split into portions of 20 to 23 ml of packed tissue (enough for one COBE run). Each portion is placed into a separate 250-ml conical tube. A large pancreas can yield enough tissue for four to five COBE runs.
24. The cells are resuspended in collecting medium and kept on ice until they are ready to be purified (islet separation from acinar tissue) using the COBE 2991 cell processor. At this time, the cells are centrifuged, and the supernant is aspirated and replaced with cold polysucrose solution (see 6.5.5, point 5).

6.5.5 Polysucrose Purification of Tissue

Although polysucrose purification of islets from the mass of nonislet pancreatic tissue can be accomplished using discontinuous gradients in 50-ml centrifuge tubes, due to the large amount of tissue, a much faster and less labor intensive method is preferred using the COBE 2991 Cell Processor. Up to 23 ml of tissue can be processed at one time and each COBE run takes about 30 min to complete.

1. COBE setup: Turn on the machine and load a processing set into place. The processing set bag is placed in the centrifuge bowl with the main tubing line through the center of the centrifuge cover (Figure 6.9). The four holes of the bag are placed over the four prongs in the centrifuge and positioned as flat as possible. The alignment blocks are placed around the inlet stem and the cover is set in place over the bag and locked. COBE is set for manual operation.
2. Typically, the feed line is placed in position V3 (yellow) and position V2 (green) is used as a pressure release. Alternatively, and in our example, the top two lines on the other side of the machine can be used V1 (red) for pressure release and collect (blue) for the feed line if it is more convenient. The machine needs to be placed close to the hood to keep the gradients and tissue sterile as they are being loaded.

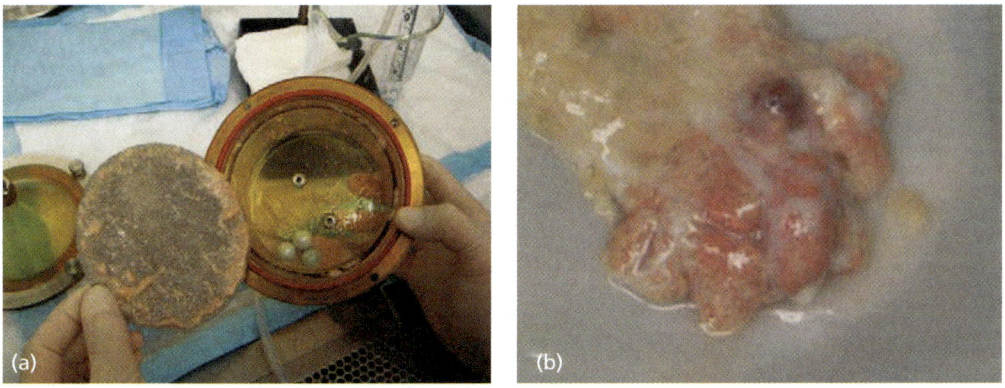

Figure 6.8 (a) Undigested tissue remaining in the Ricordi chamber after collection of digested tissue has concluded. (b) Undigested tissue that was not perfused completely to remove all blood during pancreatic harvest. Blood interferes with enzyme action and this tissue will not digest.

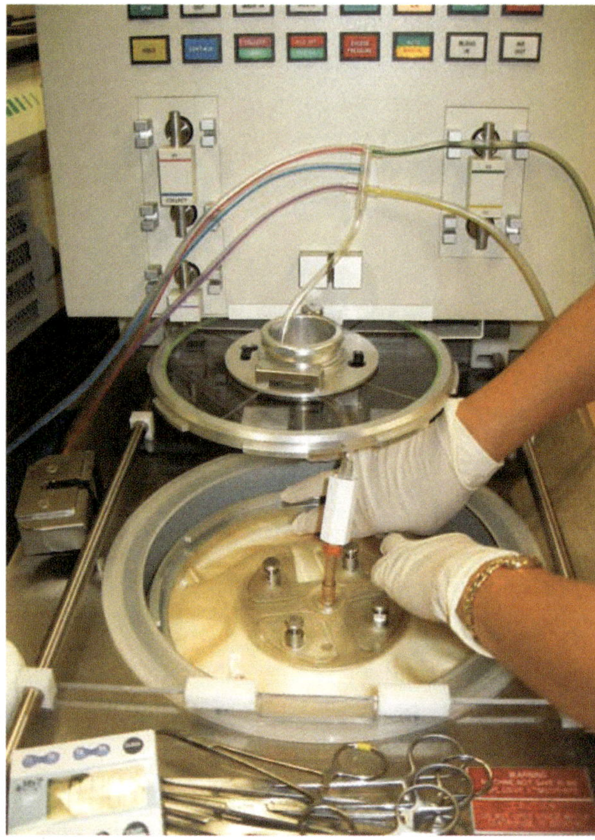

Figure 6.9 Positioning of the lines and processing set bag in the COBE centrifuge bowl.

3. The tubing lines are attached to the COBE line holders and four of the five lines are clamped with hemostats (Figure 6.10). Only the feed line tubing (blue) remains open.
4. A feed line is attached to the remaining unclamped tubing line (blue line in Figure 6.10). The other end of the feed line is attached to the conical end of a 5-ml Falcon pipette. The filter end of the pipette is broken off. This sterile pipette is kept in the hood and used to load the cells and islet gradients.
5. The feed line is hooked up to a peristaltic pump used to load the solutions. Up to 23 ml of packed tissue is washed in collecting media and centrifuged in a 250-ml conical tube at 1,100 RPM for 3 min at 4°C. The supernatant is aspirated and the cells are resuspended up to 225 ml in polysucrose stock (Euro-Collins base) solution.
6. The cells are then fed into the COBE (centrifuge not spinning) with the peristaltic pump at a rate of about 80 ml/min. or within 3 min.
7. When all cells are loaded into the bag, start to load 75 ml of the 1.108 islet gradient into the feed line. When the solution gets to the junction of the feed line (collect, blue) and the main line to the bag, clamp the line and shut off the pump.
8. Unclamp the pressure release line above the feed line (V1, red) and start the COBE centrifuge spinning at 1,000 RPM.
9. Press the SuperOut valve and set the SuperOut rate at 100 ml/min. Allow the cell solution to be pushed up to the junction with the 1.108 gradient.

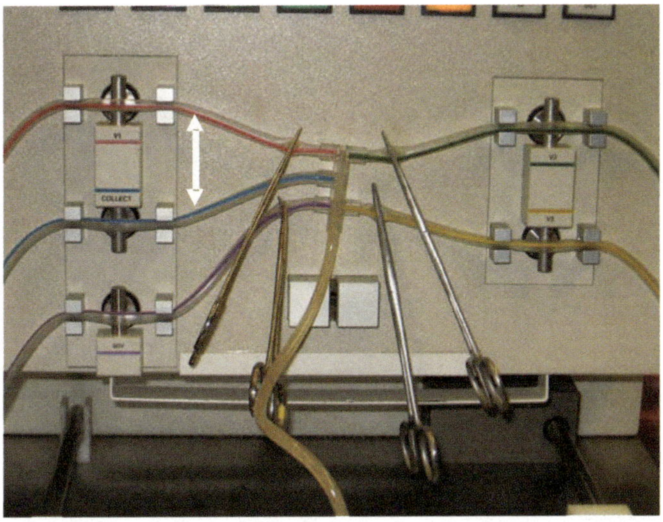

Figure 6.10 Lines attached to the COBE. All are clamped except the feed line (blue in this example). The cells are being loaded into the bag. Arrows point to the two lines that are actually used in this procedure. Upper line (red) is used for pressure release and lower (blue) is attached to the feed line.

10. Quickly clamp the open pressure release line and press the Stop/Reset button. This is done to remove any air in the system. Now the cell solution and the 1.108 gradient should meet and have no air bubbles between them in the line.
11. Unclamp the feed line and start the centrifuge spinning. The centrifuge should remain spinning until the end of the run. Turn the SuperOut rate down to zero.
12. Turn on the peristaltic pump and set it at a slower rate (50 ml/min) to load the remaining density gradients. Load, in order, 75 ml/ea (the rest of) 1.108, 1.096, and 1.037. Follow this with 50 ml of HBSS and load until the last of the fluid reaches the main line, just above the bag.
13. Clamp the feed line, turn off the peristaltic pump, and open the pump head to release the line.
14. Unclamp the feed line and press SuperOut (with the flow rate at 0) to relieve excess pressure. Centrifuge for an additional 5 min.
15. To recover cells, fractions are collected into 250-ml conical tubes at the end of the 5 min by turning the SuperOut rate up to 100 ml/min.
16. Collect 100 ml in the first tube and 75 ml/ea into five more tubes. After the last tube is filled or the flowing stops, the Stop/Reset button is pushed, the centrifuge is stopped, and the run is finished.
17. The tubes are then filled to 250 ml with collecting media and centrifuged for 3 min at 1,200 RPM (higher speed because of the polysucrose) at 4°C.
18. The supernatant is removed by aspiration leaving about 50 ml in each tube. Fill the tube up to 250 ml again with collecting media and resuspend the pellet.
19. When the tissue is well mixed, take a 1-ml representative sample from each tube and place it in a 35-mm Petri dish with 80 μl dithizone. Add 2 ml of PBS and examine each sample microscopically and estimate purity under 40X power.
20. Samples of similar purity are pooled together. Our lab typically creates a "Fraction A" of 80% purity or better (Figure 6.11) and "Fraction B" of 80% to 30% purity. Unless specifically requested, anything less than 30% purity is discarded.

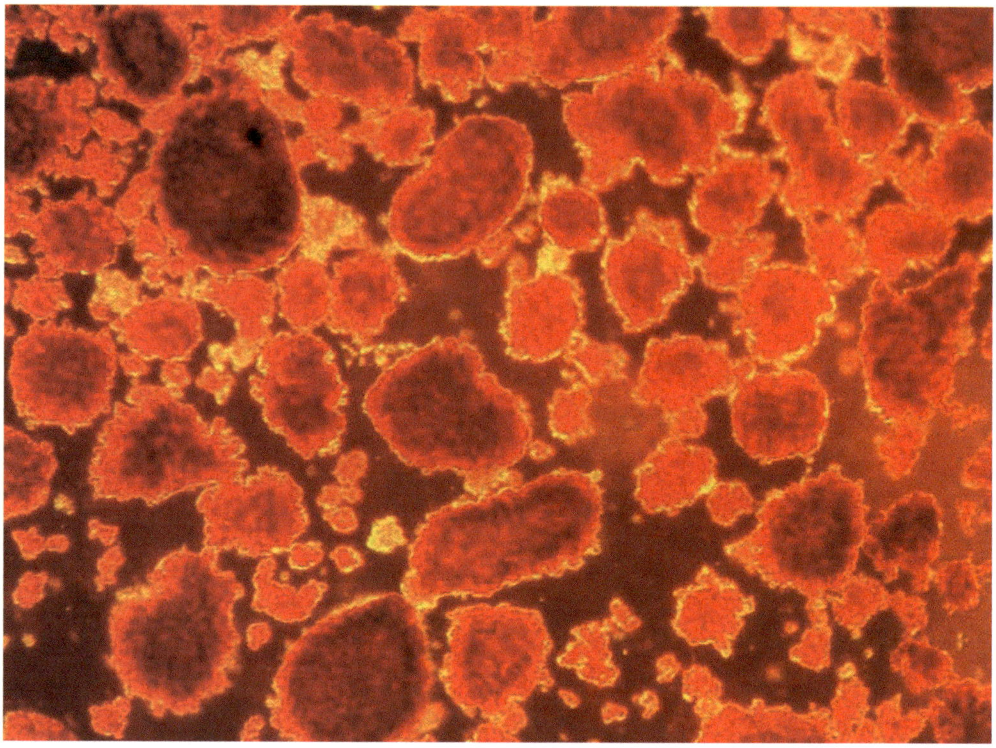

Figure 6.11 Purified pig islets after COBE discontinuous gradient separation. Islets shown are 95% pure and dithizone (red) stained. Magnification 40X.

21. The tubes are centrifuged again, supernatant aspirated, and the pellets are combined based on purity. Fractions A and B are washed once more and the pellet size of each fraction is recorded. The cells are now ready to be counted and placed in culture.

6.5.6 Counting Islet Equivalents

Islets are micro-organelles with sizes ranging on average from 50 to 350 μm in diameter. Islets of different size can be equalized based on their diameter to an islet with an arbitrary diameter of 150 μm, called an islet equivalent (IEQ). The coefficients for the conversion of a given size islet to IEQ are indicated in Table 6.1 [16]. Final counts are expressed as IEQs. Single cells or aggregates smaller than 50 μm in size are not considered for counting of adult pig islets.

1. After islets are pooled together based on purity, each fraction is resuspended in 250 ml of collecting medium.
2. A 0.5-ml representative sample is taken from each fraction and placed in a 35-mm Petri dish with grid (to aid counting) containing 80 μl dithizone and allowed to incubate at room temperature for 1 to 2 min. 2 ml PBS is then added.
3. Examine at 40X under an inverted microscope equipped with a calibrated grid in the eyepiece. Islet aggregates are stained red by the dithizone.
4. Count the stained islets by dividing them into the size categories listed in Table 6.1. Use the conversion factor for that category to determine the IEQ for that size.

Table 6.1 Conversion Factors to Calculate Islet Equivalents

Category	Islet Diameter Range (μm)	Conversion into Islets of 150 μm Diameter (IEQ Conversion Factor)
1	50–100	n/6.00
2	100–150	n/1.50
3	150–200	n×1.7
4	200–250	n×3.5
5	250–300	n×6.3
6	300–350	n×10.4
7	>350	n×15.8

5. After all islets in each sample are counted and converted to IEQs, add all the categories and determine total IEQ for that sample. Multiply that number by the volume of the fraction. (e.g., 1,000 IEQs in 0.5 ml of a 250-ml fraction = 1000×2× 250 = 500,000 IEQs in that fraction).

6.5.7 Culture

1. After the fractions of tissue are combined based on relative purity, they are washed one more time then pelleted by centrifugation for 3 min at 1,100 RPM.
2. The supernatant is aspirated and the volume of the pellet is estimated. The tissue is split into T-175 flasks (nontissue culture treated) at the rate of 50- to 100-μl tissue/flask in 75 ml culture medium.
3. The medium is changed at 24h and then every 2 days after that. Since the islets are delicate, every centrifugation will increase damage. Therefore, a preferred method of changing the medium is to let the tissue gravity settle in the flask for about 5 min, then carefully remove 20 to 30 ml without disturbing the islets and replace with fresh medium (Troubleshooting Table).

6.5.8 Assessment of Islet Quality

Assessment of islet quality is based on viability and function analysis in vitro and in vivo. For in vitro viability testing, our lab uses a calcein AM/propidium iodide assay (Molecular Probes, Carlsbad, CA) for viability.

1. For the working solution, mix 1 ml PBS, 1 μl calcein AM, and 10 μl propidium iodide (PI).
2. Place 12 to 15 islets in a small drop (20 μl) of PBS. Add 60 μl of the mix to the islets.
3. Cover the dish and place in a 37°C incubator for 25 to 30 min.
4. Fluorescent micrographs are taken immediately after incubation. Live cells appear green while dead cells will stain red from the PI (Figure 6.12).
5. Viability can be quantitated with MetaMorph® software. Typically, viability is expected to be 90% or better.

For in vitro functional testing, islet preparations are subjected to a dynamic secretagogue challenge in a perifusion apparatus [11]. The principle of this functional test focuses on measuring insulin release in response to stimuli, as a way to determine the ability of islet beta cells to recognize a secretagogue or inhibitory signal and control

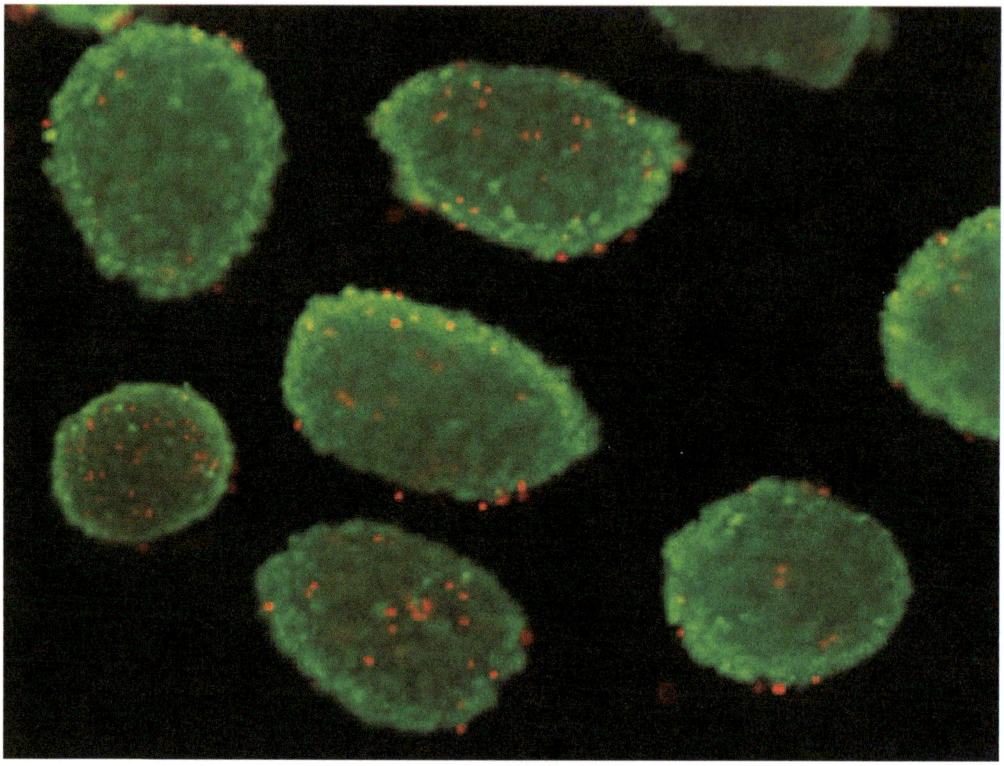

Figure 6.12 Viability assay. Green (calcein AM) shows live cells, Red (propidium iodide) stains dead cells. Viability in this example is 95% as calculated with MetaMorph® software.

insulin release accordingly. The test involves the exposure of islets to media with different glucose concentrations with and without insulin release enhancers for about 2 h in an open flow system. The media that is collected at the end of the circuit in 1-min fractions is analyzed for insulin concentration. Figure 6.13 provides an example of the porcine islet response when subjected to this assay.

1. Islets are allowed to incubate overnight at 37°C in culture medium then are loaded in the perifusion apparatus (100 handpicked islets)
2. Islets are conditioned for 30 min in Kreb's Ringer bicarbonate buffer (KRBB) supplemented with 0.5% bovine serum albumin (BSA Fraction V, Sigma-Aldrich) containing 2.8 mM glucose.
3. Experimental conditions involve exposure of islets to KRBB with 2.8 mM glucose (low concentration), 20 mM glucose (high concentration), and subsequently 2.8 mM glucose for 30 min for each glucose condition [17].
4. An additional 30 min with high glucose plus arginine (10 mM) or theophylline (5 mM) is also included, due to the prevalent response of pig islets to secretagogues other than glucose.
5. Eluates are collected every minute. Insulin concentration in the eluates is measured by ELISA (porcine insulin kit, Alpco, Salem, NH) and the data obtained are used to calculate the stimulation index [11]. The stimulation index is the ratio between insulin amounts released during stimulated (high glucose or arginine/theophyllin) over basal (low glucose) conditions [18].

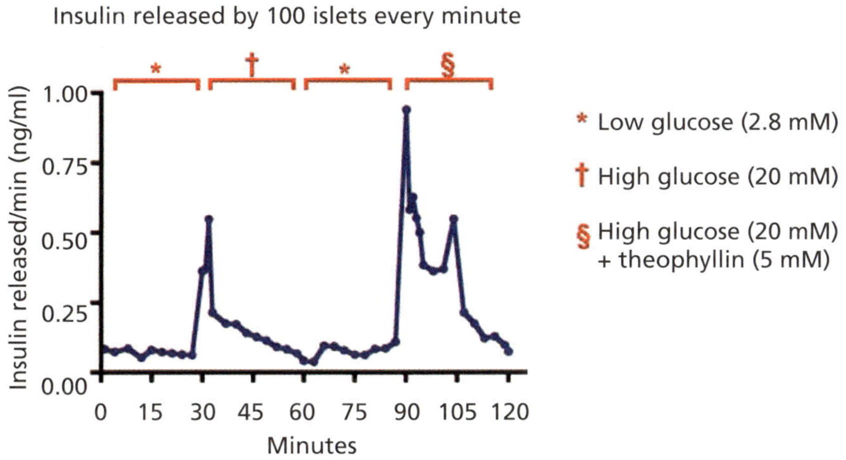

Figure 6.13 Insulin response of the pig islets to a dynamic secretagogue challenge. Islets were exposed to 30 min of low glucose (2.8 mM), followed by 30 min of high (20 mM), another 30 min of low glucose to allow for recovery, then a final challenge with high glucose coupled with theophylin (5 mM). Samples were taken of the eludate every 1 min and analyzed by porcine insulin ELISA.

In vivo analysis consists of transplantation of pig islets under the kidney capsule of immuno-deficient mice [19]. Our lab uses NOD*scid* mice but any *scid* or nude mouse used for xenotransplantation is adequate. For quality control (QC), we usually transplant four mice. Two receive 300 islets and two receive 600 islets.

1. The animals are first rendered diabetic with a single high dose of streptozotocin (200–250 mg/kg body weight).
2. Islets are usually counted by hand for transplantation into mice. When the mice are diabetic for 2 days or more, (over 300 mg/dl blood glucose) they are transplanted with differing numbers of islets for the QC assay.
3. Aseptic conditions are observed throughout surgery. The mice are anesthetized with an i.p. injection of 2.5% Avertin, 0.016 ml/g body weight. The incision site is shaved and swabbed with betadine.
4. A 1-cm incision is made in the flank over the left kidney and through the abdominal wall. The kidney is externalized and a hole is made in the capsule at one pole of the kidney.
5. The islets are gathered into a small piece of PE 50 tubing and inserted into the space between capsule and kidney. The islets are expelled as the tubing is withdrawn.
6. The kidney is then replaced in the abdominal cavity and the abdominal wall is sutured with 6-0 silk before the outer skin is closed with small wound clips. The animal is kept warm and observed throughout recovery. Wound clips are removed after 7 to 10 days.
7. The blood glucose of the grafted animals is recorded two to three times per week for 60 days. The animal should return to normoglycemia within 1 to 2 days after islet transplantation. Normal blood glucose for a mouse is 70 to 120 mg/dl.
8. At the end of 60 days, if glycemia is still normal, the graft-bearing kidney is removed and the mouse allowed to recover.

9. Nephrectomy is a similar surgery to the islet transplantation. The kidney is externalized as before and the renal artery and vein are ligated before removing the kidney.
10. If the animal returns to a diabetic state after graft removal, this is proof that the pig islets were functional. If the islets are very good, the mice with 300, as well as 600 islets, will be normal for the full 60 days. If the islets are marginally functional, only the animals receiving 600 islets will be normal. If the islets have low functionality or are damaged, the blood glucose of the mice will not normalize or will not normalize for the full 60 days.

6.6 Anticipated Results

A successful pig islet isolation is determined by the amount, purity, and function of the islets obtained. Several parameters outlined in Table 6.2 reflect the outcome of what we routinely experience with our method.

6.7 Discussion and Commentary

The future of islet transplantation for the cure of type 1 diabetes is conditional to the availability of novel islet cell sources. Swine could be the ideal islet cell donors, due to their availability, their metabolic similarities with humans, their rather large beta cell mass and the possibility of genetically engineering their tissues for greater compatibility to humans [20–22]. While xeno-rejection remains one of the major impediments to successful islet xenotransplantation programs, several possible strategies have been developed. In addition to tailoring immunosuppression to protect the graft against immune-destruction, islet encapsulation represents an attractive technology for protecting the islet graft and ensuring graft function [23]. Despite enormous progress and the availability of biocompatible materials suitable for islet encapsulation, only limited success has been achieved. Large amounts of islets are still required even after encapsulation. Whereas encapsulation seems to efficiently prevent cell mediated islet destruction, the activity of cytokines and proinflammatory molecules is only minimally modulated, and these molecules are sufficient to eventually destroy islet function. More efficient isolation approaches are therefore required. The current method for bulk isolation of porcine islets is still a work in progress, whereas standardization has not yet been fully achieved.

Table 6.2 Expected Outcomes from Pig Islet Isolations

	Young Pig	Adult Pig
Pancreas weight	200–300 g	350–500g
Islet equivalent (IEQ)	180,000–300,000	400,000–600,000
Purity (islets/whole tissue)	65%–75%	75%–90%
Stimulation index (insulin release: stimulated/basal conditions)	Glucose 1.3–2.2 Theophylline 2.0–6.5	Glucose 1.3–2.2 Theophylline 2.0–6.5
Viability	90–95%	90–95%

Troubleshooting Table

Problem	Explanation	Potential Solutions
1. COBE cell separator run failure.	Excessive fluid loads. Power failure. Faulty setup.	Wash out gradients and repeat purification steps. Manual gradient layering and conventional centrifugation.
2. Underdigestion of pancreatic tissue (more than 25% of tissue left in chamber), or most of the islets are mantled with acinar tissue.	Poor pancreas perfusion (blood remnants). Low enzyme concentration or activity. Short perfusion time. Temperature too low. Failure to maintain temperature within appropriate ranges.	This condition can only be prevented by standardization of enzyme/digestion parameters and optimization of harvesting.
3. Overdigestion of pancreatic tissue (islet fragmentation).	Prolonged warm ischemia. High enzyme concentration. Excessive exposure to digestive enzyme. Temperature too high. Failure to maintain temperature within appropriate ranges.	This condition can only be prevented by standardization of enzyme/digestion parameters and optimization of harvesting.
4. Loss of islet integrity during culture.	Acinar contamination (low purity). High density of islets/culture flask. Inadequate nutrients. Mechanical damage.	Increase medium volume/amount islets. More frequent medium changes. Centrifuge islets more slowly and for less time.

In view of a broad application of the porcine islet cells for clinical use, efforts should be undertaken to improve and automate the process of isolation and purification of islets. Pig islet manufacture should become a standardized, safe, and simple procedure that produces consistently large yields of high quality for clinical use.

The first area that needs to be optimized is the type and amount of enzyme used to digest the tissue. We have used the different enzyme blends mentioned in the methods section and are not completely satisfied [24]. We still sometimes experience a large percentage of the pancreas that remains undigested (30–40%) and feel this can be improved upon with the right blend. We also prefer to begin infusing the pancreas with prewarmed (24°C) enzyme. This results in a shorter time until the tissue begins to digest after agitation begins in the Ricordi chamber. If there is a problem with over digestion of the tissue, it may be preferable to start infusing the pancreas with cold (4°C) enzyme (Troubleshooting Table). Cold enzyme infusion will prevent the tissue from digesting too quickly and more time will be available to identify the point at which to stop digestion and begin collection. This is the point when most of the islets are free of acinar tissue and sometimes the stop point can be difficult to identify. In our experience, when we see two to four medium-to-large islets in the sample without acinar tissue attached to them (free), it is time to stop the digestion phase and begin collecting tissue.

6.8 Summary

1. Human islets are in short supply for clinical islet transplantation in diabetic patients
2. Pig islets are an ideal candidate to replace beta cell function in humans.
3. Porcine islets are difficult to isolate due to their fragility.
4. Large-scale methods for the isolation of porcine islets are available.

5. Specific enzyme blends and more gentle mechanical digestion are required to limit damage to the islets during isolation and support their engraftment and functionality.

Acknowledgments

The authors wish to thank Carmella "Candy" Knoll and Catarina Wong for their tireless dedication and excellent technical assistance.

References

[1] Berney T., L. H. Buhler, and P. Morel. "Pancreas Allocation in the Era of Islet Transplantation," *Transpl. Int.,* Vol. 18, No.7, 2005, pp. 763–7.

[2] O'Neil, J. J., et al., "The Isolation and Function of Porcine Islets from Market Weight Pigs," *Cell Transplant.,* Vol. 10, No. 3, 2001, pp. 235–46.

[3] Krickhahn, M., et al., "Highly Efficient Isolation of Porcine Islets of Langerhans for Xenotransplantation: Numbers, Purity, Yield and In Vitro Function," *Ann. Transplant.* Vol. 6, No. 3, 2001, pp. 48–54.

[4] Sabat, M., et al., "Assessment of Some Porcine Strains as Donors of Islets of Langerhans," *Transplant. Proc.,* Vol. 35, No. 6, 2003, pp. 2343–44.

[5] Kim, J. H., et al., "Influence of Strain and Age Differences on the Yields of Porcine Islet Isolation: Extremely High Islet Yields from SPF CMS Miniature Pigs," *Xenotransplantation,* Vol. 14, No. 1, 2007, pp. 60–66.

[6] Georges, P., et al., "Comparison of Size, Viability, and Function of Fetal Pig Islet-Like Cell Clusters after Digestion Using Collagenase or Liberase," *Cell Transplant.,* Vol. 11, No. 6, 2002, pp. 539–45.

[7] Rood, P. P., et al., "Pig-to-Nonhuman Primate Islet Xenotransplantation: A Review of Current Problems," *Cell Transplant.,* Vol. 15, No. 2, 2006, pp. 89–104.

[8] Cardona, K., et al., "Long-Term Survival of Neonatal Porcine Islets in Nonhuman Primates by Targeting Costimulation Pathways," *Nat. Med.,* Vol. 12, No. 3, 2006, pp. 304–6.

[9] Korbutt, G. S., et al., "Large Scale Isolation, Growth, and Function of Porcine Neonatal Islet Cells," *J. Clin. Invest.,* Vol. 97, No. 9, 1996, pp. 2119–29.

[10] Hori, H., et al., "Isolation, Culture, and Characterization of Endocrine Cells From 6-Month-Old Porcine Pancreas," *Cell Transplant.,* Vol. 10, No. 4–5, 2001, pp. 459–64.

[11] Bottino, R., et al., "Isolation Outcome and Functional Characteristics of Young and Adult Pig Pancreatic Islets for Transplantation Studies," *Xenotransplantation,* Vol. 14, No. 1, 2007, pp. 74–82.

[12] Krickhahn, M., et al., "The Morphology of Islets within the Porcine Donor Pancreas Determines the Isolation Result: Successful Isolation of Pancreatic Islets Can Now Be Achieved from Young Market Pigs," *Cell Transplant.,* Vol. 11, No. 8, 2002, pp. 827–38.

[13] Lakey, J. R. T., et al., Pancreas Procurement for Islet Isolation and Pancreas Storage Prior to Islet Isolation, Austin, TX: R.G. Landes Co., 1995.

[14] Ferrer, J., et al., "Pig Pancreas Anatomy: Implications for Pancreas Procurement, Preservation, and Islet Isolation," *Transplantation,* Vol. 86, No. 11, 2008,pp. 1503–10.

[15] Balamurugan, A. N., et al., "Flexible Management of Enzymatic Digestion Improves Human Islet Isolation Outcome from Sub-Optimal Donor Pancreata," *Am. J. Transplant.,*Vol. 3, No. 9, 2003, pp. 1135–42.

[16] Ricordi, C., Islet Isolation Assessment in Pancreatic Islet Cell Transplantation, Austin, TX: R.G. Landes Co., 1992.

[17] Davalli, A. M., et al., "Paradoxical Release of Insulin by Adult Pig Islets In Vitro. Recovery after Culture in a Defined Tissue Culture Medium," *Transplantation,* Vol. 56, No. 1, 1993, pp. 148–54.

[18] Gouin, E., et al., "Perifusion Analysis of Insulin Secretion from Specific Pathogen-Free Large-White Pig Islets Shows Satisfactory Functional Characteristics for Xenografts in Humans," *Diabetes Metab.,* Vol. 24, No. 3, 1998, pp. 208–14.

[19] Bertera S, et al. "Gene transfer of manganese superoxide dismutase extends islet graft function in a mouse model of autoimmune diabetes," *Diabetes,* Vol. 52, No. 2, 2003, pp. 387–93.

[20] Phelps, C. J., et al., "Production of Alpha 1,3-Galactosyltransferase-Deficient Pigs," *Science,* Vol. 299, No. 5605, 2003, pp. 411–14.

[21] Cooper, D. K., E. Koren, and R. Oriol, "Genetically Engineered Pigs," *Lancet,* Vol. 342, No. 8872, 1993, pp. 682–83.

[22] Kolber-Simonds, D., et al., "Production of Alpha-1,3-Galactosyltransferase Null Pigs by Means of Nuclear Transfer with Fibroblasts Bearing Loss of Heterozygosity Mutations," *Proc. Natl. Acad. Sci. USA,* Vol. 101, No. 19, 2004, pp. 7335–40.

[23] Beck, J., et al., "Islet Encapsulation: Strategies to Enhance Islet Cell Functions," *Tissue Eng.,* Vol. 13, No. 3, March 2007, pp. 589–99.

[24] Balamurugan, A. N., et al., "Harmful Delayed Effects of Exogenous Isolation Enzymes on Isolated Human Islets: Relevance to Clinical Transplantation," *Am. J. Transplant.,* Vol. 5, No. 11, 2005, pp. 2671–81.

7

Spinal Cord Injury Models and Cell Transplantation Therapy

Akio Iwanami,[1] Narihito Nagoshi,[1] Yoshiaki Toyama,[1] Masaya Nakamura,[1] and Hideyuki Okano[2*]

[1] Department of Orthopedic Surgery, Keio University School of Medicine
[2] Department of Physiology, Keio University School of Medicine
*Correspondence: 35 Shinanomachi, Shinjuku-ku, Tokyo, 160-8582 Japan
Phone: 03–5363–3747; Fax: 03–3357–5445; E-mail: hidokano@sc.itc.keio.ac.jp

Abstract

Preparation of injury models and evaluation of motor function are essential in doing research on spinal cord injury. Spinal cord injury models are broadly classified into three types: the section model, the contusion injury model, and the demyelination model. Each model has its own advantages and drawbacks. Therefore, a model suitable for the experimental design should be selected correctly. It is also of supreme importance to create reproducible models that would allow objective evaluation of the motor function. This chapter briefly describes methods for the preparation of spinal cord injury models, methods of functional evaluation using these models, and therapeutic cell transplantation strategies, while also introducing the results of studies published in recent years.

Key terms: spinal cord injury, injury models, animal models, functional analysis, cell transplantation, regeneration

7.1 Introduction

With the recent advances in the field of neurodevelopmental biology, the molecular mechanism and plasticity of axonal elongation in central nervous system (CNS) have been elucidated. Moreover, technologies developed for the identification, isolation, and expansion of stem cells (neural stem cells, embryonic stem cells, induced pluripotent stem cells, etc.) raise the enormous potential of therapeutic applications for CNS disorders such as spinal cord injury. Therefore, many researchers are currently involved in projects aiming at regeneration of injured spinal cord.

As spinal cord injuries are most often the result of trauma, no spinal cord injury mouse models can be prepared by gene manipulation. Therefore, it is really important for a researcher to select a model suitable for the purpose of experiment, create an injury model quantitatively (i.e., without individual differences) and evaluate the therapeutic response objectively when functional recovery is examined. Hence, the experimenter's anatomical knowledge and operative skill would be required, and functional evaluation by at least two observers would also be indispensable for getting reliable data. In this chapter, we briefly describe the methods for the induction of spinal cord injury models and methods that may be used for functional evaluation using these models, while also introducing the results of studies published in recent years, and review therapeutic strategies for spinal cord injury with primary reference to cell transplantation methods.

7.2 Experimental Design

7.2.1 Pathophysiologic States of Spinal Cord Injury and Therapeutic Strategies

Before proceeding to study design, one should have an understanding of the pathophysiologic states of spinal cord injuries.

In cases of spinal cord injury, neurons, astrocytes, and oligodendrocytes at the site of injury become crushed by a direct blow, along with tearing of axons involved in central-peripheral intercommunications and rupture of the microvasculature (primary injury). Tissues of regions with interrupted blood flow become hypoxic and necrotized and disruption of the blood-spinal cord barrier occurs consequent to rupture of the microvasculature, followed by infiltration of the spinal cord by inflammatory cells such as leukocytes, macrophages, and microglia, which eventually result in elevated phagocytotic activity and the release of cytokines. As a result, cells surrounding the tissues that are directly destroyed are also lost as a result of necrosis or apoptosis (secondary injury). As the inflammation subsides in due course, a cavity-like space is formed at the center of the injured region, which eventually becomes surrounded by a thick glial scar composed of proliferating astrocytes [1].

Treatment strategies for spinal cord injuries are mainly classified into two on the basis of these pathophysiologic changes in spinal cord injury: methods for preventing the so-called secondary injury, and methods to facilitate regeneration of the injured axons.

7.2.2 Induction of Models Suitable for the Study Objectives

Virtually all cases of spinal cord injury actually encountered in humans result from contusion of the spinal cord associated with dislocation/fracture of the spinal vertebrae. At the experimental level, nevertheless, it is considered to be of importance to select an injury model according to the hypothesis that needs to be corroborated. In regard to experimental models of spinal cord injury, grossly speaking, the section, contusion and demyelination injury models have been reported in the past, and among the reports are also articles dealing with microinjury models prepared using needles to explore post-traumatic cellular proliferation and migration. Experimental studies on spinal cord injury reported in the past mostly utilized laboratory rodents, but in recent years, spinal cord injury models prepared using nonhuman primates have also been documented on account of the marked differences that exist between rodents and primates, both evolutionarily and anatomically [2, 3]. For higher levels of spinal cord injury, furthermore, reports of thoracic spinal cord injury models have been overwhelmingly more frequent than those of cervical cord injury, while models of trauma to the cervical spine have also been used to examine functional recovery in respect of the deft movements of the upper extremities [4].

Thus, it is of vital importance to choose an injury model suitable for the objective of the study while designing the study.

Some examples of simple models of spinal cord injury are listed below:

- To demonstrate axonal regeneration post spinal injury → spinal cord section model (complete, partial);
- To examine regeneration of the injured spinal cord by cell transplantation → spinal cord contusion model.

7.2.3 Controls Required for the Experiments

A control must be set up as a reference for every experimental study.

When, for example, it is intended to assess the effect of some treatment in a spinal cord injury model prepared using certain knockout mice, controls with the spinal cord injury induced in the littermates in the same manner must be used for comparison. In the case where demonstration of regeneration of the injured spinal cord with the use of cell transplants is attempted, it is advisable to use controls comprising animal models with similarly induced spinal cord injury that are injected with culture medium, ground cell transplants (to rule out the possible confounding effects of humoral factors liberated from the cells rather than of the cells per se), or different cells of the same origin (e.g., transplantation of nerve cells of the cerebral cortex as a control if the study is intended to explore the effect of neural stem cell transplantation).

7.3 Materials and Methods

7.3.1 Spinal Cord Section Model

Animals and instruments/tools that need to be kept ready for use:

- For a mouse model: adult female C57BL/6J mice (20–22g).

- For a rat model: adult female Sprague-Dawley rats weighing 200–250g.
- Microscope (surgical microscope: KOM-300S; Konan Medical, Japan).
- Isodine.
- Scalpel (surgical blade).
- Cooper scissors.
- Forceps.
- Scissors for microsurgery, straight and curved.
- Forceps for microsurgery.
- Rongeur (Luer).
- Swabs.
- Normal saline.
- 27G needle.
- 5-0 or 6-0 nylon suture.
- Hemostatics (e.g., gelatin preparations).
- Staples for dermal use.

This is the method used for the preparation of an injury model wherein the vertebral arch at the level of the spinal column is resected, followed by cutting of the cord parenchyma from the dorsal aspect upon opening of the dorsal dura mater. The section model is best suited for examining the effect of therapeutic attempts to promote regeneration of the damaged axons after infliction of spinal cord injury. Establishment of a section model would consist of a technique for producing transection [5], hemisection, or dorsal hemisection (Figure 7.1) [6, 7]: each of these methods has its own advantages and drawbacks. The spinal transection model, for example, is significantly superior for the quantification of spinal cord injury, because the spinal cord parenchyma is totally transected, but is poorly suited for management of the general status. In the hemisection and dorsal hemisection models, conversely, while post-injury management of the general status is

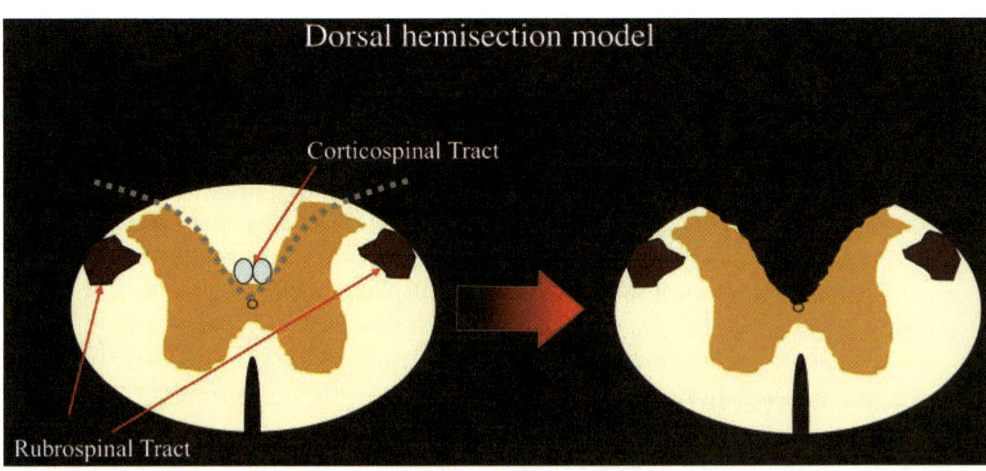

Figure 7.1 Schematic illustration of a spinal dorsal hemisection model (rat). Upon longitudinal section of the dorsal dura mater in a cephalocaudal direction, the corticospinal tract region is cut from the dorsal aspect with microscissors or some other suitable device. An incomplete section results in rather mild paralysis, causing the model to be ineffective. Paralysis induced in such a model is relatively not so severe, because the rubrospinal tract in the lateral funiculus remains intact.

relatively easy, quantification of the injury is extremely difficult. In the preparation of a dorsal hemisection model, for instance, only the posterior funiculus of a rodent (the corticospinal tract) is precisely sectioned (Figure 7.1); however, this model entails difficulty in securing an operative field due to hemorrhage as well as in scrutinizing the depth at the time of the operation. Therefore, an investigator should prepare the surgical instruments and tools for use (e.g., microscissors (Figure 7.2)), and make an effort to improve his or her technical skills beforehand.

The following operation is entirely performed under a microscope.

7.3.1.1 Anesthesia: Intraperitoneal Injection of Ketamine (100 mg/kg) and Xylazine (10 mg/kg).

1. The skin over the intended region for surgery is shaved and thoroughly disinfected with isodine. A median incision is made in the skin of the dorsal back using a scalpel or scissors. After confirming the location of the spinous processes, the dorsal muscles are severed to detach them from the spinous processes and vertebral arches with a scalpel.
2. The eighth thoracic vertebral arch is resected using a rongeur to fully expose the dorsal side of the spinal cord.

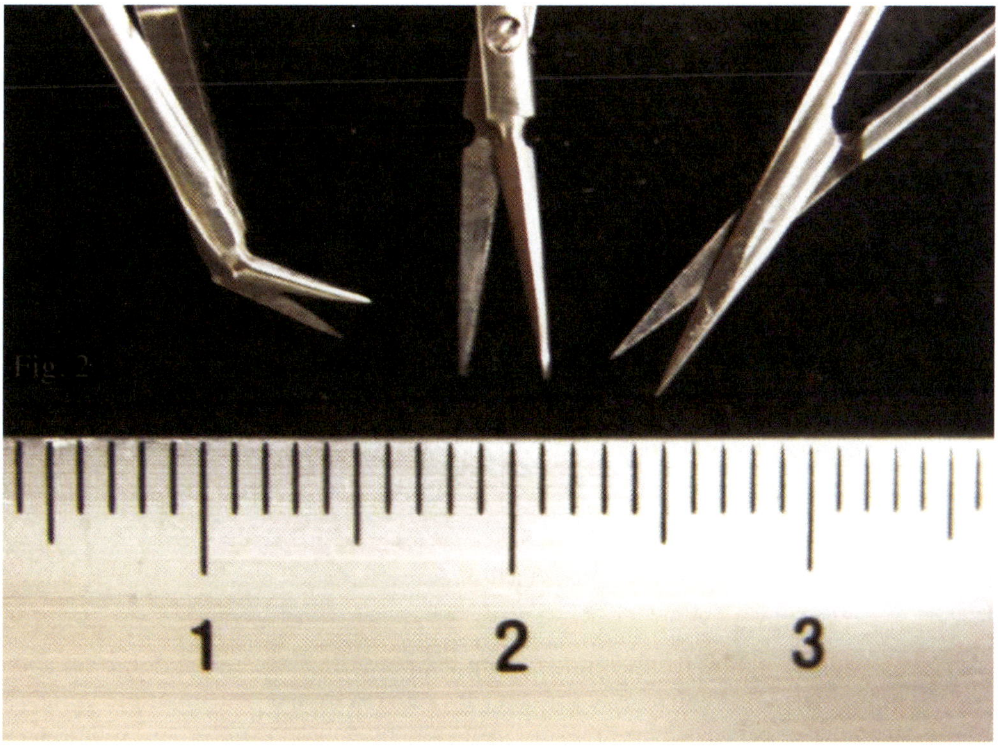

Figure 7.2 Microscissors used for spinal cord sectioning to induce injury. This tool is available from Fine Science Tools, in various shapes according to the intended use. It should even be predetermined as to how many times the same tool must be applied for the sectioning, in order to standardize the operating conditions for the preparation of a spinal section model.

3. A longitudinal slit is made in the dorsal median center of the dura mater of the spinal cord using a 27G needle. This slit is extended in the longitudinal direction using two micro forceps.
4. The spinal cord is severed with microscissors or with a surgical blade while controlling the discharge of cerebrospinal fluid and blood using swabs. In the case of transection, it should be ascertained whether the cord is entirely transected, including the lateral funiculus, while controlling bleeding with swabs and securing the visual field with normal saline. In the case of hemisection or dorsal hemisection, an effort should be made to eliminate individual differences by standardizing the method of sectioning in advance, especially in terms of the frequency and mode of sectioning and the type of micro forceps used.
5. After rinsing with normal saline, hemostasis is confirmed (if bleeding fails to cease, place a hemostatic agent such as a gelatin preparation temporarily at the site, wait until the bleeding stops, and then remove it). The muscles are then sutured using 5-0 or 6-0 nylon suture, the subcutis and skin are disinfected with isodine, and the wound is closed by stapling.
6. The animal is maintained in an incubator at 37°C until awaking from anesthesia. Analgesics and antibiotics are administered.

7.3.2 Spinal Contusion Injury Model

Animals and instruments/tools that need to be kept ready for use:

- For a mouse model: adult female C57BL/6J mice (20–22g).
- For a rat model: adult female Sprague-Dawley rats weighing 200–250g.
- Microscope (surgical microscope: KOM-300S, Konan Medical, Japan).
- Isodine.
- Scalpel (surgical blade).
- Cooper scissors.
- Forceps.
- Forceps for microsurgery.
- Rongeur (Luer).
- Various types of impactors (NYU impactor, IH impactor, etc.).
- Swabs.
- Normal saline.
- 5-0 or 6-0 nylon suture.
- Hemostatics (e.g., gelatin preparations).
- Staples for dermal use.

This is a rather reliable model from the point of view of the etiology of spinal cord injury. Previous reports describe models prepared by such means as applying a stationary weight for a specified time, fastening a microclip on the cord, or inflating a balloon placed in the vertebral canal. More recent articles have dealt with the following techniques: a weight dropping method using a NYU impactor (Figure 7.3) developed by New York University (NYU) [8], a quantitative contusion injury model developed at Ohio State University (OSU) by employing a combination of an electric magnet and con-

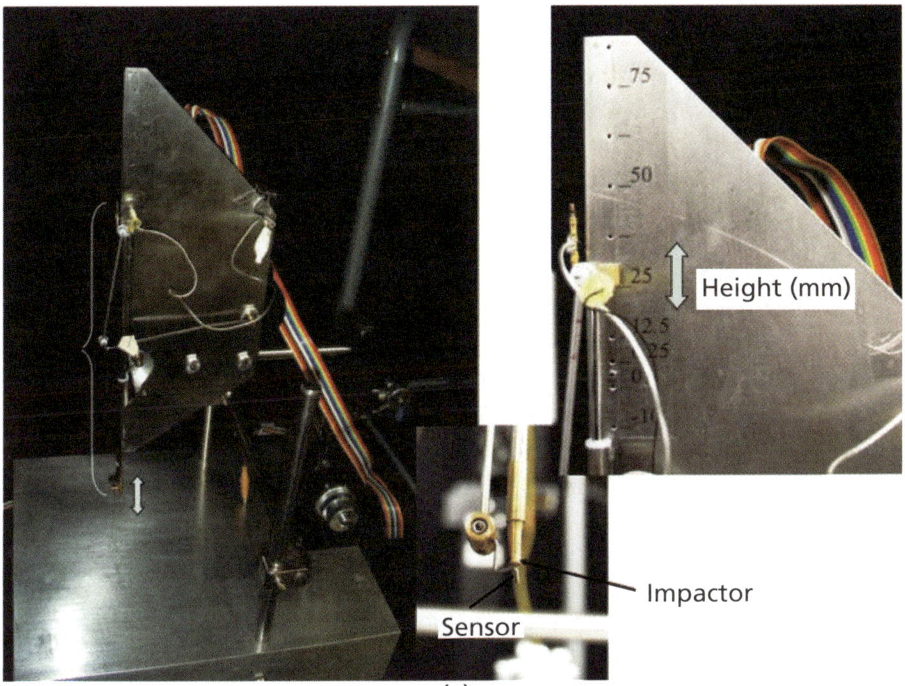

(a)

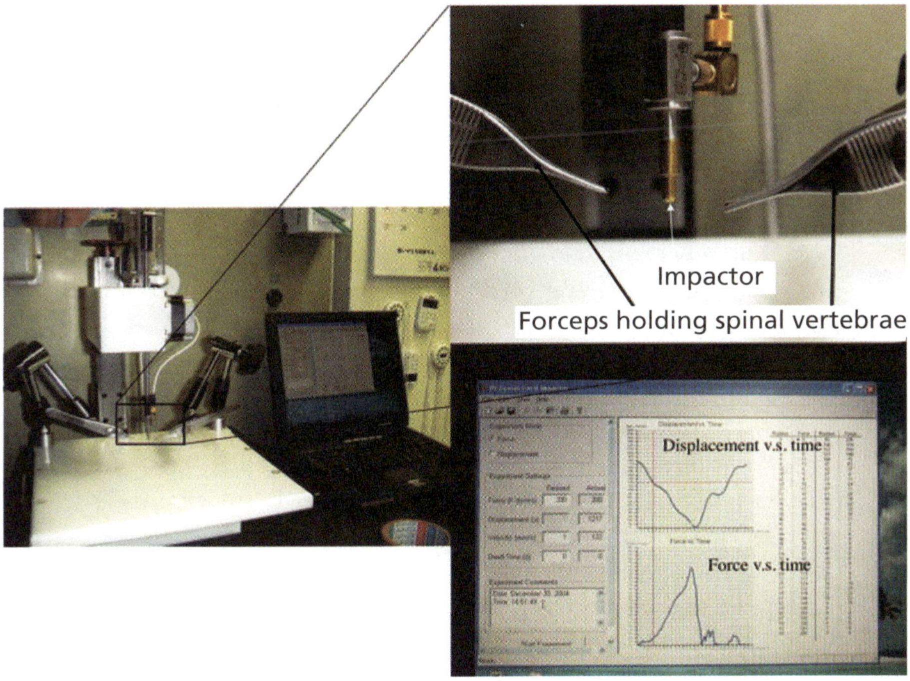

(b)

Figure 7.3 (a) NYU impactor. An apparatus designed to induce contusion injury with proven reproducibility by dropping a 10-g weight from various heights (e.g., 12.5, 25, or 50 mm). A remarkable number of investigators have used this apparatus in their studies on spinal injury. The apparatus is not equipped with a tool for holding the vertebral arch in this figure. (b)IH impactor. An apparatus designed to induce spinal injury by presetting such parameters as the intensity and velocity of the impact given to the spinal cord and the duration of the impact. This instrument is superior for reproducibility-checking performance, since the force actually applied to the spinal cord and the consequent displacement of the cord can be checked on a computer screen just after infliction of the cord injury. It also has the advantage of being applicable not to only rats, but also to mice.

denser [9], a method involving the use of a strain sensor-tipped IH impactor (Figure 7.3) developed in 2002 by Kentucky University [10, 11], and models using slightly modified instruments. All of these methods are designed for quantitative induction of spinal cord injury with an impactor set to drop on the dorsal dura mater, with the adjacent spinous processes or the vertebral body being held after resection of the vertebral arch at the prescribed level of the cord. These methods are not as dependent on the skill of the operator as the method used for establishment of a dorsal hemisection model, however, they do require some training until the operator acquires some level of skill. These models are suitable for examining the effects of cell transplantation or drug injections, because the dura mater is kept unopened, but other techniques such as immunostaining must be used in order to demonstrate the regeneration of axons.

The following operation is entirely performed under a microscope.

7.3.2.1 Anesthesia: Intraperitoneal Injection of Ketamine (100 mg/kg) and Xylazine (10 mg/kg)

1. The skin over the intended region for surgery is shaved and thoroughly disinfected with isodine. A median incision is made in the skin of the dorsal back using a scalpel or scissors.
2. After confirming the location of the spinous processes, the dorsal muscles are severed to detach them from the spinous processes and vertebral arches with a scalpel.
3. The ninth or tenth thoracic vertebral arch is resected using a rongeur to fully expose the dorsal side of the spinal cord. (The level of the cord at which injury is induced varies with the report, but is at the level of the ninth or tenth thoracic vertebral arch in most previous reports.)
4. Adjacent vertebral arches are expanded to an extent that they can be held, then the adjacent upper and lower spinous processes in the case of an NYU impactor, or the adjacent upper and lower vertebral bodies in the case of an IH impactor, are securely held with forceps and immobilized horizontally.
5. The impactor is adjusted to a position so that it drops onto the center of the exposed dorsal dura mater. The weight and height are adjusted according to the intended degree of the cord injury (Figure 7.4).

   Exercise meticulous care to ensure that the impactor drops right onto the center of the spinal cord. Be sure to observe through the microscope in two different directions (i.e., cephalocaudad and right-left direction, so that the impactor does not hit the residual vertebral arch).
6. Induction of injury. It should be ascertained whether the impactor has or has not hit the surrounding tissues such as bone at the time of the impact, or whether the forceps holding vertebral structures have come off or not. An effort should be made to eliminate individual differences by standardizing this process of the method in advance while controlling hemorrhage with swabs.
7. After rinsing with normal saline, hemostasis is confirmed (if bleeding fails to cease, place a hemostatic agent such as a gelatin preparation temporarily at the site, wait until the bleeding stops, and then remove it). The muscles are then sutured using 5-0 or 6-0 nylon suture, the subcutis and skin are disinfected with isodine, and the wound is closed by stapling.

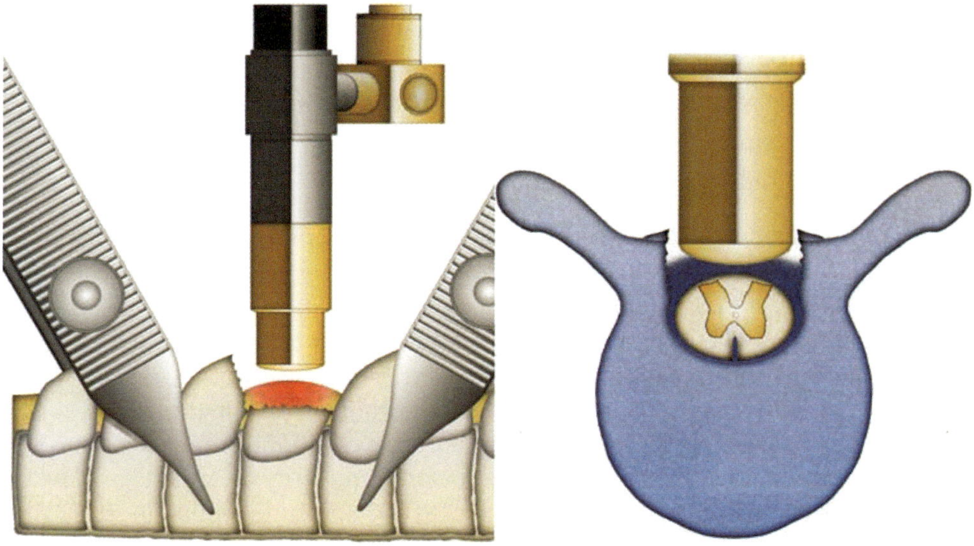

Figure 7.4 Precautions while inducing contusion injury.

8. The animal is maintained in an incubator at 37°C until awaking from anesthesia. Analgesics and antibiotics are administered.

7.3.3 Demyelination Model

Animals and instruments/tools that need to be kept ready for use:

- For a rat model: adult female Sprague-Dawley rats weighing 200–250 g.
- Microscope (surgical microscope: KOM-300S, Konan Medical, Japan).
- Isodine.
- Scalpel (surgical blade).
- Cooper scissors.
- Forceps.
- Forceps for microsurgery.
- Rongeur (Luer).
- 0.1% ethidium bromide.
- An orthovoltage X-ray emission source (Pantak Ltd, UK).
- Swabs.
- Normal saline.
- 27-G needle.
- 5-0 or 6-0 nylon suture.
- Hemostatics (e.g., gelatin preparations).
- Staples for dermal use.
- 27G needle.
- Hamilton syringe (Hamilton Company), 10 μL.
- Blunt needle for Hamilton syringe.
- Glass micropipette.

- Stereotaxic instrument and micromanipulator for the stereotaxic instrument (SM-15, Narishige, Japan).
- Stereotaxic injector (KDS 310, Muromachi, Japan).
- Glass pipette puller (PC-10, Narishige, Japan).
- Glass pipette grinder (EG-400, Narishige, Japan).

A glass pipette tapered to a point (approximate diameter: 100 μm) with a puller and a grinder is attached to the tip of a blunt needle fitted to a Hamilton syringe using a bonding agent.

It is allowed to dry overnight. Water is injected into the pipette the following day to confirm the absence of any leakage.

This model is utilized to examine the effect of any treatment of demyelinated axon in the spinal cord in an attempt to promote remyelinization of the neurons. Mainly, the procedure described by Blakemore et al. at the University of Cambridge is used, wherein the posterior funiculus is demyelinated by injecting ethidium bromide with a micropipette as well as X-ray irradiation [12]. The model is highly useful for histologic studies of demyelination/remyelinaion, but is not as suitable for examination of the pathophysiologic state of actual spinal cord injury. Functional recovery is assessed using methods such as the beam-walking test [13], in which the model animal is made to walk on a rod measuring 18 mm in diameter and 1 meter in length and the extent of hindlimb slipping is examined.

The following operation is entirely performed under a microscope.

7.3.3.1 Anesthesia: Intraperitoneal Injection of Ketamine (100 mg/kg) and Xylazine (10 mg/kg)

1. The skin over the intended region for surgery is shaven and thoroughly disinfected with Isodine. A median incision is made in the skin of the dorsal back using a scalpel or scissors. After confirming the location of the spinous processes, the dorsal muscles are severed to detach them from the spinous processes and vertebral arches with a scalpel.
2. The vertebral arches of thoracolumbar junction are resected using a rongeur to fully expose the dorsal side of the spinal cord.
3. A Hamilton syringe is filled with 5 μL of 0.1 % ethidium bromide (so that 1 μL of this solution can be injected successfully).
4. A minute fissure is opened in a cephalocaudal direction in the center of the dorsal dura mater at the site of the cord injury using a 27G needle. The Hamilton syringe is held so that its needle tip points to this site.
5. 1 μL of 0.1% ethidium bromide is slowly injected at the rate of 0.5 μL/min using a stereotaxic injector. It should be ascertained under the microscope that the solution is being securely delivered into the spinal cord.
6. Keep the needle static for 5 min, and then slowly withdraw it.
7. After rinsing with normal saline, hemostasis is confirmed (if bleeding fails to cease, place a hemostatic agent such as a gelatin preparation temporarily at the site, wait until the bleeding stops, and then remove it). The muscles are then sutured using 5-0 or 6-0 nylon suture, the subcutis and skin are disinfected with Isodine, and the wound is closed by stapling.

8. The animal is maintained in an incubator at 37°C until awaking from anesthesia. Analgesics and antibiotics are administered.
9. 24 hours later, the animal is anesthetized again. The previous skin incision was open and the injured spinal cord is exposed.
10. Using an orthovoltage source (225 kV), 40Gy x-irradiation at 4 cm section of the spinal cord is performed.
11. After rinsing with normal saline, hemostasis is confirmed. The muscles are then sutured using 5-0 or 6-0 nylon suture, the subcutis and skin are disinfected with isodine, and the wound is closed by stapling.
12. The animal is maintained in an incubator at 37°C until awaking from anesthesia. Analgesics and antibiotics are administered.

7.3.4 Cell Transplantation

Experimental investigation on nerve cell transplantation for spinal cord injury began with a peripheral nerve transplantation study conducted by Aguayo et al. in 1981 [14]. Experimental transplantation of neural stem cells was reported by McDonald et al. in 1999 [15]. McDonald et al. transplanted neural stem cells that had previously been induced to differentiate from mouse embryonic stem cells in vitro into the cord cavity of a rat thoracic spinal cord contusion injury model. They reported that the cells transplanted became engrafted and differentiated into neurons, astrocytes and oligodendrocytes and that the treated model exhibited an improvement of motor function of the hindlimbs, relative to the observations in an untreated control group. Gratifying results were also reported by Ogawa et al., who focused their attention on the changes in the micromilieu of the post-spinal injury cavity over time. These investigators transplanted fetal rat neural stem cells from spinal cord into the cord cavity of a rat model of thoracic spinal cord contusion injury on Day 9 post-injury, namely in the subacute stage between the acute inflammatory stage just after the injury and the chronic stage with progression of glial scarification. The study demonstrated differentiation of the transplanted cells into neurons, formation of synapses with the host axons, and recovery of motor function [16].

In the studies cited above, the progenitor cells were transplanted into a lesion and no section models are usually used for this purpose, because, needless to mention, the transplanted cells eventually leak out.

When cell transplantation is attempted in spinal cord section models, contrivances are needed such as cell transplant injection both cephalad and caudad to the induced injury [17] or packing of the cell transplants in a gel. Vacanti et al. documented their study, in which gel-packed adult rat neural stem cells were transplanted into a rat thoracic spinal cord transaction model with similar results [18]. Sporadic reports have dealt with transplantation by intravenous injection or intramedullary injection, however, this section introduces the most standard method of cell transplantation into the cord at the site of injury [19–21].

7.3.4.1 Instruments/Tools That Need to Be Kept Ready for Use

- Microscope (surgical microscope: KOM-300S, Konan Medical, Japan).
- Isodine.

- Scalpel (surgical blades).
- Cooper scissors.
- Forceps.
- Forceps for microsurgery.
- Swabs.
- Physiological saline.
- 5-0 or 6-0 nylon suture.
- Staples for dermal use.
- Suspensions containing various cells (for example, fetal mouse neurosphere-derived cells: 2 to 5 mL).
- 27G needle.
- Hamilton syringe (Hamilton Company), 10 μL.
- Blunt needle for Hamilton syringe.
- Glass micropipette.
- Stereotaxic instrument, and micromanipulator for the stereotaxic instrument (SM-15, Narishige, Japan).
- Stereotaxic injector (KDS 310, Muromachi, Japan).
- Glass pipette puller (PC-10, Narishige, Japan).
- Glass pipette grinder (EG-400, Narishige, Japan).

A glass pipette tapered to a point (approximate diameter: 100 μm) with a puller and a grinder is attached to the tip of a blunt needle fitted to a Hamilton syringe using a bonding agent.

It is allowed to dry overnight. Water is injected into the pipette the following day to confirm the absence of any leakage.

Thus, this device has the advantages (1) that a cell suspension placed in it can be directly seen through its glass wall and (2) that the tip diameter of the needle does not jut out from the spinal cord DSuch contrivance is essential especially when working on mice, in which the spinal cord is very small.

Preparation of cell suspension (here an example of cells is shown).

7.3.4.2 Materials

- Fetal mouse forebrain-derived neurosphere cultured cells.
- Cell culture medium
- Dulbecco's modified Eagle's medium (DMEM)/F12 medium supplemented with B27 (Gibco, Grand Island, NY).
- 20 ng/ml basic fibroblast growth factor (bFGF) (Peprotech, Rocky Hill, NJ).
- 20 ng/mL epidermal growth factor (EGF) (Invitrogen, Carlsbad, CA).
- Penicillin/streptomycin L-glutamine (Invitrogen).
- *bFGF and EGF are added twice a week.

1. Spheres are passaged into fresh media without growth factors following enzymatic dissociation with TrypLE (Invitrogen).

2. Cell suspension is prepared at the density of 500,000 to 1,000,000 cells/2 to 5 μL of fresh medium.

The following operation is entirely performed under a microscope.

7.3.4.3 Anesthesia: Intraperitoneal Injection of Ketamine (100 mg/kg) and Xylazine (10 mg/kg)

1. The skin over the intended region of surgery is shaved and totally disinfected with Isodine. The spinal cord injury animal (mouse or rat) is restrained on a stereotactic instrument under anesthesia.
2. A median incision is made in the skin of the dorsal back using a scalpel or scissors. If the animal has previously undergone an operation for infliction of a spinal injury, the previous skin incision is utilized.
3. Scars at the lesion on the spinal cord are removed with physiological saline and a cotton swab under the microscope.
4. A Hamilton syringe is filled with 2 to 5 μL of the cell suspension (containing 500,000 to 1,000,000 cells).
5. A minute fissure is opened in a cephalocaudal direction in the center of the dorsal dura mater at the site of the cord injury using a 27G needle. The Hamilton syringe is held so that its needle tip points to this site.
6. The cell suspension is slowly injected at the rate of 0.5 μL/min using a stereotaxic injector. It should be ascertained under the microscope that the cell suspension is being securely delivered into the spinal cord.
7. Keep the needle static for 5 min, and then slowly withdraw it.
8. After rinsing with physiological saline, hemostasis is confirmed (if bleeding fails to cease, place a hemostatic agent such as a gelatin preparation temporarily at the site, wait until the bleeding stops, and then remove it). The muscles are then sutured using 5-0 or 6-0 nylon suture, the subcutis and skin are disinfected with isodine, and the wound is closed by stapling.
9. The animal is maintained in an incubator at 37°C until awaking from anesthesia. Analgesics and antibiotics are administered.

7.4 Data Acquisition, Anticipated Results, and Interpretation

7.4.1 Function Assessment Methods

For assessing the therapeutic effects of measures such as cell transplantation and drug administration in spinal cord injury models, it is of supreme importance to evaluate to what extent the motor function can be restored, in other words, to what extent the induced paralysis remains, in the models. Among the variety of methods for the assessment of function that have been described, the BBB locomotor rating scale (BBB score) proposed by the OSU Group for the evaluation of hindlimb motor function in rodents, especially rats with induced thoracic spinal cord injury [22], is well-known and has been used in many previous studies. Briefly, locomotor function is assessed by parametric scoring, scores ranging from 0 to 21, in the animals focusing on the movement of the three joints (the hip joint, knee joint, and ankle joint) of the hindlimb, the presence or absence of weight support, fore- and hind leg coordination, mode of plantar stepping and tiptoe direction at hind leg toe clearance, trunk instability, tail status, and so forth. (Table 7.1). It takes time for examiners to gain skill in this procedure (training courses are provided at NYU and OSU). For a reliable assessment, the scoring and discussion must be carried out by two or more examiners, preferably with blinding, in order to eliminate the involvement of subjectivity. Such motor function assessments may be performed in mouse thoracic spinal cord injury models by using the BBB scoring scheme; however, evaluation may not always be as feasible in mice as in the rats because of their smaller body size and lower body weights. In view of this, Basso and colleagues developed and reported a new method of functional evaluation called the BMS scale in 2006 [23]. This functional assessment method is commonly employed in currently reported studies of mouse thoracic spinal cord injury models [24].

Other techniques of functional evaluation in thoracic cord injury models include the inclined plane test, in which the hindlimb function is evaluated in terms of the

Table 7.1 BBB Locomotor Rating Scale (Reference 22, with some omissions)
The right and left hindlimbs are evaluated separately. The lower (worse) score is taken as the index in the original method.

Score	
0	No hindlimb movement at all.
1	Slight movement of one or two of the hindlimb joints.
2	Extensive movement of one hindlimb joint and slight movement of another hindlimb joint.
3	Extensive movement of two of the hindlimb joints.
4	Slight movement of all three hindlimb joints.
5	Slight movement of two of the hindlimb joints and extensive movement of the third hindlimb joint.
6	Extensive movement of two of the hindlimb joints and slight movement of the third hindlimb joint.
7	Extensive movement of all three hindlimb joints.
8	Sweeping without weight support.
9	A weight support static state with the planta set on the floor.
10	Occasional weight support walk with no fore- or hindlimb coordination at all.
11	Frequent or consistent weight support walk with no fore- or hindlimb coordination at all.
12	Frequent or consistent weight support walk with occasional fore- and hindlimb coordination.
13	Frequent or consistent weight support walk with frequent fore- and hindlimb coordination.
14	Weight support walk with intact fore- and hindlimb coordination, but the hindlimbs almost consistently in an abducent position.
15	Weight support walk with intact fore- and hindlimb coordination. Hindlimbs in parallel with the trunk at initial contact, but with little or no toe clearance at liftoff.
16	Weight support walk with intact fore- and hindlimb coordination. Hindlimbs in parallel with the trunk at initial contact, and frequent toe clearance at liftoff, but with the tiptoe abducent at liftoff.
17	Weight support walk with intact fore- and hindlimb coordination. Frequent toe clearance and tiptoes in parallel with the trunk at initial contact and at liftoff.
18	Weight support walk with intact fore- and hindlimb coordination. Consistent toe clearance and tiptoes in parallel with the trunk at initial contact and at liftoff.
19	Tail status is checked. In addition to the status corresponding to score 18, tail raised occasionally during weight support walk.
20	Tail status is checked. In addition to the status corresponding to score 18, tail raised consistently during weight support walk, but the trunk is unstable.
21	No trunk instability in addition to the status corresponding to Score 20.

angle of an increasingly inclined plane at which the test rat placed on it falls off [25], and the footprints analysis, in which the right and left hind legs of a test animal are marked with ink of different colors and the animal is allowed to walk on a level ground so that the space between fore- and hind legs and the angle of the planta direction are measured [26]. For assessment of cervical cord injury models, on the other hand, documented techniques include the bar grip test, in which a test animal is made to cling on to a bar to measure the grabbing muscle force of its forelimbs [27], and the pellet retrieval test, in which a test animal that has been tethered is made to get the feed that is placed in the back of the cage [16, 28]. Other methods have also been described in which a test mouse or rat with induced spinal injury is placed in a cage equipped with a heat seeker or in a space encased with infrared ray sensors to monitor spontaneous locomotor activity or standing on hind limbs of the test animal [29]. Of importance here is to carry out integrative functional evaluation by utilizing a battery of two or more of these test methods wherever feasible. It is considerably difficult to assess the state of the induced paralysis to a necessary and sufficient extent using only a single method of evaluation insomuch as the pyramidal pathways at and below the injured level of the spinal cord are usually irregularly damaged.

The representative methods for evaluation of hindlimb function, namely, the BBB locomotor rating scale (rat) and the Basso Mouse Scale (BMS) (mouse), are introduced below.

7.4.1.1 Protocol

Materials

- Spinal cord injured mice or rats.
- Open-field (metal watering tank; smooth bottom, 30-cm sidewall height and 115-cm diameter, Behlen Manufacturing, Columbus, NE).
- At least two scorers who have the experience of scoring.
- Video (if necessary).

1. Each mouse is put on an open-field and locomotor performance is assessed for 5 min (see Table 7.1 for rats and Table 7.2 for mice below) by at least two scorers who have extensive prior experience.
2. This assessment is performed for each mouse/rat on the following days post-operation: 1, 3, 5, 7, 10, 14, 21, 28, 35, and 42.
3. Each scorer completes individual score before meeting. Careful discussion is performed and an objective reliable score is validated.

   Three joints of the hindlimb = hip joint, knee joint, and ankle joint
   Slight movement = movement with range of motion (ROM) ≤50%
   Extensive movement = movement with ROM >50%
   Sweeping = rhythmic sweeping using all three joints
   Weight support = basically, a state in which the hindlimb extensors are contracted at plantar stepping
   Occasional = frequency between 5% and ≤50%
   Frequent = frequency between 51% and 94%
   Consistent = frequency ≥95%

Table 7.2 BMS Scale

0	No ankle movement.
1	Slight ankle movement.
2	Extensive ankle movement.
3	Plantar placing of the paw with or without weight support –OR- Occasional, frequent or consistent dorsal stepping but no plantar stepping.
4	Occasional plantar stepping.
5	Frequent or consistent plantar stepping, no coordination -OR- Frequent or consistent plantar stepping, some coordination, paws rotated at initial contact and liftoff.
6	Frequent or consistent plantar stepping, some coordination, paws parallel at initial contact -OR- Frequent or consistent plantar stepping, mostly coordinated, paws rotated at initial contact and lift off.
7	Frequent or consistent plantar stepping, mostly coordinated, paws parallel at initial contact and rotated at liftoff -OR- Frequent or consistent plantar stepping, mostly coordinated, paws parallel at initial contact and lift off, and severe trunk instability.
8	Frequent or consistent plantar stepping, mostly coordinated, paws parallel at initial contact and lift-off , and mild trunk instability -OR- Frequent or consistent plantar stepping, mostly coordinated, paws parallel at initial contact and liftoff, and normal trunk stability and tail down or up and down.
9	Frequent or consistent plantar stepping, mostly coordinated, paws parallel at initial contact and lift off, and normal trunk stability and tail always up.

From: [23], with some omissions.

Toe clearance = it has been described that this may be checked by listening to the sound on hard-textured floors, such as an acrylate floor.

Objective judgment can be more readily accomplished in practice when the score is between 0 and about 14. Differences in function may be most convincing when the score is between 8 and 9 (whether capable of weight support).

In all events, it is advisable that the assessment by scoring be performed over a period of about 6 weeks, e.g., on Days 1, 3, 5, 7, 10, 14, 21, 28, and 42 after the induction of spinal cord injury. It is important that the evaluation be carried out by two or more observers whenever possible to exclude involvement of subjectivity, and that the observers are blinded as to allocation of the animal to the transplantation group versus the control group by assigning animal numbers, in order to secure objectivity of the test.

7.4.1.2 Basso Mouse Scale

The OSU Group, which first formulated the BBB scale in 2006, modified the scale for use in mouse tests. Basically, the original scoring scheme was simplified to a 0- to 9-point scale, although the points to be assessed remain consistent.

Here again, differences in function are considered as most convincing when the scores are between 3 and 4.

Immediately after infliction of spinal cord injury, in most instances, both mice and rats are in a state of spinal shock, without exhibiting any movement of either the fore- or hindlimbs, hence the score is 0. The hindlimb motor function score progressively increases spontaneously with time, from Days 3, 7, and 14 onwards after the injury. In the case of a mild injury, the animal may show recovery to the full score on the BBB or BMS rating scale. Intergroup comparison is conducted using a non-parametric method such as Mann-Whitney's U test or the Kruskal-Wallis test. A sample size of at least $n = 5$ is required in most cases to demonstrate significant differences.

7.5 Discussion and Commentary

What is of prime importance in studies of spinal cord injury is, as mentioned at the beginning, quantitative construction of a spinal cord injury model (without individual differences) suitable for the purpose of the study and, when aimed at assessing the functional recovery, is the possibility of objective evaluation of functional deficits. Needless to say, unless skilled in the operation and induction of injury, one may not be able to quantitatively induce a spinal cord injury model. In the preparation of these models, the investigator should have a thorough knowledge of the methods of operations, and thoroughly examine whether there is any inadequate sectioning in the case of section models and whether there is any asymmetric contusion injury, especially whether the impactor has touched such structures as the vertebral arch in the case of contusion model. In the case of section models, the operative technique should be standardized, for example, in terms of the use of the same scissors, specification of the site of the spinal cord, the depth of the injury, and the number of times the cord is incised for sectioning. Similarly, in the case of contusion injury induced using an impactor, the target site should be observed under the microscope not only in the transverse direction, but also in the cephalocaudad direction, in order to ensure that the injury has been inflicted at the center of the spinal cord. Although a relatively minor point, caution must be exercised in holding the adjacent vertebral arches or vertebral bodies, because overtraction of these structures would cause stretching of the spinal cord, which is liable to lead to reduction in the severity of the induced injury.

Differences between the right and left hindlimbs are often encountered in the assessment of hindlimb function if the section is inadequate or if the induced injury is not bisymmetric. It would be advisable to exclude such animals from the analysis.

When transplantation is performed, one of the key issues is how to securely inject cells into the region of injury. Toward this end, it is desirable for the laboratory staff to be capable of uniformly preparing pointed glass pipettes. Use of a puller and a grinder is recommended. Meanwhile, the tip of a glass pipette may not aspirate well enough when injection of a prepared cell suspension is attempted. In that case, the present authors inject cell suspensions from the posterior end of a syringe using Pipetman.

Troubleshooting Table

Problem	Explanation	Potential Solutions
Hindlimb function after contusion injury differs between the right and left limbs.	An asymmetric contusion injury.	Set the impactor at the center of the spinal cord by carefully observing the site crossways under the microscope.
		Exclude the animal if the induced injury is markedly asymmetric.
Milder contusion injury than planned.	Poor induction of the contusion injury.	Check whether the vertebral arches or the vertebral bodies that are held are being pulled towards each other.
		Check whether the impactor is in direct contact with such structures as the vertebral arch.
Leakage of transplanted cells.	Premature needle withdrawal.	Leave the animal as it is for at least 5 min.

7.5.1 Ethical Aspects

Animal models of spinal cord injury, in particular, must be provided with utmost care when approval is given for their use in experimental studies after full discussion of the study plan, etc. by the ethics committee, because the quality of life of these laboratory animals is badly marred due to post-injury tetraplegia and depression of the vesicorectal function [30]. Urinary retention and ileus, once developed, would directly affect the prognosis for survival of the injured animals. From an ethical standpoint, the experimenters should, for example, manually promote voiding once or twice a day following infliction of the spinal cord injury and carefully monitor the animals in an incubator maintained at 37°C until stabilization of the general condition during the immediate period after infliction of the injury. Experimenters should not spare any effort at making the animals comfortable (e.g., reforming cages to make water drinking easier and increasing the thickness of the bedding material to prevent decubitus ulcers.)

7.6 Application Notes

It would be desirable for induction of injury models to be designed so as to allow functional recovery by transplantation/drug medication (e.g., mice not initially able to support weight with their hind legs become capable of walking with weight support). To this effect, the authors consider empirically that an injury induced with an IH impactor at 50-60 kDyne for mice [11] or about 200 kDyne for rats [31] might be suitable in the case of contusion injury models.

7.7 Summary Points

A brief overview of the methods used for preparing animal models of spinal cord injury and of the techniques used for evaluating function using these models and for cell transplantation is presented. Currently, it is feasible to quantitatively create animal models of spinal cord injury using various mammals ranging from rodents to nonhuman primates, so that effective therapies for spinal cord injury will no longer remain a dream, but become applicable to humans. In conducting animal experimentation, it is considered to be of importance to pliably select an injury model suitable for the purpose of the study before starting experiments, to fully heed the ethical aspects as well, to gain sufficient anatomical knowledge and technical skill in advance, and to make it possible for objective evaluation of function to be conducted by two or more observers to assess the effects of therapies.

References

[1] Hagg, T., and M. Oudega, "Degenerative and Spontaneous Regenerative Processes After Spinal Cord Injury," *J. Neurotrauma.*, Vol. 23, No. 3–4, 2006, pp. 264–280.

[2] Fouad, K., I. Klusman, and M. E. Schwab, "Regenerating Corticospinal Fibers in the Marmoset (Callitrix Jacchus) after Spinal Cord Lesion and Treatment with the Anti-Nogo-A Antibody IN-1," *Eur. J. Neurosci.*, Vol. 20, No. 9, 2004, pp. 2479–82.

[3] Iwanami, A., et al., "Establishment of Graded Spinal Cord Injury Model in a Nonhuman Primate: The Common Marmoset," *J. Neurosci. Res.*, Vol. 80, No. 2 2005, pp. 172–81.

[4] Bradbury, E. J., et al., "Chondroitinase ABC promotes Functional Recovery after Spinal Cord Injury," *Nature,* Vol. 416, No. 6881, 2002, pp. 636–40.

[5] Kaneko, S., et al., "A Selective Sema3A Inhibitor Enhances Regenerative Responses and Functional Recovery of the Injured Spinal Cord," *Nat. Med.,* Vol. 12, No. 12, 2006, pp. 1380–89.

[6] Bareyre, F. M., et al., "The Injured Spinal Cord Spontaneously Forms a New Intraspinal Circuit in Adult Rats," *Nat. Neurosci.,* Vol. 7, No. 3, 2004, pp. 269–77.

[7] Kim, J. E., et al., "Axon Regeneration in Young Adult Mice Lacking Nogo-A/B," *Neuron,* Vol. 38, No. 2, 2003, pp. 187–99.

[8] Gruner, J. A., "A Monitored Contusion Model of Spinal Cord Injury in the Rat," *J. Neurotrauma,* Vol. 9, No. 2, 1992, pp. 123–26 (discussion, pp. 126–28).

[9] Stokes, B. T., D. H. Noyes, and D. L. Behrmann, "An Electromechanical Spinal Injury Technique with Dynamic Sensitivity," *J. Neurotrauma,* Vol. 9, No. 3, 1992, pp. 187–95.

[10] Scheff, S. W., et al., "Experimental Modeling of Spinal Cord Injury: Characterization of a Force-Defined Injury Device," *J. Neurotrauma,* Vol. 20, No. 2, 2003, pp. 179–93.

[11] Okada, S., et al., "Conditional Ablation of Stat3 or Socs3 Discloses a Dual Role for Reactive Astrocytes after Spinal Cord Injury," *Nat. Med.,* Vol. 12, No. 7, 2006, pp. 829–34.

[12] Smith, P. M., and W. F. Blakemore, "Porcine Neural Progenitors Require Commitment to the Oligodendrocyte Lineage Prior to Transplantation in Order to Achieve Significant Remyelination of Demyelinated Lesions in the Adult CNS," *Eur. J. Neurosci.,* Vol. 12, 2000, pp. 2414–424.

[13] Jeffery, N. D., and W. F. Blakemore, "Locomotor Deficits Induced by Experimental Spinal Cord Demyelination Are Abolished By Spontaneous Remyelination," *Brain,* Vol. 120, Pt. 1, 1997, pp. 27–37.

[14] Aguayo, A. J., S. David, and G. M. Bray, "Influences of the Glial Environment on the Elongation Of Axons After Injury: Transplantation Studies in Adult Rodents," *J. Exp. Biol.,* Vol. 95, 1981, pp. 231–40.

[15] McDonald, J. W., et al., "Transplanted Embryonic Stem Cells Survive, Differentiate, and Promote Recovery in Injured Rat Spinal Cord," *Nat. Med.,* Vol. 5, No. 12, 1999, pp. 1410–12.

[16] Ogawa, Y., et al., "Transplantation of In Vitro-Expanded Fetal Neural Progenitor Cells Results in Neurogenesis and Functional Recovery After Spinal Cord Contusion Injury in Adult Rats," *J. Neurosci. Res.,* Vol. 69, No. 6 ,2002, pp. 925–33.

[17] Lu, P., et al., "Olfactory Ensheathing Cells Do Not Exhibit Unique Migratory or Axonal Growth-Promoting Properties after Spinal Cord Injury," *J. Neurosci.,* Vol. 26, No. 43, 2006, pp. 11120–30.

[18] Vacanti, M. P., et al., "Tissue-Engineered Spinal Cord," *Transplant Proc.,* Vol. 33, No. 1–2, 2001, pp. 592–98.

[19] Iwanami, A., et al., "Transplantation of Human Neural Stem Cells for Spinal Cord Injury in Primates," *J. Neurosci. Res.,* Vol. 80, No. 2, 2005, pp. 182–90.

[20] Ikegami, T., et al., "Chondroitinase ABC Combined with Neural Stem/Progenitor Cell Transplantation Enhances Graft Cell Migration and Outgrowth of Growth-Associated Protein-43-Positive Fibers after Rat Spinal Cord Injury," *Eur. J. Neurosci.,* Vol. 22, No. 12, 2005, pp. 3036–46.

[21] Yamane, J., et al., "Transplantation of Galectin-1-Expressing Human Neural Stem Cells into the Injured Spinal Cord of Adult Common Marmosets," *J. Neurosci. Res,.* Vol. 88, No. 7, 2010, pp. 1394–1405.

[22] Basso, D. M., et al., "MASCIS Evaluation of Open Field Locomotor Scores: Effects of Experience and Teamwork on Reliability. Multicenter Animal Spinal Cord Injury Study," *J. Neurotrauma,* Vol. 13, No. 7, 1996, pp. 343–59.

[23] Basso, D. M., et al., "Basso Mouse Scale for Locomotion Detects Differences in Recovery after Spinal Cord Injury in Five Common Mouse Strains," *J. Neurotrauma,* Vol. 23, No. 5, 2006, pp. 635–59.

[24] Kumagai, G., et al., "Roles of ES Cell-Derived Gliogenic Neural Stem/Progenitor Cells in Functional Recovery after Spinal Cord Injury," *PLoS One,* Vol. 4, No. 11, 2009, pp. e7706.

[25] Rivlin, A. S., and C. H. Tator, "Objective Clinical Assessment of Motor Function after Experimental Spinal Cord Injury in the Rat," *J. Neurosurg.,* Vol. 47, No. 4, 1977, pp. 577–81.

[26] Kunkel-Bagden, E., and B. S. Bregman, "Spinal Cord Transplants Enhance the Recovery of Locomotor Function after Spinal Cord Injury at Birth," *Exp. Brain Res.,* Vol. 81, No. 1, 1990, pp. 25–34.

[27] Smith, J. P., et al., "Quantitative Measurement of Muscle Strength in the Mouse," *J. Neurosci. Methods,* Vol. 62, No. 1–2, 1995, pp. 15–19.

[28] Okano, H., et al., "Transplantation of Neural Stem Cells into the Spinal Cord after Injury," *Semin. Cell Dev. Biol.,* Vol. 14, No. 3, 2003, pp. 191–98.

[29] Mikami, Y, et al., "A Simple and Reliable Behavioral Analysis of Locomotor Function after Spinal Cord Injury in Mice. Technical Note," *J. Neurosurg.*, Vol. 97, No. 1, Suppl. 2002, pp. 142–47.

[30] National Research Council, *Guide for the Care and Use of Laboratory Animals*, Washington, DC: National Academy Press, 1996.

[31] Kitamura, K., et al., "Hepatocyte Growth Factor Promotes Endogenous Repair and Functional Recovery after Spinal Cord Injury," *J. Neurosci. Res.*, Vol. 85, No. 11, 2007, pp. 2332–42.

8

Cell Transplantation and Regeneration in Neurology

Takao Yasuhara and Isao Date

Department of Neurological Surgery, Okayama University Graduate School of Medicine, Dentistry and Pharmaceutical Sciences, 2-5-1, Shikata-cho, Okayama, 700-8558, Japan
Phone: 81-86-235-7336; Fax: 81-86-227-0191
E-mail: tyasu37@cc.okayama-u.ac.jp

Abstract

We describe the setup and procedures of intracerebral cell transplantation. Intracerebral cell transplantation is divided into two steps: preparation of cell suspension and stereotaxic transplantation into the brain. In addition to these two steps, an encapsulation technique that enables us to perform transplantation of xenogeneic cells or cell lines easily and safely is explained briefly. Furthermore, the essence of cell transplantation is also described with our previous studies.

Key terms behavioral test, biochemical evaluation, cell transplantation, cerebral infarct, encapsulation, immunohistochemical investigation, neurotrophic factor, Parkinson's disease, rodent

8.1 Introduction

Intracerebral cell transplantation is used for the treatment of some diseases of the central nervous system (CNS), such as Parkinson's disease and cerebral infarct in clinical trials [1–3]. However, the therapeutic effects and safety should be repeatedly confirmed in animal studies in advance. Before now, experiments using various cells from xenogeneic cells to autologous neural stem/progenitor cells have been performed [4–6]. Intravenous or intra-arterial transplantation are less invasive approaches that let the transplanted grafts survive in the cerebral tissue for a long time with subsequent functional action to the host brain, in addition to intracerebral cell transplantation, although each approach has advantages and disadvantages (Table 8.1). Intra-arterial cell transplantation might be at an intermediate position in the immediacy and invasiveness. The therapeutic effects of each administrative way remain to be solved.

How does cell transplantation exert therapeutic effects? The therapeutic effects might lie in the replacement of damaged host tissues by the transplanted grafts. However, replacement only could not explain the therapeutic effects of cell transplantation. It may be that the supply of trophic factors might dominantly contribute to the therapeutic effects. Several trophic factors including brain-derived neurotrophic factor (BDNF) [7], glial cell line-derived neurotrophic factor (GDNF) [8], or vascular endothelial growth factor (VEGF) [4] secreted from the transplanted grafts might exert neuroprotective effects through cell proliferative potencies or antiapoptotic effects [5]. Indirect effects like angiogenesis or anti-inflammatory effects also might be involved. Furthermore, these trophic factors might enhance neurogenesis with subsequent migration to the penumbra region, differentiation to the requisite phenotypes, and efficient function in the damaged brain to compensate for the lost function of the host.

In this chapter, we focus on intracerebral cell transplantation. We describe the procedures of intracerebral cell transplantation step by step with several important tips. Additionally, encapsulated cell transplantation performed in our laboratory is described. In addition, specific protocols should be followed for specific cell types.

8.2 Experimental Design

In order to make animal studies closer to clinical practice, considerable efforts have been made by great researchers. In general, several useful models of CNS diseases are used to confirm the therapeutic effects of cell transplantation in behavioral and histological aspects. When making protocols using cell transplantation, two important issues should always be considered: timing and the route of transplantation. The microenvironment and disease status are changing over time. If the disease considerably progresses and reaches irreversible conditions, the cell transplantation might not exert any therapeutic

Table 8.1 Characteristics of Intracerebral, Intra-Arterial and Intravenous Cell Transplantation

	Intracerebral Transplantation	Intra-Arterial Transplantation	Intravenous Transplantation
Advantages	Direct/secure approach	Relatively high concentration	Less invasive
	Strong therapeutic effects	Relatively less invasive	
Disadvantages	Requires surgery	Risk of embolism	Low concentration to brain
	Risk of bleeding	Relatively low cell survival	Low therapeutic effects?

effects even if the cells have great potential to ameliorate the disease. Usually, cell transplantation at an early time point might exert stronger therapeutic effects than the delayed one, although transplantation at early time points might have relatively high risk of low graft survival due to deteriorated microenvironment. As described above, the routes of cell transplantation—intravenous, intra-arterial, and intracerebral administration—are also critically important.

Intracerebral cell transplantation is divided into two steps: (1) preparation of cell suspension, and (2) stereotaxic intracerebral cell transplantation. In addition, if the researchers would like to transplant a xenogeneic graft such as a cell line, encapsulated cell transplantation is very helpful. The characteristics of encapsulated cell transplantation are summarized in Table 8.2. With the encapsulation technique, various cells with genetic manipulation can be used without immune suppression. Tailor-made medicine with potential to secrete target neurotrophic factors for each patient might also be possible. Meanwhile, real reinnervation between donor and host is prevented from the capsule itself. Thus, encapsulated cell transplantation is a unique drug delivery system.

8.3 Materials

8.3.1 Preparation for Cell Suspension

- Nutrient medium.
- 0.01M PBS (SIGMA).
- Trypsin solution (0.02-0.1 %) with EDTA (0.01%–0.02%) (Sigma).
- Trypsin inhibitor (the same dosage of Trypsin solution) (Sigma).
- Thermostatic bath.
- Bioclean bench.
- Aspirator.
- Micropipettes.
- Centrifuge.
- Centrifuge tube (15 ml, Sumitomo Bakelite, Japan).
- Eppendorf tube (1.5 mm).
- 0.4 % Trypan blue solution (Sigma).
- Cytometer.

Table 8.2 Characteristics of Encapsulated Cell Transplantation

Advantages	Disadvantages
No tumor formation	No reinnervation between donor and host
Long-term survival without immune rejection	Invasiveness at transplantation
Possibility of tailor-made medicine	Contamination of xenogeneic protein/gene from xenogeneic grants
Availability of xenogeneic grafts	
Retrievability of capsules	
Little ethical problem	
Low dose/continuous administration of target molecules	

8.3.2 Stereotaxic Intracerebral Cell Transplantation (Rat)

- Pentobarbital solution (30–50 mg/kg, Schering-Plough Animal Health).
- Lidocine hydrochloride (1%, Sigma).
- Surgical tools.
- Stereotaxic instrument (Narishige, Japan, or Kopf, CA).
- Dental drill.
- 10-μl Hamilton syringe.
- Sterilized distilled water.
- 70% ethanol solution.
- Sterilized cotton swab.
- Microscope.

8.3.3 Encapsulated Cell Production

- Dental curing light unit (Jetlite, J. Morita Corporation, Japan).
- Light curing resin (Palfique Clear, Tokuyama Dental, Japan).
- Hollow fiber with semipermeable membrane (inner diameter: 500 or 700 μm; molecular cutoff: 100 kDa, Amicon, MA).
- Hamilton syringe (blunt needle).
- Sterilized distilled water.
- Cell suspension.

8.4 Methods

8.4.1 Preparation of Cell Suspension

The cell preparation should be adjusted for each cell type and size. The protocol for genetically modified baby hamster kidney (BHK) cells is described below. Once the correct method is found, it is important to use this method each time the suspension is prepared for consistent results.

1. Efficient cells for transplantation are prepared for this procedure. On the scheduled day, the cell density should be intermediate between *confluent* to *semiconfluent*.
2. For adherent cells (e.g., BHK cells or mesenchymal stromal cells): aspirate culture medium from the dish. (For floating cells: centrifuge gently (approximately 120G× 5 minutes, although it depends on type of cells). Then aspirate culture medium with caution so as not to aspirate sediment.
3. After washing with 0.01M PBS, triturate cultured cells with warmed (37°C) trypsin solution (0.02–0.1%) for about 3 min.
4. Stop the reaction with trypsin inhibitor (adjusted at the same dose of trypsin solution). For tough cells like mesenchymal stromal cells, culture medium with serum can be used instead of trypsin inhibitor.
5. Pipette cell solution to single cell suspension using micropipetter (or transfer pipette).

6. Centrifuge the cell suspension using 15-ml centrifuge tube with subsequent aspiration of the supernatant.
7. Suspend cells to cold (4°C) culture medium (1 ml) and transfer the suspension to 1 ml Eppendorf tube. At this point, retrieve 1 μl of cell suspension and mix 0.4% trypan blue solution for cell counting. Cell viability and cell number can be evaluated.
8. Centrifuge the cell suspension (approximately 160G×3 min) using cold machine and aspirate the supernatant.
9. Adjust the cell suspension to the desired concentration from the total cell number (calculated at step 7) using medium. (Previously we used 0.01M PBS for cell solution. There was no significant difference in cell viability between cells with culture medium and those with PBS.)
10. After adjustment (e.g., 1×10^5 cells/μl), put the cell suspension on ice. Before transplantation, final concentration and cell viability should be confirmed using trypan blue solution, as described at step 7. (The appropriate cell concentration might be $0.3–3.0\times10^5$ cells/μl, although it depends on the cell type and size.)

8.4.2 Stereotaxic Intracerebral Cell Transplantation (Rat)

1. After measurement of rats' body weight, administer pentobarbital (30–50 mg/kg) intraperitoneally for general anesthesia. Other methods of anesthesia are allowed if the following procedures can be performed stably. In the case of xenogeneic cell transplantation such as human cell transplantation to rats, 10 mg/kg of cyclosporin A should be administered before anesthesia. After transplantation, daily administration of cyclosporin A is preferred for xenogeneic cell transplantation. Importantly, administration of cyclosporin A is also required for the control group, because it exerts neuroprotective effects [9].
2. Fix head of rats to the stereotaxic instrument using ear bar. For intrastriatal transplantation, we set the tooth bar at 0 mm, as reported from many laboratories [10].
3. Shave the parietal region of rats and disinfect scalp. For cleanliness, use sterilized cover and gloves.
4. Administer local anesthetic through the scalp using 0.1–0.2 ml of lidocaine hydrochloride solution (1%) (Figure 8.1(a)).
5. Make an incision approximately 1.5 cm long at the median or paramedian of the scalp.
6. Expose bone below periosteum. Bregma (intersection of the coronal suture and the sagittal suture) and lambda (intersection of the sagittal suture and lambda suture) can be confirmed under the microscope.
7. Load inner needle of 22G spinal needle to the manipulator of stereotaxic instrument. The tip of the needle should be set directly on the bregma and move the needle tip using manipulator to the target coordinates for transplantation (Figure 8.1(b)). Then perforate the cranial bone with the dental drill (Figure 8.1(c)). In order not to damage the brain surface, the thin cortical inner plate can be preserved with subsequent removal with the 22G needle. After that, dura mater is also opened with the needle.

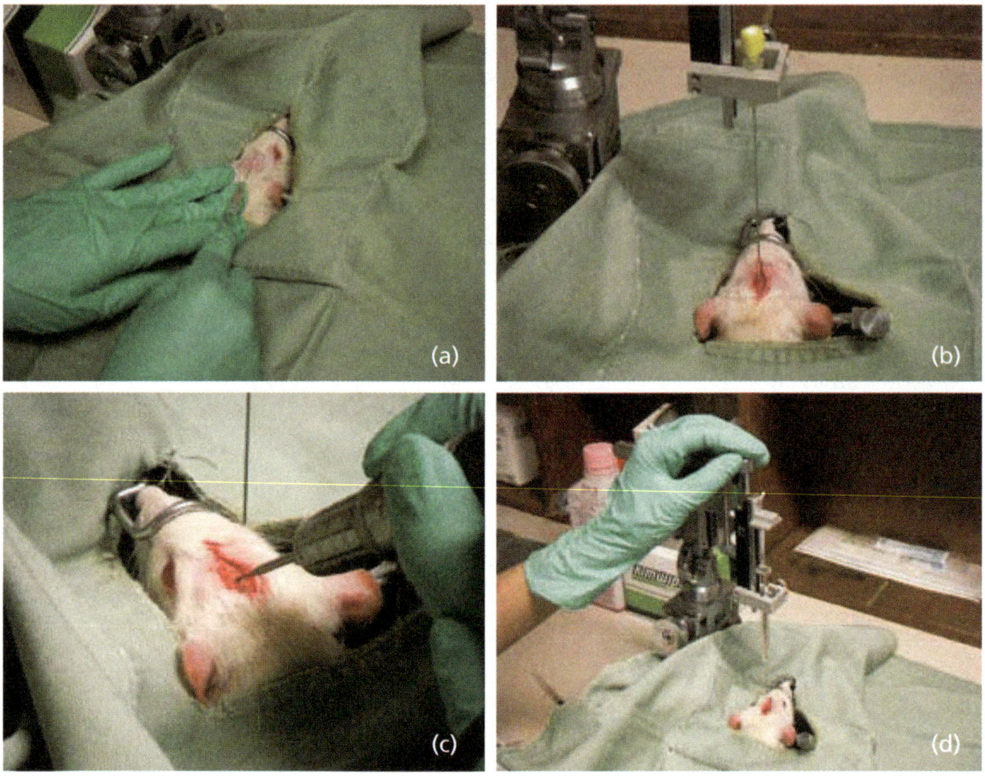

Figure 8.1 Surgical procedures of cell transplantation. (a) Local anesthesia with sterilized cover and gloves. (b) Positioning of the perforation using inner needle of 22G spinal needle. (c) Perforation of the cranial bone with the dental drill. (d) Injection of cell suspension into the brain at the speed of 1 μl/minute.

8. After disinfection of a 10 μl Hamilton syringe using 70% ethanol, aspirate sterilized distilled water twice with subsequent wash of the needle and syringe. Then slowly wash the needle and syringe with the medium. In the end, slowly aspirate cell suspension to necessary amount + 1 μl with caution so as not to aspirate air.
9. Put the 10-μl Hamilton syringe with cell suspension in the manipulator of the stereotaxic instrument. If the researcher is accustomed to this method, just fine-tune the needle tip of the syringe to the burr hole. Or reconfirm the coordinates using the Hamilton syringe from the bregma. The needle tip is just on the brain surface. We usually coordinate with the level of bregma, 3.0 mm lateral to the bregma, and 5.0 mm ventral to the brain surface for intrastriatal transplantation. However, as for intrahippocampal transplantation, the coordinates are completely different depending on age in weeks. The coordinates should be determined carefully.
10. Using slow-motion dial, put the needle tip to the targeted depth.
11. Inject cell suspension into the brain at the speed of 1 μl/minute (Figure 8.1(d)).
12. Withdraw the needle at the speed of 1 mm/min using slow-motion dial.
13. In the case of multiple transplantations, repeat this procedure.
14. Suture the wound using 3-0 silk or wound clips.
15. Carefully monitor body temperature and breathing of rats and wait for recovery.

8.4.3 Making of Encapsulated Cells

1. Under microscope, cut perpendicularly hollow fibers with blades at designed length. We usually cut at 6.5 mm.
2. Attach the light curing resin to one end of the capsule. Then, harden the resin with the dental curing light unit.
3. Wash with 70% ethanol twice.
4. Wash with sterilized distilled water.
5. Sterilize the capsules soaked in distilled water with ultraviolet light for 6 h.
6. Wash with culture medium.
7. Deeply insert blunt needle of Hamilton syringe filled with cell suspension from the open end of the capsule. Then slowly inject cell suspension inside the capsule.
8. With caution not to leak the cells outside of the capsule, pull the needle from the capsule.
9. Seal the other end of the capsule using light curing resin and dental curing light unit.
10. Culture encapsulated cells in 12 well-culture dish.
11. After confirmation that target factor is successfully secreted from the capsule without leakage at 1 month after encapsulation, the encapsulated cells can be used for transplantation.

8.5 Data Acquisition, Anticipated Results, and Interpretation

After intracerebral cell transplantation, rats can be housed freely, although xenogeneic cell transplantation (without encapsulation) requires cyclosporine A administration. Behavioral tests are important to evaluate functional recovery of transplanted rats compared to that of control rats. For example, spontaneous locomotor activity evaluation using AccuScan® is helpful to explore behavioral patterns of rodents in an open field environment, cylinder test for spontaneous forelimb usage, Rotarod test for evaluation of running/balance ability and drug-induced tests, such as amphetamine/apomorphine-induced rotational tests [4–6, 8, 9, 11]. Other than these behavioral tests, many kinds of tests are performed to reveal therapeutic effects of new therapy in animal studies. If the transplant is effective, rats might show behavioral amelioration as early as 1 week. Meanwhile, some rats can demonstrate behavioral amelioration 4 weeks after transplantation. Behavior tests should be combined correctly to evaluate functional recovery after cell transplantation. Typically, over eight rats in each group are preferred to be involved in behavioral tests for reliable data analyses.

For histological evaluation, adequate euthanasia using thorough perfusion is required. Immunohistochemical investigation should be performed for graft survival, differentiation, and migration. Neuroprotective effects of transplanted grafts can be evaluated by TTC (within 1 week after insult), Nissl, hematoxylin and eosin, NeuN, GFAP, MAP2 staining or disease-specific staining against some phenotypic marker like tyrosine hydroxylase staining for Parkinson's disease. To detect apoptotic cell death, TdT-mediated dUTP-biotin nick end labeling (TUNEL) staining is suitable using brains from rats euthanized at an early time point, usually within 1 week after insult. For graft survival, transplanted cells can be labeled by Qdot Nanocrystals with fluorescence or

GFP-gene transfer. For human cell transplantation to rodents, anti-HuNu (human nuclei) staining is very helpful. In order to clarify the involvement of neurogenesis, bromodeoxyuridine (BrdU) is administered to rats before euthanasia. BrdU is taken up by proliferative cells at DNA synthetic period. In general, for evaluation of cell proliferation, BrdU should be administered close to euthanasia, although for evaluation of differentiation and migration, prior administration of BrdU might be preferred.

For protein or mRNA assay, fresh brain tissue is required. To obtain correct data with high reproducibility, quick sampling without contamination should be preferred. The timing of euthanasia should be determined for consideration of target molecules.

8.6 Discussion and Commentary

Intracerebral cell transplantation is a direct method to demonstrate the therapeutic effects or safety of the transplanted cells in vivo. The transplanted cells can be located close to the lesion and represents an easy access with relatively strong effects compared to other administration routes. Usually many cells survive at the transplanted site. However, a number of cells can be damaged during the preparation and transplantation procedures. In addition, ischemic damage should be also considered at transplantation, although the degree of the damage might depend on the cell type and size. If we find low cell viability, we need to confirm whether the transplantation procedure is correctly performed. Additionally, after the cells are successfully transplanted with an initial survival rate, cells could die due to an adverse environment caused by secondary insult of the disease model itself or eliminated by inflammatory cells. The long-term experimental period might attenuate the fluorescence of GFP or Qdot nanocrystals with subsequent undetectable grafts. To distinguish these cases, euthanasia at early time points after transplantation might be preferred. At 3 days to 1 week after transplantation, the survival of transplanted cells can be evaluated immunohistochemically. Using encapsulated cell transplantation, the capsules can be easily retrieved at euthanasia with consequent confirmation that cells inside the capsule function as is expected as well as survive over the experiment period.

8.7 Application Notes

Parkinson's disease is one of the most explored diseases of the CNS using animal studies. Parkinson's disease is characterized by degeneration of nigrostriatal dopaminergic neuronal systems. Medication using L-DOPA or DA agonist and surgeries including deep brain stimulation are established therapies for patients with Parkinson's disease. However, these therapies are considered symptomatic therapies and a new regenerative therapy is required. 6-hydroxydopamine (6-OHDA) or 1-methyl-4 phenyl-1,2,3,6-tetrahydropyridine (MPTP) is used to make Parkinson's disease models of rodents with subsequent experiments to reveal therapeutic effects of new therapies. We performed a large number of cell transplantation experiments using Parkinson's disease model in rats. Xenogeneic cell transplantation with or without encapsulation, genetically engineered cell lines to secrete catecholamine or neurotrophic factors, neural progenitor

Troubleshooting Table

Problem	Explanation	Potential Solutions
Low viability/cell number for transplantation (before transplantation).	In appropriate preparation of cell suspension.	1. Prepare extra 30% of predicted cell number in advance. 2. Avoid long-term incubation with trypsin. Trypsin has toxicity and 5 minutes might be the limit. 3. In case of strongly attached cells, the use of cell scrapers might be considered.
Difficult to control bleeding at surgery.	Damage of the vessels.	Sterilized cotton swab might be best to control bleeding. Put the swab directly on the bleeding point and wait for a while.
Incorrect coordinates of transplantation.	Irregular shape of the bregma.	In some rats, bregma might be a line, but not a point. Be sure that the center of the bregma is considered as the true bregma and set the needle there.
Damage of the brain surface at drilling.	Clogged drill.	Check the condition of the drill. Worn drill should be replaced with new one. Drill is easily clogged by bone dust. To remove the dust, ultrasonic cleaner is usually very helpful.
Debility of rats after surgery.	Long-term administration of cyclosporin A.	Check the loss of body weight and urinary conditions. Debility might be related to infection. Dose reduction or abeyance of cyclosporin A should be considered.
Missing grafts.	Incorrect injection of cell suspension or quick withdrawal of the needle.	Be consistent with the slow injection speed and slow withdrawal of the needle.

cells and autologous neural stem/progenitor cells were transplanted with subsequent behavioral, immunohistochemical, and biochemical investigations. Other than the Parkinson's disease models, cerebral infarct models, spinal cord injury, and head trauma models are commonly used to explore the therapeutic effects of cell transplantation.

8.8 Summary Points

1. We described the setup and procedures of intracerebral cell transplantation. Intracerebral cell transplantation is a strong tool to reveal the characteristics of the cells in vivo.
2. Preparation of cell suspension. Single cell suspension is required with minimum damage to the cells. The protocol should be adjusted for each type of cell. Once the correct method is found, it is important to use this method each time the suspension is prepared for consistent results.
3. Be careful about cell number and viability before transplantation. Double checking might be preferred, because it might seriously affect cell survival and total therapeutic effects of cell transplantation.
4. Fast injection and quick movement of the needle might ruin the transplantation with consequent low cell viability.

8.9 Conclusions

Herein, we described a general procedure for intracerebral cell transplantation and several tips that might be useful to use in biotechnology, cell engineering, including induced pluripotent stem cells. We remain hopeful that techniques for intracerebral cell transplantation will be helpful in the clinical setting in the near future.

Acknowledgments

This work was supported in part by Grants-in-Aid for Scientific Research from the Ministry of Education, Culture, Sports, Science, and Technology, Japan. I also appreciate devoted the cooperation of our team, especially Drs. Kenichiro Muraoka, Masahiro Kameda, Feifei Wang, Tanefumi Baba, Akihiko Kondo, Judith Thomas Tayra, Aiko Shinko, Takaaki Wakamori, Naoki Tajiri, and Yasuyuki Miyoshi.

References

[1] Freed, C. R., et al., "Transplantation of Embryonic Dopamine Neurons for Severe Parkinson's Disease," *N. Engl. J. Med.*, Vol. 344, No. 10, 2001, pp. 710–19.

[2] Olanow, C.W., et al., "A Double-Blind Controlled Trial of Bilateral Fetal Nigral Transplantation in Parkinson's Disease," *Ann. Neurol.*, Vol. 54, No. 3, 2003, pp. 403–14.

[3] Kondziolka, D., et al., "Transplantation of Cultured Human Neuronal Cells for Patients with Stroke," *Neurology*, Vol. 55, No. 4, 2000, pp. 565–69.

[4] Yasuhara, T., et al., "Neuroprotective Effects of Vascular Endothelial Growth Factor (VEGF) upon Dopaminergic Neurons in a Rat Model of Parkinson's Disease," *Eur. J. Neurosci.*, Vol. 19, No. 6, 2004, pp. 1494–1504.

[5] Yasuhara, T., et al., "Transplantation of Human Neural Stem Cells Exerts Neuroprotection in a Rat Model of Parkinson's Disease," *J. Neurosci.*, Vol. 26, No. 48, 2006, pp. 12497–511.

[6] Muraoka, K., et al., "The High Integration and Differentiation Potential of Autologous Neural Stem Cell Transplantation Compared with Allogeneic Transplantation in Adult Rat Hippocampus," *Exp. Neurol.*, Vol. 199, No. 2, 2006, pp. 311–27.

[7] Kurozumi, K., et al., "BDNF Gene-Modified Mesenchymal Stem Cells Promote Functional Recovery and Reduce Infarct Size in The Rat Middle Cerebral Artery Occlusion Model," *Mol. Ther.*, Vol. 9, No. 2, 2004, pp. 189–97.

[8] Yasuhara, T., et al., "Early Transplantation of an Encapsulated Glial Cell Line-Derived Neurotrophic Factor-Producing Cell Demonstrating Strong Neuroprotective Effects in a Rat Model of Parkinson Disease," *J. Neurosurg.*, Vol. 102, No. 1, 2005, pp. 80–89.

[9] Borlongan, C. V., et al., "Cyclosporine-A Increases Spontaneous and Dopamine Agonist-Induced Locomotor Behavior in Normal Rats," *Cell Transplant.*, Vol. 4, No. 1, 1995, pp. 65–73.

[10] Schierle, G. S., et al., "Caspase Inhibition Reduces Apoptosis and Increases Survival of Nigral Transplants," *Nat. Med.*, Vol. 5, No. 1, 1999, pp. 97–100.

[11] Yasuhara, T., et al., "Lack of Exercise, Via Hind Limb Suspension, Impedes Endogenous Neurogenesis," *Neuroscience*, Vol. 149, No. 1, 2007, pp. 182–91.

9

Biocompatible Genetic Modification of Multipotent Mesenchymal Stromal Cells

Devang K. Thakor[1,2], Yasuhiko Tabata[3*], Jun-Ichiro Jo[3], Hideaki Obata[4], Kentaro Nagane[3], Shigeru Saito[4], Masaya Yamamoto[3], and Yang D. Teng[1,2*]

[1]Departments of Neurosurgery, and Physical Medicine and Rehabilitation, Harvard Medical School, Boston, MA
[2]Division of Spinal Cord Injury Research, Veterans Affairs Boston Healthcare System, Boston, MA
[3]Institute for Frontier Medical Sciences, Kyoto University, Japan
[4]Department of Anesthesiology, Gunma University, Japan
*Correspondence: Yasuhiko Tabata, E-mail: yasuhiko@frontier.kyoto-u.ac.jp; or Yang D. Teng, E-mail: yang_teng@hms.harvard.edu

Abstract

Cell transplantation has shown promise for treatment of spinal cord-related disorders such as neuropathic pain and traumatic spinal cord injury. Ex vivo genetic modification of cells prior to implantation has been investigated for overexpression of molecules with particular therapeutic functionality, labeling of donor cells for in vivo tracking, or even induction of the differentiation of progenitor cells to a desired phenotypic lineage. Therefore, development of suitable biocompatible gene transfer vectors is important for such biological and/or engineering strategies. This chapter describes the preparation of a polysaccharide-based nonviral vector whose unique negative surface charge permits serum compatibility and a lack of toxicity in several types of primary cells. The vector is applied to ex vivo transfection of multipotent mesenchymal stromal cells (MSCs), with subsequent minimally invasive transplantation into the rat spinal cord via a simple intrathecal lumbar puncture technique.

Key terms gene therapy, nonviral vector, pullulan, dorsal root ganglion, mesenchymal stromal/stem cell, spinal cord, neuropathic pain, anioplex

9.1 Introduction

Cell transplantation has emerged as a powerful tool for the study of spinal cord biology, with therapeutic implications leading to clinical trials currently in progress for disorders such as traumatic spinal cord injury [1–6]. Delivery of transgenes to cells prior to implantation can be used to overexpress or downregulate molecules of interest (e.g., neurotrophic factors), to label the cells with fluorescent or chemiluminescent proteins for tracking via in vivo imaging or histology, or to otherwise achieve a desired cellular phenotype (e.g., neural differentiation of progenitor cells) [7–9]. The two major technical challenges accompanying the transplantation of genetically modified cells to the spinal cord include (1) the method of ex vivo gene delivery and (2) the spinal cord implantation technique. We will address both issues in this chapter.

Traditional gene delivery techniques typically rely on a variety of viral vectors including adenovirus, adeno-associated virus (AAV), herpes simplex virus (HSV), and various retro/lentiviruses with genomic components (especially those encoding replicative machinery) replaced by an expression cassette containing the transgene [10, 11]. Infection of cells is followed by integration of the transgene into the host genome for AAV and retro/lentiviral vectors, or epigenetic expression for adenoviral and HSV vectors [10, 11]. Viral vectors typically offer the advantages of high transduction efficiency and, especially for those capable of host genome integration, prolonged transgene expression [10, 11]. However, barriers to potential clinical translation include persistent concerns of reactivation of pathogenicity via recombination, subsequent host downregulation of donor genes, and insertional mutagenesis leading to oncogenic transformation for vectors capable of host genome integration [10, 11]. Although immunogenicity of viral vectors is a concern for in vivo application, it is likely to be less of a problem for ex vivo application to nonimmune cells. However, the high cytotoxicity displayed by some viral vectors, particularly adenoviruses, could present an additional drawback for ex vivo use, especially in cells with low isolation yields.

To circumvent the disadvantages of viral gene delivery, nonviral vectors have been developed by mixing cationic polymers and lipids with plasmid DNA [12–14]. The cationic components (usually protonable amines) of these materials interact electrostatically with negatively charged phosphate groups on plasmid DNA, allowing condensation of the DNA and vector material into a micro- or nanoscale complex (polyplex or lipoplex) [12–14]. Inclusion of an excess amount of the vector material results in complexes with a positive surface charge and available protonable amines [12–14]. The positively charged complexes are thought to then interact with negatively charged cell membrane components such as proteoglycans for cell adhesion and stimulation of endocytosis; once inside the cell, the complexes are targeted to endosomes, where pH-dependent protonation of vector amines is thought to result in osmotic swelling and endosomolysis with consequent release of the complexes into the cytosol [12, 13].

Such polycationic vectors eliminate the risk of viral recombination because no viral genetic materials are used; they also greatly reduce the risk of insertional mutagenesis because the plasmid DNA should not be inserted into the host genome. However, polycationic vectors typically show greatly reduced transfection efficiency when compared to viral transduction [11, 12]. Additionally, transgene expression is usually less sustainable (more transient) than that achieved with viral gene delivery [11]. Finally, polycationic vectors are typically cytotoxic [15–17], especially at high concentrations, and often not compatible with serum in cell culture media, leading to the use of less physiologically appropriate media in their transfection protocols. The cytotoxicity and lack of serum compatibility may arise from the same positive surface charge that allows the complexes to gain entry to cells, as positively charged macromolecules are thought to destabilize cell membranes and be sequestered by negatively charged serum proteins such as albumin [15–17].

We have determined complexation conditions between plasmid DNA and a polycationic conjugate of the naturally occurring polysaccharide pullulan and polyamine spermine that permit the formation of transfectable microparticles with a negative surface charge (Figure 9.1) [18]. In terms of safety, the individual use of these materials is advantageous, as pullulan is already commercially used in oral healthcare and pharmaceutical coating applications and spermine is ubiquitously found at high concentrations in human cells, where it is thought to naturally interact with nucleic acids [18]. We hypothesized that the negative surface charge of the pullulan-spermine/DNA particles, which we have termed anioplexes, would permit decreased cytotoxicity and increased serum compatibility during transfection. This

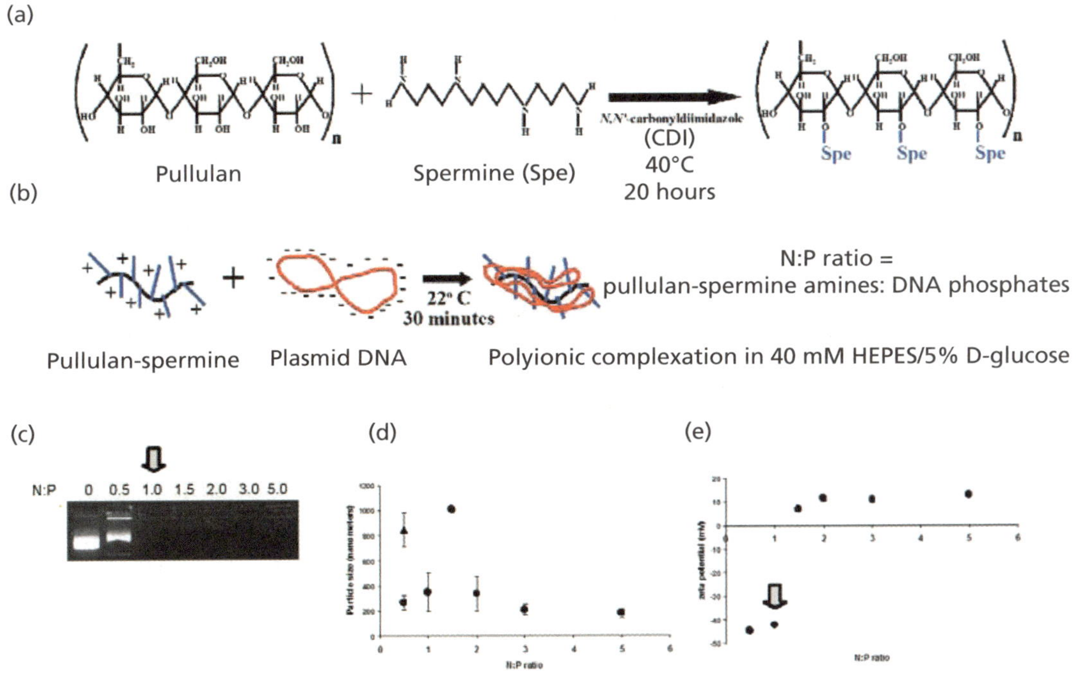

Figure 9.1 Schematic representation of pullulan-spermine/DNA anioplex strategy and physiochemical characterization of anioplexes. (a) Hydroxyl groups of pullulan, a naturally occurring and biocompatible polysaccharide, were activated by carbonyldiimidazole and then derivatized by spermine, a naturally occurring polyamine. Bottom panel: the resulting polycationic conjugate was mixed with plasmid DNA to generate pullulan-spermine/DNA complexes. (b) Electrophoresis of complexes incubated for 1 week in OPTI-MEM demonstrates complete complexation at N:P 1:1; stability is maintained at all N:P ratios. (d), (e) Complex diameter (d) and zeta potential (e) at different N:P ratios measured by dynamic and electrophoretic light scattering (DLS and ELS), respectively. Single DLS signal starting at N:P 1 suggests formation of complexes. Note the strongly negative ELS signal at N:P 1 (arrow); we have termed these negatively charged complexes "anioplexes." (Means and standard errors depicted; $n = 3$.)

hypothesis was supported by screening experiments that showed complete serum compatibility and lack of toxicity in primary sensory neurons [18].

We further applied this technology for ex vivo genetic modification of bone marrow-derived multipotent mesenchymal stromal cells (MSCs) prior to spinal cord implantation [19]. For spinal cord cell transplantation, particularly that of MSCs, three approaches are typically used: (1) intravenous injection, (2) intraspinal injection, and (3) intrathecal injection [20–22]. Intravenous injection offers the advantages of being the least invasive and allowing the largest possible injection volume. However, it is also the least specific and relies on disruption of the blood-brain barrier to permit regional access to the spinal cord, as well as on the ability of the cells (typically progenitor cells) to then home to the pathological sites. Thus, it is the least efficient method and a relatively small fraction of cells actually engrafts into the spinal cord. Intraspinal injection into the spinal cord parenchyma, on the other hand, is the most direct procedure, allowing precise localization of the implanted cells that is critical for histological evaluation of interactions between donor cells and host neural tissue. However, it is the most invasive and technically challenging of the procedures, and the most limited with respect to injection volume.

Lumbar puncture injection into the intrathecal space over the spinal cord is a minimally invasive procedure that does not require incision or laminectomy and permits relatively local delivery of MSCs (in contrast to intravenous injection) with double to quadruple the available injectate volume (and therefore cell dosage) compared to intraspinal injection. Minimal invasiveness and ease of the procedure also facilitates multiple treatments, thus further increasing the effective dose that can be delivered. We have developed a simple technique for intrathecal cell transplantation that does not require specialized or costly materials or equipment, and we present it in this chapter in the context of implantation of anioplex-transfected MSCs.

In this chapter, we describe the formation of anioplexes and the evaluation of transfection efficiency and cytotoxicity in MSCs. We also describe our simple lumbar puncture technique for spinal cord implantation of anioplex-transfected MSCs. Readers may be able to adapt these concepts and techniques for development of anioplexes from different starting materials, for anioplex-derived transfection of other cell types, or for spinal cord cell transplantation in a variety of disease models.

The experiment presented in this chapter is designed to test the hypothesis that pullulan-spermine/DNA anioplex-derived reporter gene activity can be used to track MSCs in the spinal cord after intrathecal implantation.

In this experiment, the pullulan-spermine conjugate will be synthesized and purified, and the amine content will be calculated to allow accurate mixing with plasmid DNA. Anioplexes will be formed by mixing pullulan-spermine with DNA at a nitrogen to phosphate (N:P) ratio of 1 (Figure 9.1) [18]. Anioplexes will be used to transfect MSCs with plasmid DNA in the presence of serum; luciferase expression will be chemiluminescently quantified in order to confirm transfection. Additionally, mitochondrial activity will be evaluated as a measure of cell viability after transfection. Negative control conditions for this experiment will include nontransfected cells (luciferase control), cells transfected with plasmid DNA alone (transfection control), and cells transfected with positively charged pullulan-spermine/DNA polyplexes (N:P 3; control for anioplex transfection). Cells transfected with Lipofectamine 2000/DNA lipoplexes will serve as the positive control because Lipofectamine 2000 is an established MSC transfection agent. For in vitro experiments, a sample size of six wells is deemed to have sufficient power to assess statistically significant differences.

Anioplex-transfected MSCs are then intrathecally implanted into rats, and the luciferase activity in the rat spinal cords chemiluminescently quantified. The negative control consists of rats injected with vehicle. The positive control consists of injection of 100 μg luciferase-expressing plasmid DNA because intrathecal injection of this amount of naked plasmid DNA gives rise to high levels of reporter gene expression in the spinal cord [23]. For the purpose of this experiment (to demonstrate luciferase expression in the spinal cord), three animals per group are sufficient. However, for quantitative comparisons between treated groups, especially when behavioral testing is required, at least five to eight animals per group are necessary.

9.2 Materials

9.2.1 Reagents

Unless otherwise mentioned, all reagents are obtained from Sigma, St. Louis, MO. The backbone of our gene carrier is 47.3 kDa pullulan, obtained from Hayashibara in Okayama, Japan. To our knowledge, this is the only commercial supplier of pullulan. Pullulan in powder form can be stored at room temperature. For synthesis, we also use N,N0-carbonyldiimidazole (CDI), dimethyl sulfoxide (DMSO), and spermine. CDI and spermine should be stored at 4°C. The purification step requires seamless cellulose tubing (30/32 mm, 14 kd cutoff) as well as a source of ultrapure water (i.e., MilliQ, Millipore). For the reporter gene, we use beta-actin-promoter-driven plasmid DNA expressing luciferase (pCAGGS-luciferase, a generous gift from Dr. Jyunichi Miyazaki in the Graduate School of Medicine at Osaka University); however, we have also successfully used the commercially available pGL3 luciferase-expressing plasmid from Promega Corp., Madison, WI. For plasmid preparation we use DH5α competent cells (Invitrogen, Carlsbad, CA) and the endotoxin-free Megaprep kit (Qiagen, Valencia, CA). Complex formation requires glucose and HEPES.

For MSC isolation, 3-week-old Wistar rats are used (Shimizu Laboratory Supplies). For both isolation and implantation procedures, it is essential to obtain the approval of the animal care and use committee of the respective institution prior to performing any of the experiments. Necessary reagents include phosphate-buffered saline (PBS; Invitrogen), α-MEM medium (Invitrogen), fetal bovine serum (FBS, Hyclone, Logan, UT), penicillin-streptomycin (Invitrogen), trypsin-EDTA solution (Invitrogen), black and clear 96-well plates, and 24-well TC-treated cell culture plates, as well as 25 cm^2 and 225 cm^2 TC-treated cell culture flasks. Lipofectamine 2000 (Invitrogen) and OPTI-MEM medium (Gibco, Carlsbad, CA) are required for the control. Transfection will also require a sterile-filtered solution containing 10% glucose and 40 mM HEPES (hereafter referred to as HBG). The luciferase assay requires lysis buffer and luciferase assay reagent (both from Promega), as well as a bicinchonic acid (BCA) assay kit (Pierce, Rockford, IL). For the water-soluble tetrazolium (WST-8) assay, we use the Cell Counting Kit from Nacalai Tesque of Kyoto, Japan. Similar kits are available from Dojindo and Cayman Chemical; although we have not used these kits, we expect that they would offer similar performance to the Nacalai Tesque kit. WST-8 [2-(2-methoxy-4-nitrophenyl)-3-(4-nitrophenyl)-5-(2,4-disulfophenyl)-2H tetrazolium monosodium salt] measures mitochondrial activity that can then be taken as an indicator of cellular viability or proliferation. This is done via reduction of the tetrazolium salt by NADH produced in viable mitochondria.

For implantation, 250 to 300g Sprague Dawley rats are used. The injection apparatus includes a 100-μl Hamilton syringe (MICROLITER 700 series) with needle (outer diameter [OD] 0.0283 in) as well as Tygon tubing (Cole-Parmer) with three different diameters (inner diameter [ID]/OD 0.025/0.101 in, 0.017/0.093 in, 0.09/0.166 in). To fashion the injection catheter, cut a 12-inch length of the 0.025/0.101 tubing and insert it into a 0.5 inch length of the 0.09/0.166 tubing. Insert a 1-inch length of the 0.017/0.093 tubing into the other side of the 0.5-inch length of 0.09/0.166 tubing; thus the 0.09/0.166 tubing is essentially used to splice the 0.025/0.101 and 0.017/0.093 tubes together. Because the ODs of the spliced tubes are slightly larger than the ID of the 0.09/0.166 tube, a tight seal should be formed without the use of any glue. However, if leakage is apparent or the splice comes apart, epoxy can be used to glue the tubes together. Additionally, 1-inch regular-bevel 25-gauge needles (BD) are required. Isoflurane is required for anesthesia, and ethylene oxide for sterilization of the injection apparatus.

9.2.2 Facilities/Equipment

A freeze dryer is required for lyophilization of the pullulan-spermine conjugate. Additionally, combustion elemental analysis is used to evaluate the carbon, nitrogen, and hydrogen content of the conjugate to determine the extent of spermine derivatization. We send our samples to collaborators at the Osaka Institute of Technology; however, there are many commercial services available to the reader for performing this assay. A spectrophotometer is required to read the absorbance at 260 and 280 nm for determination of DNA concentration. A luminometer that can measure luciferase chemiluminescence (Berthold) is required for the luciferase assay. A 96-well plate reader that can read absorbance at 450 and 650 nm is required for the WST-8 assay. A gas anesthesia setup with oxygen flow is required for isoflurane anesthesia; additionally, betadine, 70% ethanol, and a set of surgical tools including #11 scalpel, gripping forceps, rongeurs, needle forceps, and iris scissors are necessary for the injection procedure.

9.3 Methods

9.3.1 Preparation of Pullulan-Spermine

1. Place the entire spermine bottle in a 37°C water bath so it melts (the melting point is 28–30°C).
2. Weigh out 50 mg pullulan (this can be scaled up as needed), and estimate the number of OH groups (reaction with CDI will add spermine to OH groups on pullulan). There are 1.852×10^{-5} moles of OH groups per mg pullulan; therefore there are 9.26×10^{-4} mol OH in 50 mg pullulan. Dissolve pullulan in 5 ml DMSO; dissolution usually takes about 15 min.
3. Weigh out sufficient CDI to provide a threefold molar excess over pullulan OH groups. 50 mg pullulan will therefore require 2.78×10^{-3} moles of CDI, which weighs 450 mg. The amount of CDI controls the extent of spermine derivatization. We have used other molar ratios and found that conjugates formed at CDI/OH 1.5 provide the

best transfection performance [24]; however, we encourage further experimentation by the reader.

4. Weigh out sufficient spermine to provide a 10-fold molar excess over pullulan OH groups. For 50 mg pullulan, 9.26×10^{-3} moles of spermine are needed; this amount weighs 1.87 grams. Taking the specific gravity of the melted spermine into account, this translates to 1.97 ml. Spermine (1.97 ml) should be dissolved in 43.03 ml DMSO in a 100-ml screw-top jar and set to stir with a stir bar.
5. Add CDI to pullulan and stir for 5 min at room temperature. It is important that the reaction time is exactly 5 min; longer times will lead to over-introduction of spermine onto pullulan.
6. Using a 5- or 10-ml pipetteman or serological pipetter, add all 5 ml of the CDI/pullulan solution to the spermine solution, drop by drop with stirring. If CDI/pullulan is added too quickly, a precipitate will form, thus drop-by-drop application is critical.
7. Screw the top onto the jar and seal with parafilm. Stir the reaction solution at 35°C to 40°C for 20h. It is preferable to use a water bath to maintain the temperature.
8. Cut strips of seamless cellulose dialysis tubing to almost the length of a 3L beaker. Always wear gloves; never touch the strips with ungloved hands. Place the strips in water and bring the water to a boil, then change the water and repeat twice. After boiling, make sure that the strips always remain wet. Tie off one end of each strip.
9. Use a 35-ml syringe (without needle) to load the pullulan-spermine reaction solution into the tubes. As soon as each tube is half-filled (do not fill the tubes all the way, as some space must remain for water to enter during dialysis), squeeze out any air and seal the top with a plastic clip.
10. Place the tubes into a 3L beaker full of MilliQ water and stir for 3 days. It is best to place a stirring rod across the top of the beaker and suspend the tubes from the rod into the water using rubber bands; this will give the stirring rod at the bottom of the beaker sufficient room to rotate. For the first day, the MilliQ water should be changed every 1 to 2 h; subsequently, the water should be changed daily or twice daily. This dialysis time should be sufficient to ensure complete purification; however, completion of dialysis can be checked by comparing the conductance of the water in the beaker with that of a sample of pure MilliQ water.
11. Remove the tubes from the water and dry them using laboratory tissue. Then, empty the purified pullulan-spermine conjugate solution into an appropriate container for freeze-drying. We typically use 50-ml plastic beakers; however uncapped 50-ml Falcon tubes can also be used.
12. Freeze the samples by placing them in liquid nitrogen for 20 min. Note that liquid nitrogen should not be allowed to enter the open top of the sample containers.
13. Place a piece of laboratory tissue across the top of each sample container and fix it in place with a rubber band. Then, place the samples in the freeze dryer and freeze dry them for 3 days.
14. Remove 5 mg of the lyophilized product and submit this sample to elemental analysis. Dissolve the remaining product in MilliQ water at 4 mg/ml.
15. The degree of spermine introduction (given as the percentage of available introduction sites actually occupied by spermine molecules) can be calculated from the N and C content determined by elemental analysis as follows: % spermine introduction =

(6×N%)/(14×C%–33×N%). Divide this percentage by 100 to obtain S, the spermine introduction ratio. Under the conditions described in this chapter, an introduction ratio of about 12% to 14% can be expected [24].

9.3.2 Preparation of Plasmid DNA

1. For transformation of DH5α *E.Coli* with plasmid DNA, use the manufacturer's protocol (Invitrogen). We use ampicillin for colony selection (100 μg/ml) on LB agar plates.
2. For plasmid DNA amplification and purification, use the Qiagen endotoxin-free Megaprep system and follow the manufacturer's protocol. Use an LB broth culture volume of 500 ml to 1L with an ampicillin concentration of 100 μg/ml.
3. Prior to complexation, plasmid DNA should be concentrated and resuspended in sterile water. For this step, pipette 300 μl of the purified plasmid DNA into a 1.5-ml Eppendorf tube and add 750 μl 100% ethanol as well as 10 μl 0.22-μm-filtered 5M NaCl. Vortex well and then centrifuge at top speed for 30 min. Remove the supernatant, and resuspend the DNA pellet in 50-μl MilliQ water that has been refiltered through a 0.22-μm syringe filter. Dissolve the DNA pellet with gentle vortexing; if it is difficult to dissolve, heat to 60°C to 65°C for 5 to 10 min with intermittent vortexing.

9.3.3 Isolation and Culture of Primary Rat Bone Marrow MSCs

1. Anesthetize 3-week-old Wistar rats (Shimizu Laboratory Supplies, Kyoto, Japan) with the appropriate method dictated by the institution for a terminal procedure.
2. Remove tibia and femur bones by cutting them away from the legs and trim away all muscles and connective tissue.
3. Cut off the epiphyses (the rounded caps of the bones).
4. Flush out the bone marrow onto a Petri dish by injecting 1 ml PBS with a 21-gauge needle, and then use the needle to mince the bone marrow on the dish.
5. Disperse the bone marrow aspirate in 10-ml a-MEM medium supplemented with 15 vol% FBS and 1% penicillin–streptomycin solution (hereafter denoted as culture medium).
6. Seed the bone marrow aspirate into two 25 cm^2 culture flasks (Corning).
7. Incubate at 37°C under a 5% CO_2 atmosphere for 3 days.
8. Change the medium every 3 days for 7–10 days.
9. Wash cells with 5 ml PBS, then detach cells by adding 2 ml trypsin-EDTA solution. Trypsin incubation should not be longer than 5 min; an incubation time of 2 to 3 min is best. After about 1.5 to 2 min, the side of the culture flask can be hit with the heel of the hand to help loosen cells. Add 3 ml culture medium and then transfer the cell solution to a 15-ml Falcon tube. Centrifuge at 200×g for 10 min and 22°C to pellet the cells.
10. Resuspend the pellet in 10 ml culture medium and measure the cell concentration using a hemacytometer. Trypan blue should be used to measure the percentage of cells with intact membranes; only these cells should be included when determining the cell concentration.

11. Seed the cells in 225 cm^2 cell culture flasks at a density of 2×10^4 cells/cm^2 in a volume of 35 ml.
12. Cells should be passed 1:2 at 80% to 90% confluence and used at passage 3.

9.3.4 Preparation of MSCs for Transfection in 24-Well Plates

1. Detach MSCs from flasks as described above, and plate MSCs in 24-well plates at 1×10^4 cells/cm^3 (the surface area of each well is 24 cm^2) in 1 ml culture medium. A total of 15 wells should be seeded.

9.3.5 Pullulan-Spermine/DNA Anioplex Preparation and Anioplex Transfection

1. Measure the concentration of the plasmid DNA using the absorbance at 260 nm on a spectrophotometer (for maximum accuracy, the sample should be diluted 1:500 to 1:1000 in TE buffer prior to measurement, and the spectrophotometer should be blanked using TE buffer). At this time the purity of the plasmid DNA should be checked by taking the ratio of the absorbances at 260 nm and 280 nm; the ratio should be at least 1.85.
2. Calculate the volume of DNA solution required per well of the culture plate based on the concentration (for the in vitro experiment, we will use 5 μg DNA/well).
3. Calculate how much MilliQ H_2O is required to dilute this volume of DNA to one-fourth of the transfectate volume (for each well of a 24-well plate, the total transfectate volume is 50 μl, so one-fourth of the transfectate volume is 12.5 μl).
4. Calculate the molarity of protonable nitrogens in the pullulan-spermine conjugate solution using the equation [N]=C/Mw'×A, where [N] is the molarity of protonable nitrogens, C is the concentration (w/v) of pullulan-spermine, Mw' is the molecular weight of pullulan-spermine, and A is the amount of nitrogen in one molecule of pullulan-spermine. C=4 g/L. Mw'=(4.22×Mw x S)+Mw, where Mw is the molecular weight of pullulan (47,300 g/mol), and S is the spermine introduction ratio calculated from elemental analysis. Finally, A=0.0555×Mw×S.
5. Calculate the volume of pullulan-spermine required per well to permit N:P ratios of 1:1 or 3:1 based on the molarity of protonable nitrogens in the pullulan-spermine solution (5 μg DNA contains 1.62×10^{-8} moles of phosphate residues based on relationship given that 308 g/mol is the molecular weight of one nucleotide and each nucleotide contains one phosphate residue. If the volume is too small, dilute pullulan-spermine 1:10 in H_2O and then calculate again. Pullulan-spermine should be passed through a 0.22-μm filter prior to complexation.
6. Calculate how much H_2O is required to dilute this volume of pullulan-spermine to one-fourth of the transfectate volume.
7. Calculate how much DNA, pullulan-spermine, and H_2O will be required for all wells based on the calculations from steps 1 through 6 (for this experiment, there are six wells that will receive pullulan-spermine/DNA, i.e., three at N:P 1:1 and three at N:P 3:1; remember to account for an extra well's worth for each condition in case of pipetting errors and loss during filtering).
8. Dilute DNA in the appropriate amount of sterile filtered H_2O based on step 3.

9. Dilute pullulan-spermine in the appropriate amount of sterile filtered H_2O based on step 6.
10. Further dilute DNA and pullulan-spermine 1:1 in sterile filtered HBG (40 mM HEPES/10% glucose).
11. Mix DNA solution well by gently pipetting up and down at least 10 times.
12. Mix pullulan-spermine solution well by gently pipetting up and down at least 10 times.
13. Add pullulan-spermine solution to DNA solution in that order and immediately mix well by gently pipetting up and down at least 20 times. At high concentrations of DNA (such as those used for in vivo transfection), mixing by pipetting will generate a great amount of precipitation. In this situation, pullulan-spermine solution should instead be added dropwise to the DNA solution while gently vortexing (at 40% of maximum vortex intensity).
14. Allow complexes to incubate at room temperature for 30 min. Anioplexes should be used as soon as possible, but can be left at room temperature for several hours if necessary.
15. Prepare controls. Plasmid DNA control simply consists of diluting the same volume of DNA used for anioplex formation to a volume of 25 μl in sterile filtered H_2O, then diluting 1:1 in HBG. For preparation of Lipofectamine 2000 control, use 5 μg DNA per well as in the other conditions, and prepare lipoplexes at a ratio of 3 μl Lipofectamine 2000 per μg DNA) (i.e., 15 μl Lipofectamine 2000 per 5 μg DNA per well).
16. Replace culture medium with 500 μl fresh medium (regular culture medium containing serum should be used).
17. Add 50 μl of anioplex transfectate to the culture medium in each well (i.e., three wells for N:P 1:1, three wells for N:P 3:1) (gently pipette up and down before each addition to ensure uniform application). Add plasmid DNA control to three wells, and add Lipofectamine 2000/DNA to three wells. Leave at least three wells untreated.
18. Mix by gently shaking the plate.
19. Incubate for 6h at 37°C under a 5% CO_2 atmosphere (incubation for up to 24h does not significantly improve gene expression, but is also not toxic, so transfection time can be varied for user convenience).
20. Aspirate transfection medium and replace with 1 ml fresh medium.
21. After 24h, the cells are ready for WST-8 and luciferase assays.

9.3.6 WST-8 Assay

1. Add an amount of the WST-8 reagent equaling 10% of the culture volume (in this case, 100 μl) to each well and mix by gently shaking the plate.
2. Incubate for 4 h at 37°C under a 5% CO_2 atmosphere.
3. Transfer 150-μl medium from each well to wells of a clear 96-well plate. Avoid using the outside wells of the 96-well plate.
4. Measure the absorbance at 650 nm and 450 nm.
5. Remove the WST-8 reagent-containing medium from the cells and wash the cells three times with PBS.

6. Snap-freeze the plates by floating them on liquid nitrogen. Store the plates at –80°C; plates can be stored indefinitely prior to luciferase assay.

9.3.7 Luciferase Assay

1. Thaw the plates containing transfected MSCs and allow them to warm to room temperature. Make sure that all medium and condensation are totally removed from each well.
2. Add 100 μl lysis buffer to each well (this can be reduced to 40 μl to increase the luciferase signal) and wait 5 min.
3. Use a cell scraper to scrape the surface of each well.
4. Remove the buffer from each well and pipette it into a 1.5-ml Eppendorf tube.
5. Centrifuge at top speed and 4°C for 20 min.
6. Transfer 20 μl of buffer from each tube to wells of a black 96-well plate, keep the plate on ice until applying it to the luciferase assay (luciferase is unstable, so the assay should be performed as soon as possible after preparing the plate). Add 20 μl lysis buffer alone to an additional six wells to establish a buffer-derived baseline.
7. Subject the plate to the luciferase assay using the Berthold Technologies luminometer per the manufacturer's instructions. We use the following settings: 20 washes, 2-second delay, 10-second measuring interval, and 100-μl substrate injection.
8. For normalization of luciferase activity, determine the total protein concentration using the BCA kit from Pierce per the manufacturer's instructions. For this assay, we dilute the sample 150-fold. For the standard curve, we use serial dilutions of 200, 40, 20, 10, 5, 2.5, 1, and 0.5 μg/ml bovine serum albumin. It is important to use lysis buffer and not water as the blank to obtain an accurate concentration.

9.3.8 Anioplex-Mediated Ex Vivo Transfection of MSCs Prior to Transplantation

1. MSCs will be transfected in 225 cm^2 flasks to generate the large number required for transplantation; therefore, after passage, wait until MSCs are 70% to 80% confluent before transfecting them. Each flask will provide enough cells to implant approximately one rat.
2. Prepare pullulan-spermine/DNA anioplexes as described above; in this case, first obtain the volume of plasmid DNA required to obtain 1.125 mg DNA. Then, obtain the amount of pullulan-spermine required to complex 1.125 mg DNA at N:P 1:1. Pullulan-spermine and plasmid DNA should then each be diluted with sterile filtered H_2O to a volume of 1.41 ml. The pullulan-spermine and plasmid DNA solutions should then be diluted 1:1 in HBG to a volume of 2.82 ml; thus, when pullulan-spermine and DNA solutions are mixed, the final transfectate volume will be 5.64 ml. However, for better mixing and to prevent precipitation, we have found it better to scale the complexation reaction down to 1-ml fractions, and then combine the formed complexes together to constitute the final 5.64-ml volume.
3. Replace the culture medium with 56.25 ml of fresh medium. Add the 5.64 ml transfectate and incubate for 6h as previously described.

4. Cells can be implanted any time from 1 day after transfection up to the time of reaching confluence; if absolutely necessary, cells can be used the day after reaching confluence.

9.3.9 Lumbar Puncture Injection of MSCs

1. Sterilize the injection catheter using ethylene oxide; sterilize surgical tools by autoclaving.
2. Prepare the operating area by placing a sterile surgical drape over the operating surface, bringing the isoflurane anesthesia setup and operating lights into position.
3. Assemble the injection apparatus (Figure 9.2) by connecting the larger aperture of sterilized injection catheter to the Hamilton syringe. Take a new, sterile 25-gauge needle and break the needle off from the Luer lock support (this can be done by bending the needle in different directions at its base). Insert the broken end of needle into the narrower aperture of the injection tube.
3. Detach cells from 225 cm^2 flask, centrifuge and resuspend in PBS to a concentration of 4×10^7 cells/ml.
4. Gently pipette 35 μl of cell solution into a sterile 500-μl Eppendorf tube. Place the tube containing cells on ice. Cells can be kept on ice for several hours, but viability will decrease over time.
5. Anesthetize a 250g Sprague-Dawley rat using 5% isoflurane for induction and then 2.5% for maintenance. Shave the lumbar region of the back. Apply betadine to the shaved region, scrub for 5 min, rinse with 70% ethanol, then repeat twice.
6. Using a #11 scalpel, make a 2-cm incision along the spine, centered at the junction between the L5 and L6 vertebrae. This location can be identified by feeling for the most rostral extent of the ileac crest (hip bone) and then drawing an imaginary line between the left and right sides; the L5/L6 junction will be located close to this line.
7. Using iris scissors and needle forceps, cut and clear away the supraspinous ligament and other tissue directly over the spine between the spinous processes of the L5 and L6 vertebrae. This will leave a clear notch between the spinous processes of L5 and L6.
8. Draw sterile water into the injection catheter until it reaches the 5-μl mark on the Hamilton syringe. Then, draw air into the injection catheter until the water in the Hamilton syringe reaches the 10-μl mark.
9. Resuspend the MSCs with gentle flicking, and draw MSCs into the injection catheter until the water in the Hamilton syringe reaches the 40-μl mark; thus the injection catheter is essentially filled with a column of water separated from the 30 μl of MSCs to be injected by a small air barrier. Place the Hamilton syringe next to the rat.

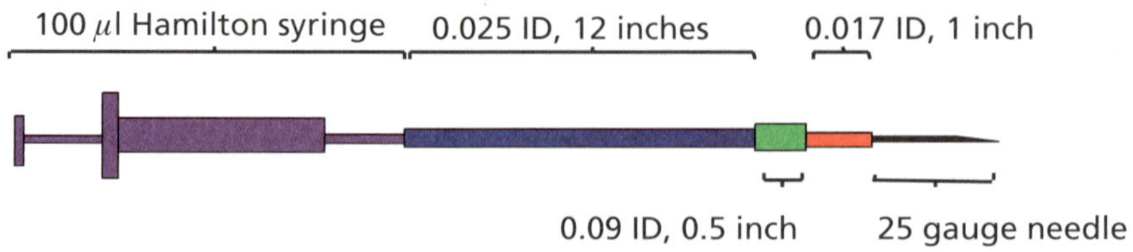

Figure 9.2 Schematic representation of injection apparatus.

10. Using the gripping forceps, grasp the spinous process of the L5 vertebra and lift slightly to widen the notch created in step 7 and expose the intraspinous ligament connecting the bases of the spinous processes of L5 and L6; the inferior vertebral notch of L5 should be clearly visible, and the intraspinous ligament should look like a smooth sheet extending out from the base of the notch.
11. Using the other hand, grasp the injection catheter where the needle meets the tubing and, holding the needle parallel to the spinal column, especially to L5, use the needle to puncture the intraspinous ligament immediately caudal to L5 with the bevel of the needle facing up. A tail-flick response is typically observed on puncture; however, successful puncture can also occur without the observation of the tail-flick.
12. Carefully advance the needle rostrally about 2.5 mm until the end of the bevel is just under the lamina of L5 in the inferior vertebral notch; essentially, the tip of the needle should appear to be underneath L5.
13. Let go of the needle and use the now-free hand to grasp the Hamilton syringe. The 12-inch length of the catheter should allow the manipulation of the Hamilton syringe without affecting needle placement. Slowly inject the MSCs over a period of 2 min. After injection, leave the needle in place for another minute, and then withdraw the needle.
14. Close the incision site with either sutures or staples, depending on the guidelines of the specific institution.
15. Wait for the animal to wake up, and monitor the animal for any evidence of motor dysfunction.
16. Replace the needle before injecting another animal.
17. For negative control injections, inject nontransfected cells as described above. For positive control injections, first calculate the volume of plasmid DNA required to obtain 100 μg DNA. Plasmid DNA should then each be diluted with sterile filtered H_2O to a volume of 15 μl, and then further diluted 1:1 in HBG to a volume of 30 μl.

9.3.10 Luciferase Assay

1. Four days after intrathecal injection, anesthetize the animal for a terminal procedure using the technique favored by the institution; we used sodium pentobarbital (60 mg/kg i.p.).
2. Expose the lumbar spinal cord from L6 to L1 by laminectomy. Using iris scissors, cut the spinal cord midway between L6 and S1. Cut the cord again midway between L4 and L3. Grasp the piece of L4-L6 cord with needle forceps, excise it using the iris scissors, place it in a 1.5-ml Eppendorf tube, and snap-freeze in liquid nitrogen. Maintenance of morphology and structure is not important; speed is the most critical factor. Next, cut the cord again midway between T12 and L1. Similarly to the procedure used for the L4-L6 segment, excise and snap-freeze the L1-L3 segment. These tissues can be stored indefinitely at –80°C prior to luciferase assay.
3. Weigh each piece of tissue without letting it thaw. Add 4 μl lysis buffer per mg tissue. Tissue should then be homogenized; this can be done using a low-volume Dounce homogenizer. If available, throughput can be increased by using a Qiagen TissueLyser for 15 min at 25 oscillations per second.

4. The subsequent luciferase assay is the same as that described for cells above, starting at step 5. However, the BCA assay does not need to be performed because all samples contain the same concentration of homogenized tissue (0.25 mg/μl).

9.4 Data Acquisition, Anticipated Results, and Interpretation

The experiments described here are designed to first demonstrate anioplex-mediated transfection and lack of cytotoxicity in MSCs in vitro, and then to use this technology to label MSCs for tracking in the spinal cord in vivo. Therefore, the outcome measures are luciferase-derived chemiluminescence as a measure of reporter gene activity (i.e., transfection efficiency/presence of labeled cells) and absorbance of formazan produced from WST-8 by mitochondrial NADH activity (i.e., viability of transfected cells).

The output of the WST-8 assay is the absorbance measured at 650 nm (A_{650}; formazan) and at 450 nm (A_{450}; background). For each well, the formazan absorbance is first normalized against that of the background by subtracting A_{450} from A_{650}. The average normalized absorbance of the untreated wells ($A_{untreated}$) is set as 100% viability. The absorbances for all wells are then divided by $A_{untreated}$ and multiplied by 100 to obtain relative measures of viability, expressed as percentages. The average percent viability for each treatment condition is then expressed as a mean accompanied by the standard error of the mean.

For the luciferase assay, the methods described above directly yield the number of relative light units (RLU) detected from 20 µl of sample lysate; to obtain RLU/μl, simply divide the total RLU by 20. The average RLU detected from wells containing lysis buffer alone serves as a baseline; any values observed to be equal to or less than this baseline should be discarded. This is an important consideration for in vivo experiments, where the observed luciferase activity is often weak.

For in vitro experiments, the final measurement of luciferase activity is given as RLU/mg total protein to provide a reference point and normalization in case the wells contain different numbers of cells; this is especially important because all wells initially receive the same volume of lysis buffer. The total protein concentration output from the BCA assay, [P], is given in μg total protein/ml and is obtained from solutions diluted 1:150. Therefore, to obtain [P′], the actual total protein concentration in mg/μl, divide [P] by 1×10^6 and then multiply by 150. Finally, to obtain RLU/mg total protein, divide RLU/μl by [P′].

For in vivo experiments, the final measurement of luciferase activity is given as RLU/g tissue to provide a reference point. In this case, additional normalization is not necessary because the amount of lysis buffer added to each tissue sample was adjusted according to the weight of the tissue. Therefore, to obtain RLU/g tissue, just multiply RLU/μl by 4×10^3. For both in vitro and in vivo experiments, the average luciferase activity for each treatment condition is then expressed as a mean accompanied by the standard error of the mean.

In the in vitro experiment described in this chapter, there is one variable (treatment type) with five conditions: (1) no treatment or treatment with (2) plasmid DNA alone, (3) positively charged pullulan-spermine/DNA complexes, (4) negatively charged pullulan-spermine/DNA anioplexes, or (5) Lipofectamine 2000/DNA. For the in vivo experiment, there is one variable with three conditions: rats receiving anioplex-transfected MSCs, nontransfected MSCs, or plasmid DNA. Thus, assuming normal distribution of the data (which should first be evaluated with the Shapiro-Wilks test) with equal variance across conditions, one-way ANOVA should be used to identify overall effects for each of the data sets (i.e., in vitro luciferase activity and MSC viability as well as in vivo luciferase activity), with Newman-Keuls post hoc tests used to identify differences between specific groups. If comparisons between the L1-L3 segments and the L4-L6 segments are added to the analysis of the in vivo data, a two-way ANOVA should be used (two variables: treatment condition and spinal cord segment).

The hypotheses of this study are that negatively charged pullulan-spermine/DNA anioplexes will permit nontoxic MSC transfection and that such technology can be used to detect transfected cells implanted in the spinal cord. Thus, in the in vitro experiment, untreated cells and cells treated with plasmid DNA or anioplexes should all show similar levels of viability, whereas viability should be reduced for MSCs transfected with positively charged pullulan-spermine/DNA or Lipofectamine 2000/DNA. Under the transfection conditions described in this chapter (but in the absence of serum), anioplex-transfected MSCs showed 89.7% (+/–10.5) viability (not significantly different from the untransfected control), whereas Lipofectamine 200/DNA-transfected MSCs showed 50.7% (+/–0.32) viability [19]. Given the positive surface charge of the lipolexes, such findings suggest that the negative surface charge of the anioplexes could be responsible for the lack of cytotoxicity, as positively charged macromolecules show local cell membrane-destabilizing effects that lead to toxicity and cell death [15–17].

We have also observed 12.3% (+/–0.75) viability for adipose-derived MSCs transfected with positively charged pullulan-spermine/DNA in the absence of serum when directly compared to anioplex-transfected MSCs [19]. This further suggests the importance of the negative surface charge for biocompatibility, as the same pullulan-spermine vector was used in both conditions; the primary difference between the particles formed was the surface charge. However, it remains possible that some other toxic epitope is preferentially displayed on the surface of positively charged pullulan-spermine DNA, or even that the threefold increase in the amount of pullulan-spermine (N:P 3:1 vs. N:P 1:1) may play a role. The primary interpretation of the WST-8 results should thus be that anioplex-derived transfection is nontoxic to MSCs, and that this lack of toxicity is likely (but not conclusively) derived from the negative surface charge. It is important to remember, however, that the WST-8 assay only measures NADH activity in the mitochondria; it should not be considered to directly indicate cell death. Thus, decreased WST-8 activity may also be interpreted as decreased proliferation (which can still be considered decreased viability), although this is highly unlikely in the described experiments given their short time course.

Regarding transfection efficency, luciferase reporter gene activity for wells treated with anioplexes should be higher than that for untreated wells, or wells treated with plasmid DNA or positively charged pullulan-spermine/DNA complexes (Figure 9.3). MSCs do not natively express luciferase; thus there should be no measurable luciferase activity in these cells above the baseline values of the luminometer. Plasmid DNA alone does not give rise to efficient gene delivery in most nonphagocytosing cell types; this is why it is typically necessary to utilize gene delivery vectors such as pullulan-spermine for transfection. We have found that positively charged pullulan-spermine works quite well in serum-free transfection conditions [24–26]; however, its performance is

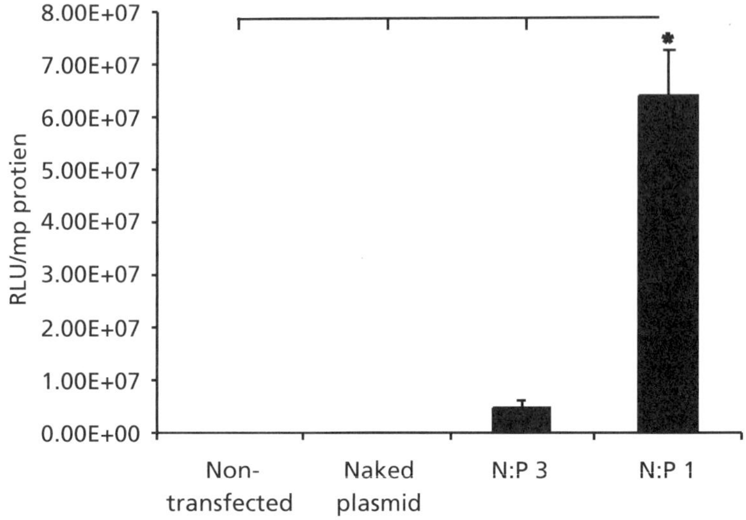

Figure 9.3 Luciferase reporter gene activity in MSCs 24h after transfection with naked plasmid DNA, positively charged pullulan-spermine/DNA complexes (N:P 3), and anioplexes (N:P 1) in the presence of serum. Activity levels were higher for anioplexes than in any of the other conditions. ($p = 0.002$, $p = 0.001$, $p = 0.001$ vs. positively charged pullulan-spermine, naked plasmid, and nontransfected, respectively; one-way ANOVA with Newman-Keuls post-hoc; * indicates significance; means and standard errors depicted; $n = 3$.)

decreased in the presence of serum [18], likely due to sequestration by negatively charged serum molecules such as albumin. On the other hand, negatively charged pullulan-spermine/DNA anioplexes permit efficient transfection of several different cell types, including MSCs, even in the presence of serum [18, 19]; this is likely due to their negative surface charge as well as receptor-mediated interactions that may involve members of the galectin or asialoglycoprotein receptor families. We also typically observe moderately increased transfection with anioplexes compared to Lipofectamine 2000 [18, 19], the commercially available gold standard, although results obtained with Lipofectamine 2000 can vary (Figure 9.4).

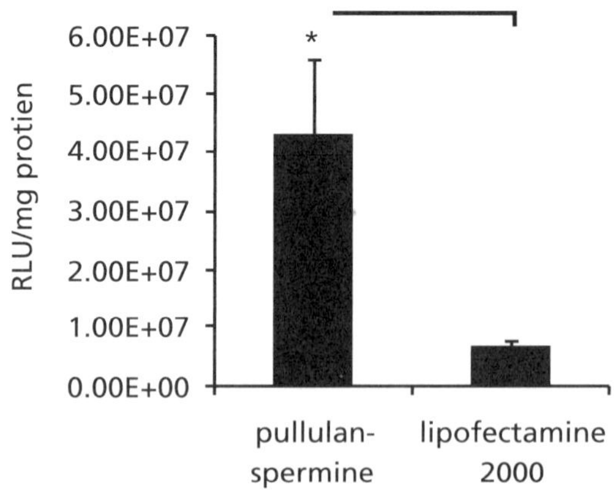

Figure 9.4 Luciferase reporter gene activity in MSCs 24h after transfection with pullulan-spermine/DNA anioplexes (N:P 1) and Lipofectamine 2000/DNA in the presence of serum. Activity levels were higher for anioplexes. ($p = 0.033$, t-test; * indicates significance; means and standard errors depicted; $n = 3$.)

The results of the in vitro luciferase assay described in this chapter should thus be interpreted as a proof that anioplexes can indeed be used to transfect MSCs under optimal culture conditions (i.e., in the presence of serum). This is important given the biological inertness (i.e., lack of toxicity) of this carrier, and also because thus far, polycationic gene delivery vectors have typically relied on a net positive charge for cell adsorption and uptake. The advantage of the luciferase assay used here is that it quantitatively measures overall functional reporter gene activity for a given amount of cultured cells. It should be noted, however, this assay cannot determine the percentage of cells transfected or identify specific subtypes of cells transfected. Fluorescence microscopy of reporter gene expression or fluorescence-activated cell sorting is typically used to evaluate these parameters [27], but these techniques are less favorable for quantification of overall expression levels. Thus, we recommend that the luciferase assay should be used to determine transfection efficiency in applications where the total production of transfected gene product is important (e.g., growth factor overexpression), whereas fluorescence microscopy should be used when the proportion of transfected cells is important (e.g., induction of progenitor cell differentiation).

In the in vivo experiment described in this chapter, spinal cord luciferase activity should be observed for rats intrathecally injected with anioplex-transfected MSCs as well as those injected with naked plasmid DNA, but not for those injected with vehicle, because rat tissues do not natively express luciferase. Intrathecal injection is an established technique for introducing plasmid DNA to the spinal cord [23]; recent evidence suggests that the DNA plasmids are taken up in host macrophages, probably via phagocytosis [28]. Thus, observation of spinal cord luciferase activity after naked plasmid administration will confirm that the lumbar puncture injection technique is successful. Luciferase activity should be higher in the L4-L6 region of the spinal cord near the injection site; however, it should still be detected in the L1-L3 region because of the relatively high volume of the injection.

The greatest advantage of the whole-tissue luciferase assay used here is that quantitative and specific comparisons of functional transgene activity can be made across tissue samples. Additionally, the use of luciferase introduces the subsequent possibility of noninvasive in vivo imaging of transplanted cells. Observation of spinal cord luciferase activity at 4 days after transplantation (Figure 9.5) likely indicates the presence of viable MSCs, as the half-life of the luciferase protein is only ~3h [29]. However, it remains to be examined systematically whether plasmid DNA in dead MSCs could be consumed and expressed by resident or infiltrating macrophages. Lastly, measurement of total luciferase activity in a spinal cord segment does not reveal the specific location of the transfected cells or allow the evaluation of their relationship to host cells and structures. Immunohistochemical staining should therefore be used to confirm reporter gene expression by viable MSCs and to better describe MSC localization and host interactions.

9.5 Discussion and Commentary

Transplantation of genetically modified cells has great therapeutic and investigative potential in the central nervous system, particularly in the spinal cord. However, successful implementation relies upon the development of effective gene transfer vectors and transplantation techniques.

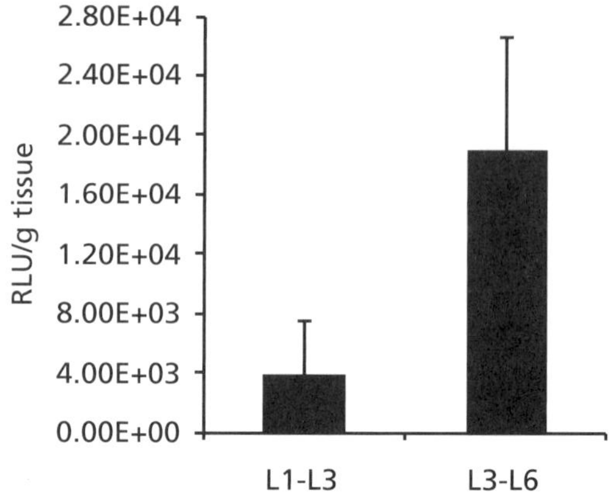

Figure 9.5 Spinal cord implantation of anioplex-transfected MSCs. At 4 days after minimally invasive lumbar intrathecal injection with 1.2×10^6 MSCs that had been transfected ex vivo with pullulan-spermine/DNA anioplexes carrying approximately 40 μg of DNA, luciferase reporter gene activity was detected in spinal cord tissue both in the vicinity of and rostral to the injection site (L3-L6 and L1-L3 zones, respectively). (Means and standard errors depicted in all graphs; $n = 4$.)

Nonviral vectors are conventionally thought to be safer, easier to manipulate, and more easily manufactured than viral vectors. However, they are typically characterized by much lower transfection efficiency, and polycationic nonviral vectors are often toxic. This could potentially reduce the yield and alter the biological properties of the transfected cells; such considerations are especially important for progenitor cells or primary cells that can only be isolated or prepared in small numbers. We therefore have envisioned the pullulan-spermine anioplex vector described in this chapter as a tool for ex vivo gene delivery in cell transplantation because it is easy to prepare and use, serum-compatible, and nontoxic even at high concentrations and over long incubation times.

Although the preparation of pullulan-spermine is relatively simple and resistant to procedural error, there are several points where problems can occur. Overaddition of spermine to pullulan will occur if the initial reaction of CDI with pullulan is allowed to proceed for too long before addition of spermine; it is particularly easy to make this mistake when simultaneously preparing multiple batches of pullulan-spermine. Thus, to ensure repeatable synthesis, it is absolutely necessary to keep the initial CDI-pullulan reaction to only 5 min; if possible, only one batch of pullulan-spermine should be prepared at a time. Moreover, the dropping rate of the spermine solution into the CDI-pullulan solution is important; if the spermine is added too quickly, precipitation will form, but if it is added too slowly, the CDI reaction with pullulan may proceed too far. This issue is best addressed by practice; however, one drop at a time from a 10-ml Eppendorf pipetter with not more than a 1-second interval between drops is typically sufficient.

When complexing pullulan-spermine with plasmid DNA, precipitation can occur; this usually does not happen at the relatively low concentrations described here for in vitro transfection, but it should still be looked for. If precipitation is observed, start again with a new sample, and this time add the pullulan-spermine solution dropwise to the DNA solution while gently vortexing (at 40% of maximum vortex intensity); precipitation will be greatly reduced, although it may still not be eliminated.

In the described study, the WST-8 and luciferase assays are performed on the same wells; this allows double the amount of data to be collected from the same experiment and also allows direct correlation of the viability of the cells in each well with their reporter gene expression level. We have not observed any downstream effects on the luciferase assay after performing the WST-8 assay and recommend this technique.

When performing the luciferase assay, it is critical to ensure that all liquid is removed from each well prior to adding the lysis buffer because such a small volume of lysis buffer is used; remaining liquids can affect luciferase concentration and activity. It is also important to keep the samples on ice because luciferase is relatively unstable. Although patterns of luciferase activity (i.e., comparisons between different groups) are usually consistent, the actual RLU levels can vary across trials. It is therefore preferable to run all samples for a given set of comparisons at the same time. If the samples cannot be run at the same time, then the data should be carefully interpreted, and it may be useful to present different comparisons in different graphs.

In addition to the use of pullulan-spermine for ex vivo gene delivery, we have presented here a simple and minimally invasive technique for spinal cord implantation of the transfected cells. When practicing this method, any concerns about injection success or localization can be addressed by applying the same technique to injection of 1% methylene blue dye [30] followed by dissection and visual inspection. It is notable that our technique utilizes an incision in the back of the animals to expose the injection site, as lumbar puncture is often performed without such an incision; please see the excellent review article by Fairbanks [30] for a thorough description of this technique in multiple species.

Lumbar puncture without surgical incision, which is colloquially known as a spinal tap and is a frequently used clinical procedure, is advantageous in that it is truly minimally invasive [30]. This is especially critical for studies involving injection of short-acting drugs. However, this technique requires practice and skill to achieve accuracy and reproducibility. In this chapter we described how to use the incision-introduced spinal tap because it provides easier access, manipulation, and visualization of the injection site, thus allowing increased reproducibility and faster mastery of the technique. The animals show fast recovery from the incision without impairment, and given that the time course of cell transplantation-derived effects is likely much longer than that of short-acting drugs, the use of anesthesia and incision should not affect the results. We therefore recommend this technique for researchers, especially those who will not be frequent users of intrathecal cell transplantation. Conversely, for frequent users, the nonincision technique should be considered, especially for conducting studies with large treatment groups.

The 30-μl injection volume described here is at the upper limit of the acceptable intrathecal injection volume in rats, which results in fairly wide dispersal of cells throughout the spinal cord. In addition, the cells are administered to the intrathecal space; they must migrate to the spinal cord parenchyma and penetrate it to engraft within the spinal cord, although intraspinal engraftment may not be necessary if the desired effect of the cells is derived from secreted molecules. Thus, although MSCs are known to home to sites of injury and inflammation [20–22], this technique may not be suitable for studies requiring the localized administration of cells within the tissue. In such cirumstances, direct intraparenchymal injection of cells can be used [20, 31]. However, this method is technically challenging and invasive (the injection itself may dam-

Troubleshooting Table

Problem	Explanation	Potential Solutions
Spermine introduction too high.	CDI-pullulan reaction time too long.	Keep reaction to 5 min.
Precipitation observed on addition of spermine to CDI-pullulan.	Spermine added too fast.	Be sure to add spermine drop-by-drop.
Precipitation observed during complexation of pullulan-spermine with DNA.	Pullulan-spermine added too fast.	Add pullulan-spermine drop-by-drop with gentle vortexing.
Extreme variability in RLU measurements of wells from the same treatment condition.	Insufficient removal of liquids from well prior to addition of lysis buffer.	Carefully remove all liquid from each well prior to addtion of lysis buffer.

age the spinal cord, especially with larger volumes), and fewer cells can be implanted (typically around 300,000 [31]). Additionally, repeated injections (which can be used to further increase the cell dose or to evaluate effects at different time points) are more difficult than when using the lumbar puncture technique. Therefore, for most studies involving genetically modified MSCs, especially those engineered to overexpress soluble factors, we recommend intrathecal injection.

9.6 Application Notes

Thus far, NGF-, bFGF-, BDNF-, and NT-3-overexpressing fibroblasts [32, 33] and met-enkephalin-overexpressing adrenal chromaffin cells [34] have been parenchymally injected in rat models of spinal cord injury and inflammatory pain, respectively. Retroviral overexpression of BDNF [35] and IGF [36] in MSCs prior to parenchymal transplantation resulted in enhanced axonogenesis and neuroprotection, respectively, in rat models of spinal cord injury. Lumbar puncture injection has been used to apply nontransfected MSCs to rat spinal cord injury. However, to our knowledge the technique described in this chapter is the first to combine nonviral genetic modification with lumbar puncture injection. We have applied this technology to the spinal nerve ligation (SNL) model of neuropathic pain, where spinal cord neuroinflammation is thought to contribute to the maintenance of pain. Using the technique described in this chapter, we injected 1.2×10^6 rat MSCs transfected with anioplexes containing plasmid DNA expressing a variant of IL-10, an anti-inflammatory cytokine (generously provided by Avigen and Erin Milligan of the University of New Mexico) under the conditions described above into SNL rats. After injection, treated rats showed a significant decrease in mechanical hypersensitivity on the side of the lesion compared to vehicle-injected controls, suggesting partial relief of neuropathic pain (Figure 9.6). This supports previous reports of antinociception in rodent neuropathic pain after lumbar puncture of naked plasmid DNA expressing IL-10 [23, 37, 38], suggesting that transplantation of ex vivo anioplex-transfected cells with autologous potential could provide an alternative method for gene therapy of the spinal cord.

9.7 Summary Points

1. Pullulan-spermine/DNA anioplexes can be used for nontoxic genetic modification of MSCs in the presence of serum.

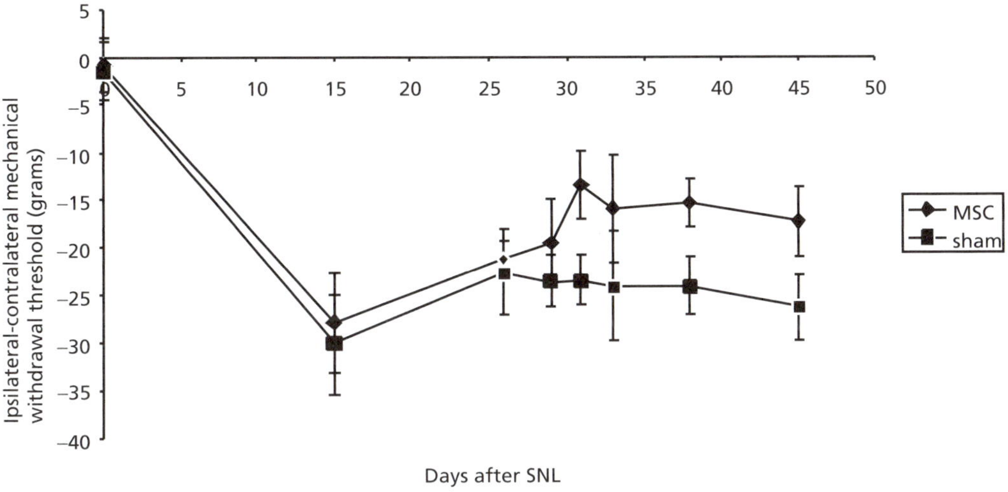

Figure 9.6 Mechanical withdrawal thresholds of SNL-injured animals after intrathecal injection of 1.2× 10^6 MSCs genetically engineered to express an IL-10 variant using pullulan-spermine/DNA anioplexes. Significant antinociception ($p = 0.025$, *t*-test) was observed 8 days after injection. Animals tended to show continued analgesic effects at 15 and 22 days after injection. ($p = 0.06$ for both time points, *t*-test) ($n = 7$ for MSC treatment, $n = 6$ for vehicle treatment.)

2. Minimally invasive lumbar puncture can be used to inject transfected MSCs into the intrathecal space of the rat spinal cord.
3. Reporter gene activity can be used to identify the transplanted cells in spinal cord tissue.

Acknowledgments

The work of DKT and YDT is currently supported by VARR&D, CIMIT, and DOD. DKT has been supported by the Japan Society for the Promotion of Science, the Clinical Implant Society of Japan, and the Takanawakai. We gratefully acknowledge Erin Milligan of the University of New Mexico and Avigen for the IL-10 plasmid as well as Dr. Jyunichi Miyazaki of Osaka University for the pCAGGS plasmid DNA.

References

[1] Chernykh, E. R., et al., "Application of Autologous Bone Marrow Stem Cells in the Therapy of Spinal Cord Injury Patients," *Bull. Exp. Biol. Med.*, Vol. 143, No. 4, 2007, pp. 543–47.

[2] Saito, F., et al., "Spinal Cord Injury Treatment with Intrathecal Autologous Bone Marrow Stromal Cell Transplantation: The First Clinical Trial Case Report," *J. Trauma*, Vol. 64, No. 1, 2008, pp. 53–59.

[3] Yoon SH, et al., "Complete Spinal Cord Injury Treatment Using Autologous Bone Marrow Cell Transplantation and Bone Marrow Stimulation with Granulocyte Macrophage-Colony Stimulating Factor: Phase I/II Clinical Trial," *Stem Cells*, Vol. 25, No. 8, 2007, pp. 2066–73.

[4] Syková, E., et al., "Autologous Bone Marrow Transplantation in Patients with Subacute and Chronic Spinal Cord Injury," *Cell Transplant.*, Vol. 15, No. 8–9, 2006, pp. 675–87.

[5] Callera, F., and R. X. do Nascimento, "Delivery of Autologous Bone Marrow Precursor Cells into the Spinal Cord Via Lumbar Puncture Technique in Patients with Spinal Cord Injury: A Preliminary Safety Study," *Exp. Hematol.*, Vol. 34, No. 2, 2006, pp. 130–31.

[6] Park, H. C., et al., "Treatment of Complete Spinal Cord Injury Patients by Autologous Bone Marrow Cell Transplantation and Administration of Granulocyte-Macrophage Colony Stimulating Factor," *Tissue Eng.,* Vol. 11, No. 5–6, 2005, pp. 913–22.

[7] Giudice, A., and A. Trounson, "Genetic Modification of Human Embryonic Stem Cells for Derivation of Target Cells," *Cell Stem Cell,* Vol. 2, No. 5, 2008, pp. 422–33.

[8] Giudice, A., and A. Trounson, "Genetic Modification of Stem Cells for Transplantation," *Adv. Drug Deliv. Rev.,* Vol. 60, No. 2, 2008, pp. 160–72.

[9] Tey, S. K., and M. K. Brenner, "The Continuing Contribution of Gene Marking to Cell and Gene Therapy," *Mol. Ther.,* Vol. 15, No. 4, 2007, pp. 666–76.

[10] Kimelman, N., et al., "Review: Gene- and Stem Cell-Based Therapeutics for Bone Regeneration and Repair," *Tissue Eng.,* Vol. 13, No. 6, 2007, pp. 1135–50.

[11] Papapetrou, E.P., N. C. Zoumbos, and A. Athanassiadou," "Genetic Modification of Hematopoietic Stem Cells with Nonviral Systems: Past Progress and Future Prospects," *Gene Ther.,* Vol. 12, Suppl. 1, 2005, pp. S118–30.

[12] Godbey, W. T., and A. G. Mikos, "Recent Progress in Gene Delivery Using Non-Viral Transfer Complexes," *J. Control Release,* Vol. 72, 2001, pp. 115–25.

[13] Lungwitz, U., et al., "Polyethylenimine-Based Non-Viral Gene Delivery Systems," *Eur. J. Pharm. Biopharm.,* Vol. 60, 2005, pp. 247–66.

[14] Thakor, D., et al., "Subcutaneous Peripheral Injection of Cationized Gelatin/DNA Polyplexes as a Platform for Non-Viral Gene Transfer to Sensory Neurons," *Mol. Ther.,* Vol. 15, No. 12, 2007, pp. 2124–31.

[15] Werkmeister, J. A., et al., "Sequence Requirements for the Activity of Membrane-Active Peptides," *J. Pept. Res.,* Vol. 60, 2002, pp. 232–38.

[16] Dokka, S., et al., "Oxygen Radical-Mediated Pulmonary Toxicity Induced by Some Cationic Liposomes," *Pharm. Res.,* Vol. 17, 2000, pp. 521–25.

[17] Hunter, A. C., "Molecular Hurdles in Polyfectin Design and Mechanistic Background to Polycation Induced Cytotoxicity," *Adv. Drug Deliv. Rev.,* Vol. 2, 2006, pp. 1523–31.

[18] Thakor, D. K., Y. D. Teng, and Y. Tabata, "Neuronal Gene Delivery by Negatively Charged Pullulan-Spermine/DNA Anioplexes," *Biomaterials,* Vol. 30, No. 9, 2009, pp. 1815–26.

[19] Thakor, D. K., et al., Program No. 774.16, 2008 Neuroscience Meeting Planner, Chicago, IL: Society for Neuroscience, 2008.

[20] Vaquero, J., et al., Cell Therapy Using Bone Marrow Stromal Cells in Chronic Paraplegic Rats: Systemic or Local Administration?" *Neurosci. Lett.,* Vol. 398, No. 1–2, 2006, pp. 129–34.

[21] Ohta, M., et al., "Bone Marrow Stromal Cells Infused into the Cerebrospinal Fluid Promote Functional Recovery of the Injured Rat Spinal Cord with Reduced Cavity Formation," *Exp. Neurol.,* Vol. 187, No. 2, 2004, pp. 266–78.

[22] Shi, E., et al., "Therapeutic Benefit of Intrathecal Injection of Marrow Stromal Cells on Ischemia-Injured Spinal Cord," *Ann. Thorac. Surg.,* Vol. 83, No. 4, 2007, pp. 1484–90.

[23] Milligan, E. D., et al., "Repeated Intrathecal Injections of Plasmid DNA Encoding Interleukin-10 Produce Prolonged Reversal of Neuropathic Pain," *Pain,* Vol. 126, No. 1–3, 2006, pp. 294–308.

[24] Jo, J., Ikai, T., et al., "Expression Profile of Plasmid DNA by Spermine Derivatives of Pullulan with Different Extents of Spermine Introduced," *J. Control Release,* Vol. 118, No. 3, 2007, pp. 389–98.

[25] Kanatani, I., et al., "Efficient Gene Transfer by Pullulan-Spermine Occurs Through both Clathrin- and Raft/Caveolae-Dependent Mechanisms." *J. Control Release,* Vol. 116, No. 1, 2006, pp. 75–82.

[26] Jo, J., et al., "Expression Profile of Plasmid DNA Obtained Using Spermine Derivatives of Pullulan with Different Molecular Weights," *J. Biomater. Sci. Polym. Ed.,* Vol. 18, No. 7, 2007, pp. 883–99.

[27] Yang F, et al., "Gene Delivery to Human Adult and Embryonic Cell-Derived Stem Cells Using Biodegradable Nanoparticulate Polymeric Vectors," *Gene Ther.,* Vol. 16, No. 4, 2009, pp. 533–46.

[28] Hughes, T.S., et al., "Immunogenicity of Intrathecal Plasmid Gene Delivery: Cytokine Release and Effects on Transgene Expression," *J. Gene Med.,* Vol. 11, No. 9, 2009, pp. 782–90.

[29] Corish, P., and C. Tyler-Smith, "Attenuation of green fluorescent protein half-life in mammalian cells," *Protein Eng.,* Vol. 12, No. 12, 1999, pp. 1035–40.

[30] Fairbanks, C. A. "Spinal Delivery of Analgesics in Experimental Models of Pain and Analgesia," *Adv. Drug Deliv. Rev.,* Vol. 28, No. 8, 2003, pp. 1007–41.

[31] Hofstetter, C. Pet al., "Marrow Stromal Cells Form Guiding Strands in the Injured Spinal Cord and Promote Recovery," *Proc. Natl. Acad. Sci. USA,* Vol. 99, No. 4, 2002, pp. 2199–04.

[32] Mitsui, T., et al., "Transplants of Fibroblasts Expressing BDNF and NT-3 Promote Recovery of Bladder and Hindlimb Function Following Spinal Contusion Injury in Rats," *Exp. Neurol.,* Vol. 194, No. 2, 2005, pp. 410–31.

[33] Nakahara, Y., F. H. Gage, and M. H. Tuszynski, "Grafts of Fibroblasts Genetically Modified to Secrete NGF, BDNF, NT-3, or Basic FGF Elicit Differential Responses in the Adult Spinal Cord," *Cell Transplant.,* Vol. 5, No. 2, 1996, pp. 191–204.

[34] Duplan, H., et al., "Grafts of Immortalized Chromaffin Cells Bio-Engineered to Improve Met-Enkephalin Release Also Reduce Formalin-Evoked c-Fos Expression in Rat Spinal Cord," *Neurosci. Lett.,* Vol. 370, No. 1, 2004, pp. 1–6.

[35] Lu, P., L. L. Jones, and M. H. Tuszynski, "BDNF-Expressing Marrow Stromal Cells Support Extensive Axonal Growth at Sites of Spinal Cord Injury," *Exp. Neurol.,* Vol. 191, No. 2. 2005, pp. 344–60.

[36] Hollis, E.R., 2nd, et al., "IGF-I Gene Delivery Promotes Corticospinal Neuronal Survival But Not Regeneration after Adult CNS Injury," *Exp. Neurol.,* Vol. 215, No. 1, 2009, pp. 53–59.

[37] Sloane, E. M., et al., "Long-Term Control of Neuropathic Pain in a Non-Viral Gene Therapy Paradigm," *Gene Ther.,* Vol. 16, No. 4, 2009, pp. 470–75.

[38] Sloane, E., et al., "Immunological Priming Potentiates Non-Viral Anti-Inflammatory Gene Therapy Treatment of Neuropathic Pain," *Gene Ther.,* Vol. 16, No. 10, 2009, pp. 1210–22.

10

Adipose Tissue as a Source of Mesenchymal Stem Cells for Hepatocyte Transplantation

Tetsuya Ishikawa, Keitaro Hagiwara, Laura Gómez-Santos, Hideki Iwaguro, and Takahiro Ochiya*

Division of Molecular and Cellular Medicine, National Cancer Center Research Institute, Tokyo 104-0045, Japan
*Correspondence: Division of Molecular and Cellular Medicine, National Cancer Center Research Institute 5-1-1, Tsukiji, Chuo-ku, Tokyo 104-0045, Japan
E-mail: tochiya@ncc.go.jp

Abstract

Adipose tissue-derived mesenchymal stem cells (AT-MSCs) can be isolated, expanded ex vivo, and reintroduced into the patient as an autologous transplant. These attributes overcome the limitations imposed by ethical issues and immune system rejection. Currently, adipose tissue has received a great deal of attention as a source of MSCs for cell therapy. A sufficient number of cells for therapeutic use can be obtained from adipose tissue with minimal invasiveness and without affecting the health of the patient. AT-MSCs, also referred to as processed lipoaspirate (PLA) cells, adipose-derived stromal cells (ADSCs), adipose-derived adherent stromal cells/adipose-derived adult stem cells (ADASs), or adipose-tissue-derived stromal cells (ATSCs), are considered to be a multipotent population of adherent cells. After isolation of the adipose stromal vascular fraction (SVF), the cells attach to plastic culture dishes and remain there as a heterogeneous population of fibroblast-like cells. Although AT-MSCs are more heterogeneous than bone marrow mesenchymal stem cells (BM-MSCs), they express a surface antigen marker profile and differentiation potential similar to BM-MSCs. Cultured AT-MSCs are characterized as $CD45^- CD34^+ CD105^+$. Because they are easy to procure in large numbers, they are attractive for use in clinical studies focused on the establishment of alternative therapies for the treatment of liver dysfunction. Freshly isolated or frozen stocks of uncultured AT-MSCs could provide a future therapeutic option for regenerative medicine in a clinical setting.

Key terms: adipose tissue-derived mesenchymal stem cells (AT-MSCs), regenerative cells, hepatic regeneration, hepatocyte differentiation, cell transplantation

10.1 Introduction

The direct differentiation of mouse embryonic stem (ES) cells into functional hepatocytes was established in adherent monoculture using a hepatocyte induction factor cocktail (HIFC) [1, 2] (Figure 10.1). Growth factors that direct differentiation into hepatocytes have been identified by in vivo transplantation of ES cells into liver-injured animals [3]. In addition, ES cell-derived hepatocytes have been maintained for long periods of time in three-dimensional cultures with a porous hyaluronate (3D-HA) sponge [4]. In this review, we suggest that AT-MSCs hold promise for the development of cell-based therapy for the treatment of liver dysfunction [5, 6]. Adipose tissue-derived mesenchymal stem cells (AT-MSCs) can be isolated ex vivo, expanded, and transplanted back into the patient as an autologous transplant. This overcomes the limitations imposed by ethical issues and immune system rejection. Currently, adipose tissue has received great attention as a source of MSCs for cell therapy [7, 8]. A sufficient number of cells for therapy can be easily obtained from adipose tissue with minimal invasiveness and without causing damage to a patient's health.

AT-MSCs are very similar to BM-MSCs. Despite the fact that AT-MSCs are a more heterogeneous population of cells, they exhibit a surface antigen marker profile and differentiation potential similar to BM-MSCs. Cultured AT-MSCs are characterized as $CD45^-$ $CD34^{\pm}$ $CD105^+$. One of the markers that defines MSCs is CD105 (endoglin), which is a component of the transforming growth factor (TGF)-β receptor complex. TGF-β is a pleiotropic cytokine involved in cellular proliferation, differentiation, and migration. $CD105^+$ cells from bone marrow contain more colony-forming unit-fibroblasts (CFU-Fs), and possess the capacity to differentiate into a chondrogenic lineage and form bone in vivo. The adipogenic and myogenic differentiation potential of $CD105^+$ BM-MSCs is not influenced by the age of the donor; however, the fraction of MSCs obtained from BM

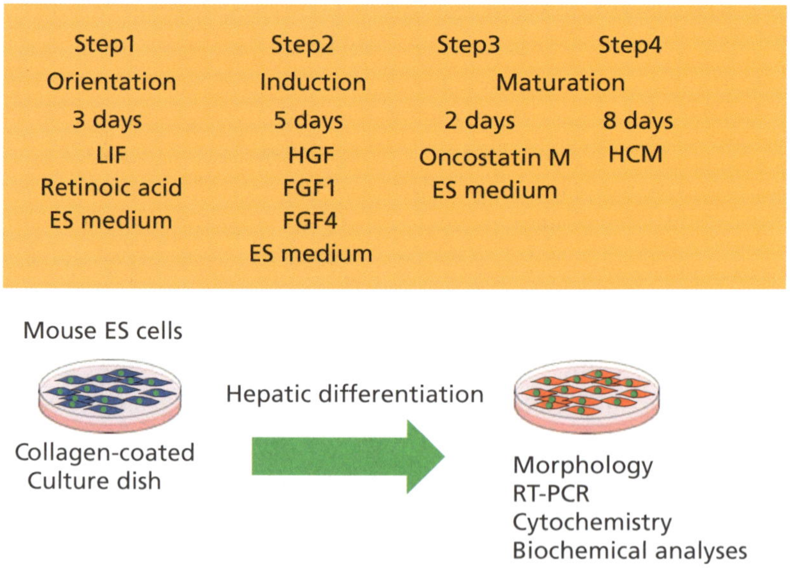

Figure 10.1 Schematic representation of the differentiation protocol for the induction of hepatocytes from mouse ES cells in monolayer culture, steps 1 to 4. ES cells = embryonic stem cells; LIF = leukemia inhibitory factor; RA = retinoic acid; FGF = fibroblast growth factor; HGF = hepatocyte growth factor; OsM = oncostatin M; HCM = hepatocyte culture medium; ESM = ES medium.

samples tends to decrease in older patients. Whether there is a correlation between the number of $CD105^+$ MSCs in adipose-tissue and the age of the donor is not clear at this point.

10.2 Materials

- Subcutaneous abdominal adipose tissue, approximately 5g.
- Surgical tools (scissors, scalpels, etc.).
- Phosphate buffered saline (PBS).
- Collagenase type 1.
- Water bath shaker (used for adipose tissue dissociation).
- Dulbecco's Modified Eagle Media (DMEM) (Gibco).
- Fetal bovine serum (FBS).
- NH_4Cl.
- 40 μm syringe filters (Millipore).
- 75 cm^2 flask.
- 60-mm cell culture plates.
- Trypsin-EDTA (Gibco).
- CD105-coupled magnetic microbeads (Miltenyi Biotec).
- Magnetic-activated cell sorting (MACS) buffer (PBS, 0.5% bovine serum albumin, 2 mM EDTA).
- Magnetic columns (Miltenyi Biotec).
- MACS Separator (Miltenyi Biotec)
- Fluorescent-activated cell sorting (FACS) (EPICS XL, Software Expo 32, Beckman Coulter).
- Mouse anti-human CD29 antibody (BD Bioscience Pharmingen).
- Mouse anti-human CD31 antibody (eBioscience).
- Mouse anti-human CD45 antibody (eBioscience).
- Mouse anti-human CD34 antibody (DacoCytomation).
- Mouse anti-human CD105 antibody (Ancell, Bayport, MN).
- Human bone marrow-mesenchymal stem cells (BM-MSCs,Lonza).
- Mesenchymal stem cell-growth medium (MSC-GM, Lonza).
- Hematopoietic $CD34^+$ stem cells (Lonza).
- Hematopoietic progenitor growth medium (HPGM, Lonza).
- HepG2 cell line.
- Collagen-coated dishes (ASAHI GLASS, Tokyo, Japan).
- Primary human hepatocytes (TaKaRa Bio Kyoto, Japan).
- Hepatic culture medium (HCM, Lonza).
- Hepatic induction media (HCM media supplemented with: transferrin (5 μg/ml), hydrocortisone-21-hemisuccinate (10^{-6}M), bovine serum albumin (0.5 mg/ml), ascorbic acid (2 mM), epidermal growth factor (20 ng/ml), insulin (5 μg/ml) gentamicin (50 μg/ml; Cambrex) and dexamethasone (10^{-8}M), supplemented with

HIFC that contained HGF (150 ng/ml), FGF1 (300 ng/ml), and FGF4 (25 ng/ml) (PeproTech, London).

- Oncostatin M (OsM)(Pepro Tech).
- Dexamethasone.
- SuperScript III reverse transcriptase (Invitrogen).
- Ampli Taq Gold kit (Applied Biosystems).
- Formaldehyde.
- Protein block (DakoCytomation).
- Transthyretin antibody (TTR).
- Albumin antibody (ALB; Sigma).
- Mouse anti-human Cytokeratin-18 antibody (CK-18, Cosmo Bio, Japan).
- Mouse anti-human CYP3A4 antibody (Cosmo Bio, Japan).
- Rhodamine-conjugated secondary antibodies.
- FITC-conjugated secondary antibodies.
- Mouse anti-human vimentin antibody (Sigma).
- Mouse anti-human CD105 antibody (Cosmo Bio).
- Mammalian protein extraction reagent (Pierce).
- SDS-polyacrylamide gels.
- PVDF (polyvinylidine difluoride) membranes (BIORAD).
- Tris-buffered saline with 0.1% tween-20 (TBS-T).
- Rabbit anti-human CYP3A4 polyclonal antibody (Cosmo Bio).
- Goat NADPH-cytochrome P-450 reductase antibody (CPR, Daiichi Pure Chemicals, Tokyo)
- Mouse anti-human CYP1A1 antibody (Santa Cruz Biotechnology).
- Rabbit anti-human CYP2C9 antibody (Fitzgerald).
- Sheep anti-rabbit IgG-HRP-linked whole antibodies (GE Healthcare Bio Sciences).
- Sheep anti-mouse IgG-HRP-linked whole antibodies (GE Healthcare Bio Sciences).
- Rabbit anti-goat IgG (H+L) HRP-linked antibodies (Southern Biotechnology Associates, Inc., Birmingham)
- Mouse anit-human GAPDH monoclonal antibody (Santa Cruz).
- Acetylated LDL labeled with 1,1-dioctadecyl-3,3,3,3-tetramethylindocarbocyanine perchlorate (Dil-Ac-LDL, Biomedical Technologies, Inc., Stoughton).
- Periodic acid.
- Schiff's reagent.
- Hematoxylin.
- Ammonia-test Wako (Wako Pure Chemicals, Tokyo).
- Bromocresol green (BCG).
- Olive oil.
- CCl_4.
- BALB/c nude mice (SLC, Tokyo).
- Collagenase/dispase.

10.3 Methods

10.3.1 Isolation and Culture of AT-MSCs

10.3.1.1 Isolation and Plating of AT-MSCs

1. Prepare adipose tissue for isolation of AT-MSCs by mincing the tissue with scissors and/or scalpels into pieces measuring less than 3 mm.
2. Combine the minced tissue with an equal volume of PBS and shake gently until the mixture separates into two phases.
3. Wash the upper phase (containing stem cells, adipocytes, and blood) once with PBS.
4. Enzymatically dissociate the tissue by incubation with 0.075% type I collagenase in PBS for 1h at 37°C with gentle shaking.
5. Mix the dissociated tissue with an equal volume of DMEM supplemented with 10% FBS, and incubate the solution for 10 min at room temperature, until the solution separates into two phases.
6. Centrifuge the lower phase at 1,500 rpm for 5 min at 20°C.
7. Resuspend the cellular pellet in 160 mM NH_4Cl to eliminate erythrocytes, and pass it through a 40-μm mesh filter into a new tube.
8. Resuspend the cells in an equal volume of DMEM/10% FBS and then centrifuge.

Isolation of AT-MSCs from 5g of adipose tissue should result in a primary culture of 7.7×10^6 adherent cells (approximately 1.0×10^5 to 4.6×10^6/g) after 7 to 10 days in culture.

10.3.1.2 Separation of $CD105^+$ AT-MSCs by MACS

1. Resuspend the cells in DMEM/10% FBS and plate at a density of 1–5×10^6 cells/75cm^2.
2. When the cells reach 70% to 80% confluence, harvest them with 0.25% trypsin-EDTA.
3. Suspend the cells in MACS buffer (0.5–1.0×10^8 cells) and incubate them with CD105-coupled magnetic microbeads at 8°C for 15 min.
4. Rinse the cells with MACS buffer and centrifuge at 200×g for 10 min.
5. Apply the cells to a magnetic column and separate the $CD105^+$ cells with a MACS separator. Discard the flow through as the fraction of $CD105^-$ cells.
6. Remove the magnetic column from the MACS separator and elute the retained cells as the $CD105^+$ fraction.

After separation, approximately 2.0–8.0×10^5 $CD105^+$ should be obtained, which can be expanded for experimental analysis.

10.3.2 Flow Cytometry

10.3.2.1 Phenotype Evaluation of AT-MSCs

1. Add AT-MSCs to clean tube and incubate with blocking solution for 30 min at 4°C.
2. Add the appropriate antibody (CD29, CD31, CD45, CD34, or CD105 coupled to either PE or FITC) to the tube and incubate for 30 min at 4°C.
3. Wash cells twice with FACS buffer and resuspend.

Analyze the cells as soon as possible by flow cytometry analysis.

10.3.3 In Vitro Culture and Expansion of BM-MSC, BM-HSC, and HepG2 Cell Line

10.3.3.1 Culture of Human BM-MSCs, BM-HSCs, and HepG2 Cells

1. Seed human BM-MSCs and BM-HSCs at a density of 5–6×10^3 cells/cm^2 and maintain the cells in MSC-GM or serum-free HPGM at 37°C, 5% CO_2, and 90% humidity.
2. Plate and maintain HepG2 cells in type I collagen-coated dishes under the same conditions described for culture of AT-MSCs.
3. Seed human cryo-preserved primary hepatocytes at a density of X cells/cm^2 and maintain in HCM at 37°C, 5% CO_2, and 90% humidity.

10.3.3.2 Hepatic Differentiation by HIFC Method

1. At passage 5–9, plate 2–3×10^5 $CD105^+$ AT-MSCs onto 60mm type I collagen-coated dishes.
2. After the cells have adhered to the plate, induce hepatocyte differentiation by culturing in hepatocyte induction medium for three weeks (Figure 10.2).
3. For the following 2 weeks, culture the cells in the presence of oncostatin M (30 ng/ml) and dexamethasone (2×10^{-5}mol/l).
4. Finally, culture the induced AT-MSCs in HCM alone for an additional 2 to 5 weeks.

Hepatic differentiation can be confirmed by examination of cell morphology, immunohistochemistry, RT-PCR analysis, Western blot, and analysis of biochemical functions. Human primary hepatocytes or HepG2 cells should be included as positive controls.

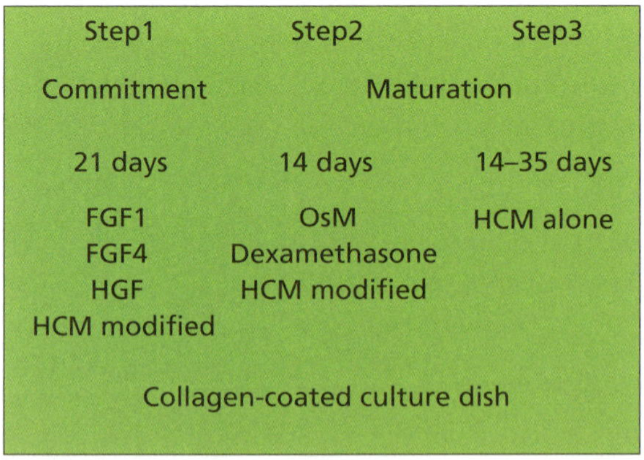

Figure 10.2 Human AT-MSCs were cultured on collagen for 21 days in the presence of a hepatic induction growth factor cocktail (HIFC: FGF1, FGF4, and HGF), followed by 14 days of culture in the presence of OsM/dexamethasone. The cells were then maintained in HCM alone for 2 to 5 weeks.

10.3.4 RT-PCR

1. Isolate total RNA from the cells, and reverse-transcribe 0.5 μg of RNA using SuperScript III Reverse Transcriptase according to the manufacturer's guidelines.
2. Use PCR analysis to identify the expression of hepatocyte specific cell markers. A kit such as the Ampli Taq Gold kit from Applied Biosystems is suitable for this application. The PCR primer sequences and conditions that were used by our laboratory are listed in Table 10.1.

10.3.5 Immunofluorescence

1. Fix cells in 4% formaldehyde for 10 min at room temperature, followed by incubation with Protein Block for 30 min with a wash step after fixation or blocking.
2. Incubate undifferentiated $CD105^+$ AT-MSCs and $CD105^+$ AT-MSC-derived hepatocytes with appropriate dilutions of the desired antibodies overnight at 4°C. The antibodies used in our laboratory were transthyretin (1:300), albumin (1:250), cytokeratin-18 (1:200), and CYP3A4 (1:2000).
3. After washing, incubate the cells with rhodamine (1:1000) or FITC (1:1000)-conjugated secondary antibodies for 30 min.

Undifferentiated AT-MSCs were stained according to the same protocol used for mouse anti-human vimentin (1:50) and mouse anti-human CD105 (1:40) antibodies.

10.3.6 Western Blot Analysis

The presence of CYP3A4, CYP1A1, CYP2C9, and NADPH P450 reductase can be detected by Western blot analysis of AT-MSC-derived hepatocytes.

1. Lyse the cells using a buffer such as mammalian protein extraction reagent.
2. Determine the protein concentration of the lysate and separate protein on an SDS-polyacrylamide gel.
3. Transfer the proteins to a PVDF (polyvinylidine difluoride) membrane and incubate the blots in a blocking solution of 5% skim milk in TBS-T for 1h at room temperature.
4. Incubate the blots overnight at 4°C with the appropriate antibody. The antibodies used by our laboratory were anti-human rabbit polyclonal CYP3A4, goat NADPH-cytochrome P-450 reductase, mouse CYP1A1, or rabbit CYP2C9. Monoclonal human GAPDH antibodies were used to confirm equal protein loading.
5. Wash the membranes in TBS-T, and incubate them for 30 min with sheep anti-rabbit or anti-mouse IgG-HRP-linked whole antibodies or rabbit anti-goat IgG (H+L) HRP-linked antibodies.
6. Visualize the proteins by chemiluminescence.

10.3.7 LDL Uptake

LDL uptake can be assessed by incubating the cells for 4h at 37°C with 10 μg/ml acetylated LDL labeled with 1,1-dioctadecyl-3,3,3,3-tetramethylindocarbocyanine perchlorate (Dil-Ac-LDL). The cells are then washed twice with PBS, fixed in 10% formalin, and analyzed by fluorescence microscopy.

Table 10.1 Primers and Conditions for RT-PCR

GenBankAccession No.	Primers Sequence*	Annealing Temperature (°C)	Cycles	PCR Product Size (bp)
CD10 NM_000902	F: TTGTAAGCAGCCTCAGCCG R: TTGTCCACCTTTTCTCGGAG	60	35	459
CD13 NM_001150	F: GCCGTGTGCACAATCATCGCACT R: CACCAGGGAGCCCTTGAGGTG	60	31	449
CD44 M59040	F: GATCCACCCCAATTCCATCTGTGC R: AACCGCGAGAATCAAAGCCAAGGCC	60	30	586
CD59 NM_203330	F: ACACTCTACTACATGTGACTG R: TGCAAAAGTCAGCCTATGCC	58	31	439
CD105 NM_000118	F: TGTCTCACTTCATGCCTCAGCT R: AGGCTGTCCATGTTGAGGAGT	58	33	377
CD166 NM_001627	F: AGATACCATTATCATCATACCTTGCCGACT R: TGTCTTTGTATTCGTGTACATCGTCG	60	35	157
CD49d NM_000885	F: GAGTGCAATGCAGACCTTGA R: AAAGAAGCCAGCCTTCCACA	58	33	417
SH3 NM_001009555	F: CTCAAGATGTGTTGCTCGGT R: CGGTACTATGGCCAACATAC	58	35	474
CD11b NM_000632	F: TCAGAGAATACCAGTCGGGT R: AAGAAGCCGAGCTTGTACAG	58	40	571
CD29 NM_002211	F: ACCAAGGTAGAAAGTCGGGA R: TGACCACAGTTGTTACGGCA	58	31	366
CD71 NM_003234	F: TCGCTGGTCAGTTCGTGATT R: CAGCTCCCTGAATAGTCCAA	58	31	461
CD90 NM_006288	F: GCTCTCCTGCTAACAGTCTT R: TCACAGGGACATGAAATCCG	58	31	467
CD106 NM_001078	F: GAGATACAACCGTCTTGGTC R: GATGGTATAGGCGCCATCTA	58	40	433
CD117 NM_000222	F: AACGCTCGACTACCTGTGAA R: GACAGAATTGATCCGCACAG	58	33	401
CD120a NM_001065	F: TCTCTTCTTGCACAGTGGAC R: CACCGTTGGTAGCGATACAT	58	31	378
CD124 NM_000418	F: ACCAAGTGGCACAACTCCTA R: GCCTTGTGAGGATCTTCATC	58	35	328
CD135 NM_004119	F: GGCTCGAGATATCATGAGTG R: CTACGAATCTTCGACCTGAG	58	33	489
CD48 BCO16182	F: GGAAGATCAAGCTGCAAGTG R: CACTAGCCAACTTGCAATCC	58	33	334
CD45 NM_002838	F: TTCACTGCAGGGATGGATCT R: GACTTGCAGGACCATTGACA	58	33	401
CD14 NM_000591	F: AGACCTGTCTGACAATCCTG R: AGATCGAGCACTCTGAGCTT	58	31	317
CD31 NM_000442	F: CAGTGGTTATCATCGGAGTG R: GCCTTCCGTTCTAGAGTATC	58	31	388
CD34 NM_001025109	F: GTTCAGGCATCAGAGAAGTG R: TCAGGAAATAGCCAGTGATG	58	35	407
ALB NM_000477	F: GTCACCAAATGCTGCACAGA R: ACGAGCTCAACAAGTGCAGT	56	38	180
TDO2 NM_005651	F: GTGTGCATGGTGCACAGAAT R: GGGTTCATCTTCGGTATCCA	56	35	166
HNF4β X76930	F: CCGGATCAGCTACTCGAA R: AGCTCGTCAAGGATGCGTATG	58	36	411
AFP NM_001134	F: CATTTCTCCAACAGGCCTGA R: TCCCTCCTGCATTCTCTGAT	58	33	182
CYP7A1 NM_000780	F: AGGACGGTTCCTACAACATC R: CGATCCAAAGGGCATGTAGT	56	45	196
TTR NM_000371	F: TACTGGAAGGCACTTGGCAT R: TTCCTTGGGATTGGTGACGA	56	40	152
GAPDH BC025925	F: GACTTCAACAGCGACACCC R: GCCAAATTCGTTGTCATACCA	60	30	105

F = forward; R = reverse.

10.3.8 Periodic Acid-Schiff Staining

Glycogen levels in undifferentiated AT-MSCs and HIFC-induced AT-MSC hepatocytes can be determined by periodic acid-Schiff (PAS) staining.

1. Fix cells with 10% formaldehyde oxidized in 1% periodic acid for 10 min. Rinse twice with distilled water.
2. Incubate the cells with Schiff's reagent for 10 min, rinse well with tap water.
3. Stain the cells with hematoxylin for 1 min, then rinse well with distilled water.
4. Dehydrate the cells with alcohol, apply a coverslip and observe under a light microscope.

10.3.9 Ammonia Clearance

To measure the ammonia clearing function of AT-MSCs, culture the cells in the presence of 2.5 mM NH_4Cl for 30h. At the 9th, 19th, and 30th hours of incubation, collect the medium and determine the concentration of NH_4Cl using Ammonia-Test Wako.

10.3.10 Albumin Production

The albumin level in the culture medium can be estimated using the BCG method.

1. Mix BCG with culture medium from undifferentiated AT-MSCs or HIFC induced AT-MSC hepatocytes.
2. Mix the samples gently and incubate for 5 min.
3. Read the samples in a spectrophotometer at 630 nm to determine albumin concentration.

10.3.11 $CD105^+$ AT-MSC-Derived Hepatocyte Transplantation into Mice with CCl_4-Induced Injury

To assess the therapeutic capabilities of $CD105^+$ AT-MSC derived hepatocytes, the effect of transplantation of in vitro generated $CD105^+$ AT-MSC-derived hepatocytes on CCl_4-induced liver injury can be assessed.

1. Nonlethal liver injury is induced in BALB/c nude mice by intraperitoneal injection of 100 μL/20g body weight of olive oil containing 10 μL of CCl_4.
2. $CD105^+$ AT-MSC-derived hepatocytes are collagenase-dispase-treated, while undifferentiated AT-MSCs are trypsinized. After centrifugation, the cell pellet is rinsed in PBS and resuspended for transplantation.
3. $CD105^+$ AT-MSC-derived hepatocytes can then be transplanted into the mice by injection of 5×10^5 cells per mouse into the tail vein. CCl_4-treated mice without hepatocyte transplantation and untreated (olive oil) mice should be included as controls.

Histological analysis of liver tissues can be performed by serial tissue section 1 day after cell transplantation and either stained with hematoxylin and eosin or immunohistochemically examined for human specific albumin expression.

10.4 Anticipated Results

10.4.1 Characterization of AT-MSCs

Banas et al. [9] characterized AT-MSCs by flow cytometry, immunofluorescence, and RT-PCR analysis. Flow cytometry revealed that the cells from all patients ($n = 6$) were positive for: CD31 (12.9–30%), CD34 (5.3–9.9%), CD45 (0.06–8.4%), CD29 (96–99%), and CD105 (63.9–98.5%). The number of $CD105^+$ cells varied between donors. AT-MSCs were identified by comparing the expression profiles of stem cell markers (mesenchymal: CD10, CD13, CD59, CD105, CD166, CD49d, SH3, CD29, CD44, CD71, CD90, and CD106; and hematopoietic: CD120a, CD124, CD11b, CD14, CD31, CD34, CD45, CD48, CD135, and CD117) with mRNA from BM-MSCs and BM-$CD34^+$ stem cells (Table 10.2) [9]. AT-MSCs and BM-MSCs share similarities in morphology and marker profile, however, AT-MSCs were also found to express CD34, CD31, and CD14 mRNA. The expression of CD106, CD14, CD31, and CD117 varied according to the donor. In addition, immunostaining revealed that the expression of CD105 and vimentin was lower in unfractionated AT-MSCs than in BM-MSCs. The $CD105^+$ fraction of AT-MSCs obtained by MACS sorting revealed a population of cells with a homogeneous morphol-

Table 10.2 Expression of Stem Cell Markers in AT-MSCs, BM-MSCs, and BM CD34+ Stem Cells, by RT-PCR Analysis

	Bone Marrow-Derived Mesenchymal Stem Cells	Adipose Tissue-Derived Mesenchymal Stem Cells	Bone Marrow-Derived CD34+ Stem Cells
CD10	+	+	–
CD13	+	+	+
CD59	+	+	+
CD105	+	+	+
CD166	+	+	–
CD49d	+	+	+
SH3	+	+	–
CD29	+	+	+
CD44	+	+	+
CD71	+	+	+
CD90	+	+	–
CD106	+	+/–	–
CD120a	+	+	+
CD124	+	+	+
CD11b	–	–	+
CD14	–	+/–	+
CD31	–	+/–	+
CD34	–	+	+
CD45	–	–	+
CD48	–	–	+
CD135	–	–	+
CD117	+	+/–	+

+/– means that mRNA expression varies depending on the patient.

ogy and high proliferation ability (doubling time of early passages = 34–36h, doubling time of the unfractionated population = 36–40h). The efficiency of MACS sorting was evaluated by CD105 staining and flow cytometry analysis before and after MACS sorting. Immunostaining of the $CD105^+$ fraction of AT-MSCs revealed expression of vimentin as well. The adipogenic differentiation potential of unfractionated AT-MSCs and sorted $CD105^+$ AT-MSCs was determined by inducing adipogenic differentiation. Oil Red O staining revealed that $CD105^+$ AT-MSCs had a higher adipogenic induction ratio (60%–82%) than unfractionated AT-MSCs (21%–67%). Oil Red O staining of noninduced AT-MCSs showed no lipid droplets. Similar results were obtained from other donors of different ages. Chondrogenic induction of $CD105^+$AT-MSCs also resulted in a higher induction ratio than the induction of unfractionated AT-MSCs. In summary, the paper reported that $CD105^+$ cells exhibit higher homogeneity, proliferation ability and potentiality than unfractionated AT-MSCs.

10.4.1.1 Hepatic Differentiation of AT-MSCs

It was also reported that the hepatic-differentiation potential of $CD105^+$ AT-MSCs [9]. During the initial stages of hepatic differentiation, $CD105^+$cells derived from three independent donors showed a remarkable transition from a bipolar fibroblast-like morphology to a round epithelial-like shape. The contraction of the cytoplasm progressed further during maturation, and most of the treated cells adopted a dense and round shape with clear or double nuclei in the late stage of differentiation. The cells eventually became small round or oval-shaped cells that were grouped together into a polyhedral structure with tight cell-to-cell interactions and visible bile canaliculi in a configuration similar to primary human hepatocytes. The $CD105^+$ AT-MSC-derived hepatocyte-like cells from a 45-year-old male (Donor #2) were used for further analysis.

10.4.1.2 RT-PCR Analysis of Hepatocyte-Specific Markers

The temporal expression pattern of a number of hepatocyte-specific genes, such as ALB, AFP, TTR, TDO2, and CYP7A1, and the HNF-4-alpha were analyzed by RT-PCR [9]. The expression of the early endoderm differentiation marker, AFP, decreased when the cells underwent maturation, whereas the expression of mature hepatocyte markers, such as ALB, TDO2, and CYP7A1 appeared at around day 6 of treatment with HIFC and persisted until day 70. HNF-4-alpha, which is essential for morphological and functional hepatocyte differentiation, was detected at day 41. Therefore, HIFC treatment induced $CD105^+$ AT-MSCs to differentiate into hepatocyte-like cells that exhibit a gene expression pattern of hepatocyte-specific markers and transcription factors similar to human hepatocytes.

10.4.1.3 Immunostaining and Western Blot Confirmation of Hepatic Differentiation

It was examined whether $CD105^+$ AT-MSCs also expressed hepatocyte-specific proteins following hepatic induction [9]. Markers such as ALB, TTR, CYP3A4, and CK-18 were positively identified by immunofluorescence staining. Cells were doubly positive for ALB/CYP3A4, CK-18/TTR, and ALB/TTR at day 40. Undifferentiated cells were not stained with any of these four markers. To further confirm hepatic induction, the protein expression of the hepatocyte-specific enzymes CYP1A1, CYP3A4, CYP2C9, and

NAPDH P-450 reductase was analyzed by Western blot analysis. A clear band was detected for each of the four proteins in $CD105^+$ AT-MSC-derived hepatocyte-like cells. Undifferentiated $CD105^+$ AT-MSCs also express a low, detectable level of CYP1A1, as had been indicated by RT-PCR analysis.

10.4.1.4 Functional Characterization of AT-MSC-Derived Hepatocytes

Afterwards, it was examined whether hepatocytes derived from $CD105^+$ AT-MSCs were functionally competent [9]. After 2 weeks of hepatic induction, almost 20% of the cells were able to incorporate LDL, and by maturity (40 days), almost all of the induced cells were capable of LDL uptake. At 35 days, $CD105^+$ AT-MSCs were analyzed by PAS staining for glycogen-storage ability. Almost all $CD105^+$ AT-MSC-derived hepatocytes exhibited strong PAS staining, while undifferentiated $CD105^+$ AT-MSCs were only weakly stained. Moreover, it was examined whether $CD105^+$ AT-MSCs produced albumin and detoxify ammonia. AT-MSC-derived hepatocytes had the capacity to secrete albumin and clear ammonia from culture media. This data indicates that $CD105^+$ AT-MSC-derived hepatocyte-like cells are capable of liver-specific functions similar to normal mature human hepatocytes.

10.4.1.5 Transplantation of $CD105^+$ AT-MSC-Derived Hepatocytes into Mice with CCl_4 Injury

It was also examined whether HIFC induced $CD105^+$ AT-MSC-derived hepatocytes were therapeutically applicable [9]. Transplantation of 5×10^5 hepatocytes was performed into BALB/c mice that had been injected with CCl_4. Prior to transplantation, hematoxylin-eosin staining of liver sections clearly indicated damage to the liver following CCl_4 injection. Twenty-four hours after transplantation of undifferentiated $CD105^+$ AT-MSCs and $CD105^+$ AT-MSC-derived hepatocytes, staining of liver sections with human specific albumin immunostaining revealed that $CD105^+$ AT-MSC-derived hepatocytes had incorporated into host livers. Measurement of parameters such as ammonia concentration level and levels of alanine aminotransferase (ALT) in the peripheral blood indicated that at least some liver functions had improved in mice transplanted with $CD105^+$ AT-MSC-derived hepatocytes. Some human albumin-positive cells were found in liver sections from mice implanted with undifferentiated $CD105^+$

Troubleshooting Table

Problem	Cause	Solution
No viable cells after thawing stock.	Lab-made stock not viable.	Freeze cells at a density of $1–2\times10^6$ viable cells/ml.
No viable cells after thawing stock.	Lab-made stock not viable.	Use low-passage cells to make your own stocks.
No viable cells after thawing stock.	Lab-made stock not viable.	Slow freezing and fast thawing is the key.
Cells grow slowly.	Cells too diluted.	Generally we recommend thawing one vial at a density of $1–5X10^6$ cells/75 cm^2.
Cells grow slowly.	Cells too old.	Use healthy ADSCs, do not overgrow and use low-passage cells.
Cells differentiated.	Cells too old.	Use healthy ADSCs, do not overgrow and use low-passage cells.
Cell aging.	The passaging time is not appropriate.	The cells should be subcultured when reaching 80% to 90% confluence.

AT-MSCs; however, these cells did not exhibit a typical hepatocyte morphology. In summary, the paper reported that transplantation of AT-MSC-derived hepatocyte-like cells is applicable for the treatment of liver dysfunction.

10.5 Discussion and Commentary

10.5.1 Hepatocyte Differentiation and Hepatic Regeneration

Human $CD105^+$ AT-MSCs are readily induced into the hepatic lineage, and the resulting hepatocyte-like cells express liver markers, proteins, and enzymes, and exhibit some of the functions of primary human hepatocytes [9]. In addition, signaling pathway analysis has identified genes that are significantly upregulated in hepatocyte-like cells, and these signals are also upregulated in whole liver [10]. Following in vivo transplantation into immunodeficient mice, $CD105^+$ AT-MSC-derived hepatocytes were incorporated as hepatocyte-like cells into CCl_4-injured livers. Human AT-MSCs, which can be obtained in large numbers, represent a very attractive source of stem cells for therapy of liver diseases. Such therapy would be comparatively easier, more efficient, and safer than whole organ transplantation in patients suffering from end-stage liver dysfunction. Considering the positive attributes of AT-MSC cells, such as a minimally invasive collection procedure, an abundant yield of isolated cells, and the nominal effects to donor health, AT-MSCs are an ideal tool for stem cell therapy for liver dysfunction. MSCs from other sources, such as bone marrow, exhibit a lower abundance and proliferation capacity than AT-MSCs, and are influenced by age, gender, the presence of osteoporosis, and prior exposure to high-dose chemotherapy or radiation. It is possible to obtain approximately 40-fold more MSCs from 200 ml of lipoaspirate than from 40 ml of bone marrow. Even though adipose tissue provides a more heterogeneous population of stem cells, the potential number of AT-MSCs is sufficiently large to allow for the selection of multipotential AT-MSCs.

$CD105^+$ BM-MSCs have been reported to have greater homogeneity and a greater ability to form CFU-Fs. As age does not affect the $CD105^+$ fraction, isolation of $CD105^+$ AT-MSCs for transplantation into elderly patients with liver failure or hepatocellular carcinoma may be possible. Large amounts of fat are discarded during liposuction, and in the future, this valuable resource could be used for the isolation of AT-MSCs, which might then be stored for later use. Hepatic differentiation of $CD105^+$ AT-MSCs produced functional and transplantable hepatocytes that could be used in the treatment of liver disease.

Unfractionated AT-MSCs also possess some hepatogenic potential. However, the ratio of differentiation was relatively lower than in $CD105^+$ AT-MSCs (20%–60% vs. 60%–85%). The hepatogenic potential of AT-MSCs was confirmed by the detection of liver specific markers and biochemical functions. Albumin, the major protein produced by hepatocytes, was synthesized and secreted into the medium, as measured on days 30 and 50 (5 and 14 mg/dl/4 days/10^6 cells, respectively). The expression and activity of microsome CYP enzymes, which are involved in drug and xenobiotic metabolism as well as in sterol and bile acid synthesis, are indicators of hepatocyte specificity. Hepatocyte-like cells generated from $CD105^+$ AT-MSCs expressed CYP7A1, CYP1A1, CYP2C9, CYP3A4, and NADPH-cytochrome P450 reductase. These findings suggest that these cells may be valuable for in vitro preclinical drug investigations.

Following the transplantation of AT-MSC-generated hepatocytes into CCl_4-injured nude mice, direct incorporation of the cells into the liver was revealed by human albumin immunostaining. Induction with HIFC does not require coculturing, thus, undesired cell–cell interactions are reduced. Nevertheless, further studies that define the in vivo mechanism of homing, engraftment, and liver regeneration need to be performed in order to eliminate post-transplantation complications. Some in vivo parameters, such as serum ammonia and ALT levels, decreased after transplantation of undifferentiated AT-MSCs, indicating the possibility of in vivo differentiation of AT-MSCs in a regeneration microenvironment. The subsequent interaction of transplanted undifferentiated MSCs with liver parenchyma cells needs to be further evaluated in the context of the promotion of fibrosis, cancer, or liver dysfunction. The correlation between factors such as donor age, history of cancer, and the differentiation potential of stem cells should be further evaluated as well. Three independent donor-derived lines of $CD105^+$ AT-MSCs showed similar hepatic differentiation abilities. Furthermore, induction of human AT-MSCs into hepatocyte-like cells occurs within a very short time. A period as short as 13 days may be sufficient for hepatic differentiation of unfractionated AT-MSCs if cells are pretreated with activin A and fibroblast growth factor 4 (FGF4) [11]. Additionally, the components of the growth factor cocktail used in this study were modified to include dimethyl sulfoxide (DMSO), nicotinamide, and insulin-transferrin-selenium. Using this revised protocol, functional hepatocyte-like cells were obtained in a much shorter period of time. When these AT-MSC-derived hepatocyte-like cells were transplanted into immunodeficient mice with nonsevere acute liver injury, significant decreases in levels of ammonia, aspartate aminotransferase (AST), ALT, and uric acid (UA) were observed in the blood plasma of the mice after transplantation. The mechanisms by which the transplanted cells improve liver function are not yet known and need to be further investigated. Future clinical applications will require refinement of these techniques, such as reducing ex vivo manipulations including in vitro expansion and hepatic differentiation as much as possible. In the context of clinical applications, AT-MSCs can be obtained from a patient's own tissue, in contrast to ES cells, which eliminates many obstacles to stem-cell-based treatments, such as ethical issues, the probability of immune system rejection, and the risk of teratoma and tumor formation. The outlook for the future application of AT-MSCs in the treatment of liver diseases is very promising.

10.5.2 Uncultured AT-MSCs

The use of uncommitted AT-MSCs, which are minimally manipulated ex vivo and ready for transplantation within a short period of time, such simplified procedure is crucial for future clinical usage, quality control of cells, and safety issues. The therapeutic potential of uncommitted human AT-MSCs and their potential future clinical use for the treatment of hepatic dysfunctions has been previously addressed [12]. Because uncommitted human AT-MSCs produce a variety of growth factors and cytokines, their use in the treatment of liver dysfunction offers many potential benefits. Undifferentiated AT-MSCs can produce trophic factors that may inhibit fibrosis and apoptosis, promote angiogenesis, and stimulate host regenerative cells to proliferate and differentiate into functional mature cells in vivo. The trophic activity of these uncultured AT-MSCs offers great potential for the restoration of liver function. The efficacy of these cells in vivo needs to be confirmed and analyzed by long-term preclinical and clinical studies. At present, the detailed mechanisms by which the cells restore liver function remain unclear, and further characterization will be required to in order to utilize these cells as a reliable therapy. Freshly isolated or frozen stocks of uncultured AT-MSCs could provide a future therapeutic option for hepatic dysfunctions in a clinical setting (Figure 10.3).

10.6 Clinical Applications of Uncultured AT-MSCs

10.6.1 Breast Reconstruction

Breast surgery as a treatment for breast cancer typically results in postoperative breast asymmetry. Reconstructive surgery is often difficult, because whole breast irradiation commonly results in severe fibrosis and atrophy of the breast tissue and underlying chest wall. Furthermore, until new blood vessels grow into the graft, the supply of oxygen and nutrients to target tissues is limited. A new reconstructive technique was recently developed using a combination of autologous adipose tissue and concentrated autologous AT-MSC from a patient who received partial mastectomy for breast cancer in 2006 [13]. This procedure utilized the automated Celution System (Cytori Therapeutic Inc., CA) for rapid extraction and concentration of uncultured AT-MSCs, and has been subsequently performed in 21 patients who had postoperative follow-up for 6 months or

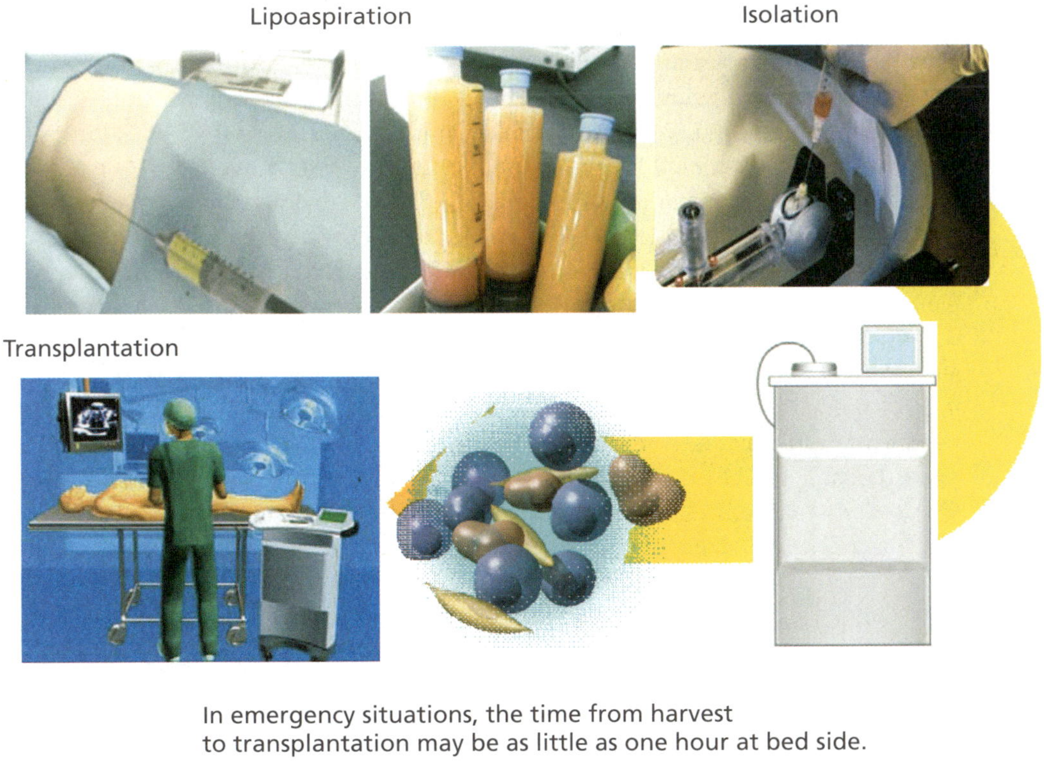

Figure 10.3 Diagram of the autologous cell transplantation procedure. Adipose tissue is the most abundant source of MSCs and is easily harvested by liposuction. The resulting lipoaspirate is rich in regenerative cells (ADRCs). ADRCs, such as uncultured AT-MSCs, are prepared from the lipoaspirate within 1h, either manually or with an automated system. Uncultured AT-MSCs are morphologically fibroblast-like cells that can be processed and transplanted within the time frame of a single surgical procedure.

longer. Patients for whom 2 years had passed since partial mastectomy for breast cancer and who had not had recurrence irrespective of postoperative radiation were selected. Participants were enrolled into this study based on criteria such as the age of the patient, stage of breast carcinoma, skin condition at recipient site, general condition, ability to undergo liposuction, and so forth. However, patients with a severe past history, skin adhesion at the recipient site, recurrence of active tumor, pregnancy or nursing, and so forth were excluded from the study. Adipose tissue can be extracted from abdomen, thighs, hips, or lower back of patients. Only one surgical procedure is required for this breast reconstruction technique. In this Japanese study, patients were followed for up to 18 months after the procedure, with no loss of tissue volume and no recurrence of breast cancer. Using ultrasound, investigators measured the average tissue thickness from the most prominent indentation of the scar to the pectoral muscle. Results show that thickness increased from 6 mm at baseline to 17 mm after 1 month ($P < .05$). Furthermore, the thickness was 15 mm at 6 months, demonstrating that there had not been any significant loss of tissue from the 1-month measurement. This procedure has been used on 21 patients, 79% of which said they were pleased with the results.

10.6.2 Cardiovascular Disease

Myocardial infarction (MI) is a leading cause of morbidity and mortality in industrialized nations. Current therapy has significantly improved patient outcome; however, this disease remains a major medical issue. Researchers are in the early stages of clinical investigation into the early administration of stem cells post MI in an effort to reduce scarring of the myocardium and thus improve myocardial performance. In patients for whom scarring has already occurred, there is no current treatment available to replace the dead myocardium. Investigation into the use of uncultured AT-MSCs was instigated to determine if they could differentiate into cardiomyocytes and effectively replace the dead/dying myocardium and restore cardiac function. Previous studies have focused on the isolation, characterization, and preclinical and clinical application of uncultured AT-MSCs in the cardiovascular field [14, 15]. The PRECISE trial is a prospective, double-blind, randomized, placebo-controlled, sequential dose-escalation trial that will randomize up to 36 patients with end-stage coronary artery disease and moderate to severe left ventricular dysfunction to receive freshly isolated uncultured AT-MSCs or a placebo in a 3:1 ratio. This study is focused on the safety and feasibility of uncultured AT-MSC transplantation in patients with acute MI. Finally, the APOLLO trial is designed to investigate the same source of uncultured AT-MSCs in patients with acute MI and LV ejection fraction impairment, and will be a prospective, double-blind, randomized, placebo-controlled, sequential dose-escalation trial that will include up to 48 patients.

10.7 Summary Points

1. Stem cells from several sources can differentiate into hepatocytes; thus, cell-based therapy is a potential alternative to liver transplantation.
2. The goal of the present study was to examine the in vitro hepatic differentiation potential of AT-MSCs.

3. AT-MSCs were used from patients of various ages and regardless of age, the CD105+ fraction of AT-MSCs exhibited high hepatic differentiation ability in an adherent monoculture condition in response to incubation with specific growth factors (HGF, FGF1, and FGF4).
4. CD105+ AT-MSC-derived hepatocyte-like cells expressed several liver-specific markers and exhibited hepatocyte-like functions, such as albumin production, low-density lipoprotein uptake, and ammonia detoxification.
5. CD105+ AT-MSC-derived hepatocyte-like cells incorporated into the parenchyma of the liver after transplantation into immunodeficient mice.
6. Adipose tissue is a source of multipotent stem cells that can be easily isolated, selected, and induced into mature, transplantable hepatocytes.
7. Because AT-MSCs are easy to procure ex vivo in large numbers, they are an attractive tool for clinical studies designed to establish an alternative therapy for liver dysfunction.
8. Freshly isolated or frozen stocks of uncultured AT-MSCs could provide a future therapeutic option for regenerative medicine in a clinical setting.

Acknowledgments

This work was supported in part by a grant-in-aid from the Third-Term Comprehensive 10-Year Strategy for Cancer Control, and by Health Science Research Grants for Research on the Human Genome and Regenerative Medicine from the Ministry of Health, Labor, and Welfare of Japan, and a grant from the Program for Promotion of Fundamental Studies in Health Sciences of the National Institute of Biomedical Innovation (NiBio).

References

[1] Teratani, T., et al., "Direct Hepatic Fate Specification from Mouse Embryonic Stem Cells, *Hepatology,* Vol. 41, 2005, pp. 836–46.

[2] Yamamoto Y, et al., "Recapitulation of In Vivo Gene Expression During Hepatic Differentiation from Embryonic Stem Cells," *Hepatology,* Vol. 42, 2005, pp. 558–67.

[3] Yamamoto, H., et al., "Differentiation of Embryonic Stem Cells into Hepatocytes: Biological Functions and Therapeutic Application," *Hepatology,* Vol. 37, 2003, pp. 983–93.

[4] Teratani, T., et al., "Long-Term Maintenance of Liver-Specific Functions in Cultured ES-Derived Hepatocytes with Hyaluronic Acid Sponge," *Cell Trasplant.,* Vol. 14, 2005, pp. 629–35.

[5] Ishikawa, T., et al., "Regenerative Cells for Transplantation in Hepatic Failure," *Cell Transplant.,* in press, 2011.

[6] Ishikawa, T, et al., "Stem Cells for Hepatic Regeneration: The Role of Adipose Tissue Derived Mesenchymal Stem Cells," *Curr. Stem Cell Res. Ther.,* Vol. 5, 2010, pp. 182–89.

[7] Zuk, P. A., et al., "Multilineage Cells from Human Adipose Tissue: Implications for Cell-Based Therapies," *Tissue Eng.,* Vol. 7, 2001, 211–28.

[8] Zuk, P. A., Zhu, M., Ashjian, P., et al., "Human Adipose Tissue is a Source of Multipotent Stem Cells," *Mol. Biol. Cell,* Vol. 13, 2002, pp. 4279–95.

[9] Banas, A., et al., "Adipose Tissue-Derived Mesenchymal Stem Cells as a Source of Human Hepatocytes," *Hepatology,* Vol. 46, 2007, pp. 219–28.

[10] Yamamoto, Y., et al., "A Comparative Analysis of The Transcriptome and Signal Pathways in Hepatic Differentiation of Human Adipose Mesenchymal Stem Cells," *FEBS J.,* Vol. 275, 2008, pp. 1260–73.

[11] Banas A, et al., "Rapid Hepatic Fate Specification of Adipose-Derived Stem Cells and Their Therapeutic Potential for Liver Failure," *J. Gastroenterol. Hepatol.,* Vol. 24, 2008, pp. 70–77.

[12] Banas, A., et al., "IFATS Collection: In Vivo Therapeutic Potential of Human Adipose Tissue Mesenchymal Stem Cells after Transplantation into Mice with Liver Injury," *Stem Cells*, Vol. 26, 2008, pp. 2705–12.

[13] Kitamura, K., et al., "Stem Cell Augmented Reconstruction: A New Hope for Reconstruction after Breast Conservation Therapy," *Breast Cancer Res. Treat.*, Vol. 106 (Suppl. 1), 2007, Abstract 4071.

[14] Sanz-Ruiz, R., et al., "Early Translation of Adipose-Derived Cell Therapy for Cardiovascular Disease," *Cell Transplant.*, Vol. 18, 2009, pp. 245–54,

[15] Meliga, E., et al., "Adipose-Derived Cells," *Cell Transplant.*, Vol. 16, 2007, pp. 963–70.

11

Culture Protocols for Producing Definitive Endoderm and Pancreatic Lineage Cells from Human ES or iPS Cells

Xiaofang Xu, Victoria L. Browning, and Jon S. Odorico

Division of Transplantation, Department of Surgery, University of Wisconsin-Madison School of Medicine and Public Health and WiCell Research Institute, Madison, WI, 53711

Abstract

The genes and intercellular signaling molecules controlling definitive endoderm and pancreas development in vertebrates are rapidly being elucidated. This knowledge has informed the establishment of methods for generating such cells from human ES and iPS cells in vitro. Based on the defined roles of nodal/activin signaling in definitive endoderm development we have identified a protocol using Activin A, BMP4, and bFGF in chemically defined, serum-free media which efficiently directs human pluripotent stem cells into an enriched population of definitive endoderm that can be differentiated in EB suspension cultures to produce a homogeneous population of foregut and pancreatic progenitors, including PDX1- and Sox9-expressing cells. Under defined conditions, a proportion of progenitors ultimately give rise to insulin^{+}C-peptide^{+}PDX1^{+} betalike cells as well as cells expressing other endocrine hormones in vitro. As more knowledge of developmental processes is realized, information can be assimilated to progressively refine and improve protocols, with the ultimate goal of generating large quantities of physiologically responsive beta cells that can be used to restore normoglycemia in patients with diabetes.

Key terms pluripotent stem cells, endoderm, pancreas, diabetes, activin, bone morphogenetic protein, fibroblast growth factor, pancreatic duodenal homeobox 1, insulin, beta cells

11.1 Introduction

Insulin–secreting beta cells of the islets of Langerhans arise by a sequence of developmental events beginning with the formation of definitive endoderm during gastrulation, eventually giving rise to the pancreas and other organs of the aerodigestive tract. Until recently, little was known about the extrinsic signaling events controlling the formation of definitive endoderm and regional specification of the pancreas, and even less about how to manipulate embryonic stem (ES) or induced pluripotent stem (iPS) cells in culture to form this germ layer. Once studies in lower organisms began to provide hints as to the molecules regulating the process of definitive endoderm formation and pancreas specification, it became possible to apply this knowledge to the in vitro culture of human pluripotent stem cells. Owing to the fact that the achievement of functional beta cell differentiation from pluripotent stem cells would provide a nearly limitless source of cells for transplantation for millions of patients with diabetes, this field has received intense experimentation in recent years. Although this highly relevant public health goal remains a work in progress, significant advances have been made. And, as we accumulate greater knowledge of embryonic signaling events and progressively assimilate this information to the culture of human pluripotent stem cells in a temporally and spatially specific manner, we will be able to achieve the efficient production of functional endocrine cells.

As early as 2001, the capacity of mouse and human pluripotent stem cells for spontaneously differentiating into pancreatic duodenal homeobox 1 $(PDX1)^+$ foregut progenitors and rare insulin-expressing cells was established [1–4]. However, spontaneously differentiating cultures of pluripotent stem cells contain a mixture of lineages, the majority of which are not definitive endoderm or pancreas. Thus, untreated, unselected cultures have limited therapeutic potential, and retain the possibility of unwanted side effects, such as teratoma formation. Based on studies in lower organisms identifying a role for nodal/activin A/TGFβ family signaling in definitive endoderm development, this pathway has been explored to drive differentiation of human ES (hES) cells toward definitive endoderm, as defined by coexpression of Sox17, Foxa2, CXCR4, and E-cadherin [5–8]. These studies were among the first to demonstrate that exposure of hES cells to Activin A 100ng/ml with low dose serum resulted in a highly enriched population of definitive endoderm. A subsequent refinement employed activation of Wnt signaling with Wnt3a agonist to inhibit PI3K activation [9]. This endoderm could be differentiated using retinoic acid and other growth factors into cells expressing PDX1, a marker of pancreatic and foregut endoderm, and other pancreas lineage markers [10]. However, insulin-expressing cells that were ultimately derived showed a phenotype inconsistent with mature beta cells and appeared to be poorly responsive to glucose. Subsequent studies have also demonstrated the importance of nodal/activin A signaling in differentiating ES cells to definitive endoderm and derivative pancreatic and hepatic cell types [11–19]. Although there is now convincing evidence that an Activin A-based differentiation protocol can promote differentiation of hES cells to definitive endoderm and a pancreatic progenitor-like state, highly functional glucose-responsive beta cells have not been reproducibly achieved through in vitro differentiation protocols [20, 21]. This finding could indicate that activin A alone does not sufficiently induce competent definitive endoderm, and/or that another inductive cell type is necessary, and/or that later stage culture steps require modification in order to efficiently produce endocrine progenitors and mature endocrine cells. Furthermore, culture protocols suitable for clinical use would ideally use chemically-defined media free from contamination by xenoproteins and problems with lot-to-lot variation which can affect standardization of scaled-up production methods.

Varying combinations and concentrations of a small number of developmentally relevant growth factors regulate the formation of many different tissues during gastrulation and organogenesis. Based on this we tested various combinations and concentrations of activin A, BMP4, and bFGF for their ability to differentiate hES cells into definitive endoderm and prospective foregut endoderm. We found that activin A alone is capable of directing differentiation to definitive endoderm in monolayer culture, but these cells do not survive subsequent differentiation and growth as embryoid bodies (EBs) in suspension culture. BMP4- and bFGF- treated hES cells similarly differentiate into definitive endoderm in the presence of MEF-conditioned media, which is known to contain TGFβ signaling effects, and so treated, cells survive suspension culture and differentiate into foregut progenitors. However, cells induced to differentiate in the presence of BMP4, bFGF, and mouse embryonic fibroblast (MEF)-conditioned media not only generated endoderm but also trophoblast and mesoderm lineages. Thus, we reasoned that the addition of activin A to BMP4- and bFGF-treated hES cells in chemically defined media would not only provide the signals for definitive endoderm cells to survive in suspension culture but would also yield a more enriched endoderm population than activin A treatment alone or combined BMP4/bFGF treatment. We found this to be the case; in addition, bFGF, activin A, and BMP4-derived endoderm allow cells to readily differentiate into homogeneous clusters of PDX1$^+$Sox9$^+$ pancreatic progenitors to a greater degree. Furthermore, this method appears to be equally effective with several different hES cell lines and three different human iPS (hiPS) cell lines and is highly reproducible. The protocol described herein produces a population of cells that can be used to derive endocrine progenitors and an abundant supply of betalike cells.

11.2 Experimental Design

11.2.1 Materials

11.2.1.1 Tools and Plasticware

- Sterile tissue culture hood.
- Humidified 37°C incubator with 5% CO_2.
- Inverted phase microscope for viewing live cultures.
- Aspirator.
- Pipette aid.
- 1-ml, 5-ml, and 10-ml glass pipettes.
- T75 flasks.
- Culture plates: 6-well and 24-well.
- Suspension culture dish, 100 m×20 mm, Corning Cat #430591.
- Coverslips.

11.2.1.2 Reagents

- DMEM/F12 media.
- PBS/BSA.
- L-Glutamine.
- 2-Mercaptoethanol.
- Nonessential amino acids.
- Dispase.
- bFGF.
- Activin A.
- BMP4.
- Serum replacement.
- Matrigel.

Dissolve dispase in DMEM/F12 media and bFGF in 0.1% BSA in PBS. Reagents 1–5 are stored at 4°C. Reagents 6–11 are all stored frozen in small aliquots. Avoid repeated freeze-thaw cycles.

11.2.2 Protocols

1. Culture undifferentiated hES cells or hiPS cells. Most-often used cell lines in this lab are NIH approved H1 (WA01) and H9 (WA09) between passage 18 and 42. This method also yields similar quantities of definitive endoderm gene expression and pancreatic progenitor cells from hiPS cell lines, IMR-90, Foreskin, and DF 19-9-7T. Using a similar bFGF/activin A/BMP4–based differentiation protocol, definitive endoderm is readily generated from mouse ES cells. Media is comprised of 80% DMEM/F12 and 20% knockout serum replacement supplemented with 1 mM L-glutamine, 1% nonessential amino acids, 0.1 mM 2-mercaptoethanol and 4 ng/ml bFGF (all from Invitrogen). Culture hES cells in 6-well plates on a feeder layer of irradiated MEFs and refresh media daily.
2. Split and passage cells when nearly confluent. Detach ES cell colonies from the feeder layer with 2 mg/ml dispase (Invitrogen), then rinse cells off the plates and filter through a 70-μm cell strainer to remove MEFs. The colonies to be retained remain in the strainer. Rinse cells retained in strainer and pipette into small pieces and plate on growth factor-depleted Matrigel (BD Biosciences) at a density of 70% to 80% confluence in MEF-conditioned media.
3. Matrigel-coated plates are generally prepared the day before splitting. Thaw one aliquot of Matrigel (2 mg/tube) and dilute with 6 ml cold DMEM/F12. Add 1 ml of diluted Matrigel to each well of a 6-well plate. Place plate at 4°C overnight. On the day of splitting, allow plate to sit at room temperature for 1h before use.
4. Stage 1. On the day after splitting, initiate differentiation by replacing media with media containing 100 ng/ml bFGF, 100 ng/ml activinA and 50 ng/ml BMP4 (FAB) in DMEM/F12 supplemented with 2% BSA, 1 mM L-glutamine, 1% nonessential amino acids and 0.1 mM 2-mercaptoethanol for 3 to 4 days. Change media daily.
5. Stage 2. Detach FAB-treated cells from plate with 2 mg/ml dispase and grow in suspension as cell aggregates in MEF-conditioned media for 2 days. On the third day switch media to DMEM/F12 with ITS supplement (BD, 5 μg/ml insulin + 5 μg/ml

transferrin + 5 ng/ml selenous acid) and 20 to 100 ng/ml bFGF for 12 days in suspension culture dishes. Change media every 2 days.

6. Stage 3. Transfer cell aggregates to a 15 ml centrifuge tube and allow settling by gravity. Aspirate the media and place the pellet-containing tube on ice. Dilute 2 mg Matrigel with 3 ml cold DMEM/F12. Add diluted Matrigel to the pellet at a density of 50 to 100 cell aggregates/ml. Add 0.5 ml of this diluted Matrigel/cell aggregate mixture onto a coverslip previously loaded into wells of a 24-well plate. Place plate in incubator overnight.
7. Carefully aspirate nongelled Matrigel and add ITSFINE media, a serum-free media comprised of DMEM/F12 with ITS, 10 ng/ml FGF7 (R&D), 200 nM INGAP (PSN-4765), 10 μM nicotinamide (Sigma), 10 nM exendin-4 (Sigma), 4 μg/ml insulin (Gibco) and 2 g/L BSA (Sigma) for 21 to 28 days. Change media every other day.
8. Stage 4. Add 0.2 ml of 2 mg/ml dispase to each well of a 24-well plate to dislodge the sphere-structured cell masses. Rinse cells off plate and place in suspension culture dishes in DMEM/F12 supplemented with 1X B27 and 10 μM nicotinamide for 1 to 2 weeks.

11.3 Data Acquisition, Anticipated Results, and Interpretation

The cell differentiation process is monitored at key time points: (1) after 3 days of FAB treatment (Stage 1); (2) after 14 days of suspension culture with bFGF+ITS following FAB treatment (Stage 2, EB formation); (3) 21 to 28 days after initiating EB adherent culture in ITSFINE media (Stage 3); and (4) 1 to 2 weeks after cell aggregates are further grown in suspension culture (Stage 4).

Gene and protein expression are assessed at the indicated time points, most commonly by quantitative RT-PCR and immunostaining. In stage 1, cells are expected to differentiate towards mesendoderm/primitive streak and definitive endoderm fates. Compared to the undifferentiated state, FAB treatment induced significant upregulation of genes associated with primitive streak (*mixl1*, 363 fold; *gsc*, 335 fold), mesendoderm (*T*, 335 fold), and definitive endoderm (*sox17*, 2485 fold; *foxa2*, 471 fold). Immunostaining showed the intermingling of Brachyury$^+$ cells with SOX17$^+$ / FOXA2$^+$ costained cells. While treatment of hES cells with higher concentrations of BMP4 alone for 7 days is known to yield a high proportion of trophoblast differentiation as manifested by human chorionic gonadotropin and CDX2 expression ([22] and data not shown), lower concentrations and shorter exposure of BMP4 in combination with bFGF and activin A leads to increased mesendoderm and definitive endoderm formation.

At Stage 2, one begins to see widespread FOXA2$^+$ staining and the appearance of PDX1$^+$ cell clusters within EBs. Compared to untreated cells (i.e., cells treated with the same Stage 1 media except *without* FAB and the same Stage 2 media), FAB-treated Stage 2 cells have an induction of *pdx1* gene expression of approximately 176-fold.

By 2 weeks into Stage 3, ductlike structures begin to appear, gradually becoming larger and forming into spheroids. These spheroids become more numerous and eventually are found in all areas of the cultures by the end of this stage (Figure 11.1). More than 80% of cells in each spheroid stain with PDX1 antibody. In many spheres, this percentage approaches 100%. At this stage, gene expression studies show *pdx1* gene expression is comparable to that found in 50% pure human islets. Essentially all PDX1$^+$ cells costain with HNF1, HNF6, and SOX9. Some of the cells within the spheroids also stain positively with antibodies to carboxypeptidase A (CPA), NKX6.1, NGN3, and P48 (aka Ptf1a). Scattered insulin$^+$/C-peptide$^+$ cells also appear at the end of Stage 3.

Suspension culture in Stage 4 promotes an approximately 50-fold increase in insulin gene expression and a nearly 10-fold increase in the number of insulin$^+$/C-peptide$^+$ cells over that found at Stage 3. The majority of these cells do not coexpress glucagon or somatostatin, yet glucagon$^+$ and somatostatin$^+$ cells are also readily detected within these endocrine cell clusters. All of the insulin$^+$ cells coexpress PDX1.

11.4 Discussion and Commentary

The procedure described above is one we have developed for obtaining enriched pancreatic progenitors and isletlike cell clusters from hES cells. This protocol was developed by first taking a rational approach based on known developmental signals, followed by empirical testing to refine it. The stepwise differentiation process should be monitored at different time points for expression of stage-specific markers at both gene and protein levels. In our hands this method appears to work equally well for human ES cells and iPS cells.

An important aspect of this procedure is to begin with healthy colonies of low passage, normal-appearing undifferentiated pluripotent stem cells, ideally at 80% confluence. If nonoptimally maintained, some stem cell colonies exhibit spontaneous differentiation at the periphery of colonies. In our experience, even mild amounts of spontaneous differentiation within stem cell colonies can lead to significantly altered responses to growth factors and result in heterogeneity at later stages of differentiation. Differentiated cells or colonies can be physically removed. To accomplish this, one can

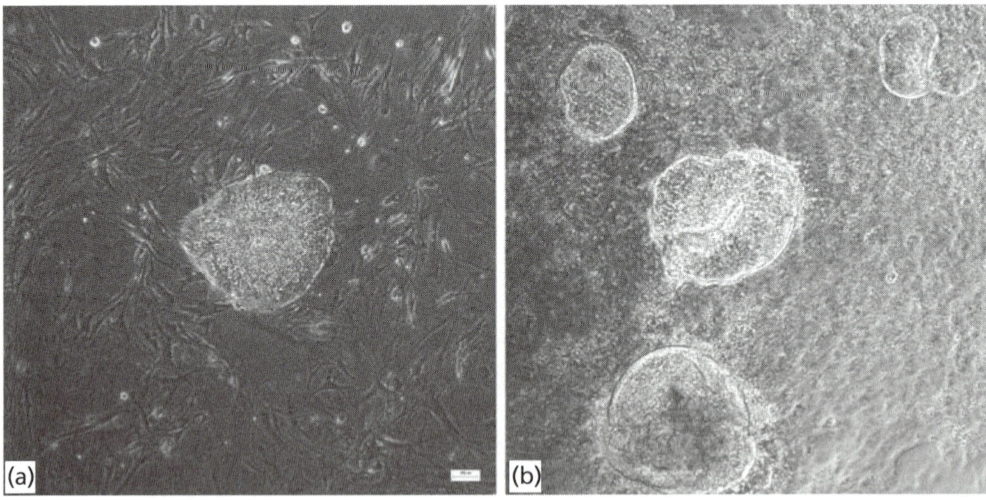

Figure 11.1 (a) Single colony of undifferentiated WA09 hES cells growing on a fibroblast feeder layer. (b) Phase-contrast image of multiple hES cell-derived "pancreas spheres" at Stage 3. Undifferentiated hES cells were grown as described: Stage 1—treated with bFGF, activin A, and BMP4 for 3 to 4 days; Stage 2—EBs in bFGF and ITS for 14 days; Stage 3—ITSFINE media and BSA for 21 to 28 days.

first mark the cells or colonies using the colony marker with a 2.5× or low power objective of microscope, and then scratch them off with a pipette ("pick-to-remove"). Cell lines should also be periodically karotyped. We generally use low passage stem cells (passage 18–42).

During Stage 1, we treat stem cells with a combination of growth factors bFGF, activin A and BMP4 (FAB) in chemically defined media for 3 to 4 days and initiate differentiation towards mesendoderm. This specific combination of growth factors and concentrations is essential to achieve these results. Other studies have shown activin A alone or activin A and Wnt3a to be adequate to achieve subsequent $PDX1^+$ cells if differentiation occurs entirely in adherent monolayer culture. However, when using a second stage of EB suspension culture, as in this protocol, EBs generated from Stage 1 cells treated only with activin A do not survive. In contrast, cells treated at Stage 1 with FAB are able to survive EB suspension culture and initiate a developmental pathway leading to $PDX1^+$ pancreas-spheres. The signaling within EBs responsible for this differentiation is not known presently. Thus, we believe this triple growth factor protocol may represent an optimal system for differentiating ES cells into pancreatic progenitors.

During the process that FAB-treated hES cells are detached and gently pipetted to small pieces to initiate suspension cultures, it is very important to avoid centrifuging the cells. Cells at this stage stick together very easily and become clumped. If they are centrifuged, even at relatively low speeds (i.e., 1,000 rpm) and for short periods of time (i.e., 2 min), large cell clumps will likely form and be very difficult to subsequently break up into small clusters. If this should occur, there will be very few cell aggregates, and thus EBs, formed and subsequent experimental yields will be reduced.

Matrigel is used in this protocol at multiple stages of differentiation. The main objective when working with Matrigel is to keep it cold. As Matrigel warms to room temperature, it begins to solidify, making it more difficult to work with and potentially leading to a nonhomogeneous growth matrix. After plates are coated with Matrigel, the coated plate is stored at 4°C overnight. This should "re-melt" any possible chunks and make Matrigel into an even layer. During the process that Matrigel and EBs are mixed and plated together (i.e., initiation of Stage 3), make sure EBs are cold beforehand. Otherwise, they do not mix well.

11.5 Application Notes

A major goal of the clinical treatment of patients with insulin-deficient diabetes is to replace damaged or dysfunctional cells with new beta cells from an abundant source such as human pluripotent stem cells. Whether such a beta cell replacement therapy entails use of functionally mature insulin-secreting beta cells either in purified form or as a mixture of endocrine cells types, or a population of endocrine progenitors, or both, remains to be tested. It is certain, however, that to get to that point requires production of a homogeneous or highly enriched population of definitive endoderm and pancreatic progenitors such as can be achieved by the described protocol. This protocol can form the basis for further exploration and refinement of methods to efficiently convert these enriched populations of pancreatic progenitors into endocrine cell types.

Troubleshooting Table

Problem	Explanation	Potential Solutions
Stem cells show spontaneous differentiation.	Density of MEFs is too high or too low.	Optimal density is 0.75×10^5 cells/ml; plate 2.5ml into each well of a 6-well plate.
	Intrinsic cell abnormalities resulting in autonomous differentiation.	"Pick-to-remove" differentiated colonies, may need to repeat for several passages before initiating growth factor directed differentiation. If problem is still not corrected, discard the cells. Consider karyotyping cells or obtaining lower passage cells.
Few EBs are formed.	Cells are centrifuged during collection of FAB-treated cells.	Let cells settle by gravity for 30–45 min.
	Cells are pipetted too vigorously.	Treat cells more gently with less pipetting.
	Few stem cells are present prior to Stage 1.	Starting colony density should optimally be 70% to 80% confluent. FAB treatment results in significant cell loss. Higher initial cell densities will improve yields of endoderm cells at the end of Stage 1.
EBs form incompletely or become attached during Stage 2 culture.	Incorrect EB formation method used.	Use suspension culture dishes; alternatives for producing EBs include hanging drop method and embedding into methylcellulose semisolid media. These methods are untested with this protocol. Consider plate rocker in incubator.
EBs are not embedded well in Matrigel after plating.	Matrigel warmed during mixing with cells.	Keep Matrigel and EBs cold during mixing process and plating.
	Matrigel is too concentrated.	Dilute Matrigel slightly.

11.6 Summary Points

1. Competent techniques in hES cell culture methods are essential for successfully performing this protocol.
2. Excellent quality, karyotypically normal cells are critical to achieving success.
3. There is some inherent minor variability that occurs during in vitro differentiation of pluripotent stem cells. This protocol leads to highly reproducible results because of use of chemically defined conditions and high-quality growth factors.
4. Expertise in methods of quantitative RT-PCR and immunofluorescence staining and microscopy are necessary for analysis.

References

[1] Assady, S., et al., "Insulin Production by Human Embryonic Stem Cells," *Diabetes*, Vol. 50, 2001, pp. 1691–97.

[2] Brolen, G. K., et al., "Signals from the Embryonic Mouse Pancreas Induce Differentiation of Human Embryonic Stem Cells into Insulin-Producing Beta-Cell-Like Cells," *Diabetes,* Vol. 54, 2005, pp. 2867–74.

[3] Kahan, B. W., et al., "Pancreatic Precursors and Differentiated Islet Cell Types from Murine Embryonic Stem Cells: An In Vitro Model to Study Islet Differentiation," *Diabetes*, Vol. 52, 2003, pp. 2016–24.

[4] Xu, X., et al., "Endoderm and Pancreatic Islet Lineage Differentiation from Human Embryonic Stem Cells," *Cloning Stem Cells*, Vol. 8, 2006, pp. 96–107.

[5] D'Amour, K. A., et al., "Efficient Differentiation of Human Embryonic Stem Cells to Definitive Endoderm," *Nat. Biotechnol.*, Vol. 23, 2005, pp. 1534–41.

This landmark paper is the first to report that activin A promotes definitive endoderm formation from hES cells in vitro.

[6] Jiang, J., et al., "Generation of Insulin-Producing Islet-Like Clusters from Human Embryonic Stem Cells," *Stem Cells,* Vol. 25, 2007, pp. 1940–53.

[7] Phillips, B. W., et al., "Directed Differentiation of Human Embryonic Stem Cells into the Pancreatic Endocrine Lineage," *Stem Cells Dev.,* Vol. 16, 2007, pp. 561–78.

[8] Shim, J. H., et al., "Directed Differentiation of Human Embryonic Stem Cells Towards a Pancreatic Cell Fate," *Diabetologia,* Vol. 50, 2007, pp. 1228–38.

[9] McLean, A. B., et al., "Activin A Efficiently Specifies Definitive Endoderm from Human Embryonic Stem Cells Only When Phosphatidylinositol 3-Kinase Signaling Is Suppressed," *Stem Cells*, Vol. 25, 2007, pp.29–38.

[10] D'Amour, K. A., et al. "Production of Pancreatic Hormone-Expressing Endocrine Cells from Human Embryonic Stem Cells," *Nat. Biotechnol.*, Vol. 24, 2006, pp. 1392–1401.

This important paper describes a multistep differentiation protocol applying lessons from developental biology studies to activate or inhibit different signaling pathways to derive pancreatic progenitors and insulin+ cells.

[11] Touboul, T., et al. "Generation of Functional Hepatocytes from Human Embryonic Stem Cells under Chemically Defined Conditions that Recapitulate Liver Development," *Hepatology,* Vol. 51, 2010, pp. 1754–65.

[12] Shi, Y., "Generation of Functional Insulin-Producing Cells from Human Embryonic Stem Cells In Vitro," *Methods. Mol. Biol.,* Vol. 636, 2010, pp. 79–85.

[13] Cai, J., et al., "Generation of Homogeneous PDX1(+) Pancreatic Progenitors from Human ES Cell-Derived Endoderm Cells," *J. Mol. Cell Biol.,* Vol. 2, 2010, pp. 50–60.

[14] Ameri, J., et al., "FGF2 Specifies hESC-Derived Definitive Endoderm into Foregut/Midgut Cell Lineages in a Concentration-Dependent Manner," *Stem Cells*, Vol. 28, 2010, pp. 45–56.

[15] Vallier, L., et al., "Early Cell Fate Decisions of Human Embryonic Stem Cells and Mouse Epiblast Stem Cells Are Controlled by the Same Signalling Pathways," *PLoS ONE*, Vol. 4, 2009, p. e6082.

This comprehensive study analyzes the differentiation effects of different doses of, and combinatorial actions of Activin A, BMPs, and FGFs in hES cell cultures.

[16] Sulzbacher, S., et al., "Activin A-Induced Differentiation of Embryonic Stem Cells into Endoderm and Pancreatic Progenitors—The Influence of Differentiation Factors and Culture Conditions," *Stem Cell Rev.,* Vol. 5, 2009, pp. 159–73.

[17] Johannesson, M., et al., "FGF4 and Retinoic Acid Direct Differentiation of hESCs into PDX1-Expressing Foregut Endoderm in a Time- and Concentration-Dependent Manner," *PLoS ONE*, Vol. 4, 2009, p. e4794.

[18] Hay, D. C., et al., "Highly Efficient Differentiation of hESCs to Functional Hepatic Endoderm Requires ActivinA and Wnt3a Signaling," *Proc. Natl. Acad. Sci. USA*, Vol. 105, 2008, pp. 12301–306.

[19] Vallier, L., et al., "Signaling Pathways Controlling Pluripotency and Early Cell Fate Decisions of Human Induced Pluripotent Stem Cells," *Stem Cells,* Vol. 27, 2009, pp. 2655–66.

[20] Kroon, E., et al. "Pancreatic Endoderm Derived from Human Embryonic Stem Cells Generates Glucose-Responsive Insulin-Secreting Cells In Vivo" *Nat. Biotechnol.*, Vol. 26, 2008, pp. 443–52.

[21] Matveyenko, A. V., et al., "Inconsistent Formation and Nonfunction of Insulin-Positive Cells from Pancreatic Endoderm Derived from Human Embryonic Stem Cells in Athymic Nude Rats," *Am. J. Physiol. Endocrinol. Metab.*, Vol. 299, 2010, pp. E713–E720.

This recent study casts some doubt upon the claimed in vivo maturation of hES cell-derived pancreatic progenitors.

[22] Xu, R. H., et al., "BMP4 Initiates Human Embryonic Stem Cell Differentiation to Trophoblast," *Nat. Biotechnol.* Vol. 20, 2002, pp. 1261–64.

12

Human Placenta Cell Isolation

Roberto Gramignoli[1], Fabio Marongiu[2], Toshio Miki[3], Kristen J. Skvorak[1], Veysel Tahan[1], Kenneth Dorko[1], Marc C. Hansel[1], Julio C. Davila[4], Ewa C.S. Ellis[5], and Stephen C. Strom[1*]

[1] Department of Pathology, School of Medicine, University of Pittsburgh, Pittsburgh, PA
[2] Department of Biomedical Sciences and Technologies, University of Cagliari, Cagliari, 09124, Italy
[3] Department of Developmental Biology, School of Medicine, University of Pittsburgh, Pittsburgh, PA
[4] Pfizer Inc., St. Louis, MO
[5] Department of Clinical Science, Intervention and Technology (CLINTEC), Division of Transplantation Surgery, Liver Cell Lab, Karolinska University Hospital Huddinge, Stockholm, Sweden
* Correspondence: Department of Pathology, University of Pittsburgh, 200 Lothrop Street, BST S-450, Pittsburgh, PA 15261; E-mail: strom@pitt.edu

Abstract

In this chapter, we present procedures to isolate amniotic epithelial cells and amniotic mesenchymal cells from a human term placenta. Human term placenta is readily available and easily procured tissue and it is gaining importance as a reserve of different types of progenitor/stem cells. The placenta-derived cells can be easily isolated from different compartments and their properties suggest they may become a useful, noncontroversial, extensive, and unlimited source for cells for cell therapy-based clinical applications.

Key terms amniotic epithelial cell, amniotic mesenchymal cells, chorion

12.1 Introduction

In the search for stem cells, many sources have been used, such as umbilical cord, embryos, bone marrow, fat, and blood. Term placenta contains cells with characteristics of embryonic stem (ES) cells and and has been shown to be a valuable source of cells for cell-based therapies. Four regions can be distinguished in the full-term placenta: amniotic epithelial, amniotic mesenchymal, chorionic mesenchymal, and chorionic trophoblastic cells. Amniotic epithelial and mesenchymal regions are of major importance for cell-based therapy and tissue engineering. Due to the complexity of the structure of the placenta, many diferent techniques and culture conditions and have been described for the isolation of cells from amniotic epithelial and mesenchymal membranes. Through these studies, a consensus on the minimum criteria for cell isolation from both of these membranes has been reached.

In this chapter we provide a preferred method to isolate human amniotic epithelial and amniotic mesenchymal cells.

12.2 Placenta Embryological Development

During the implantation process, approximately 1 week after fertilization, the human blastocyst implants and placenta development begins. The blastocyst is flattened and composed of an outer wall (trophoblast) that surrounds the blastocystic cavity. A small group of larger cells, known as the inner cell mass (ICM), are opposed to the inner surface of the trophoblastic vesicle. It is from the ICM that the embryo, the umbilical cord, and the amniotic epithelium originate. As the blastocyst adheres to the endometrium, invading trophoblasts erode the decidua, facilitating implantation of the blastocyst. In the following days, trophoblastic cells at the implanting pole of the blastocyst proliferate robustly, forming a bilayered trophoblast. The outer of the two layers becomes the syncytiotrophoblast by fusion of neighboring trophoblast cells, whereas the inner cells (i.e., the cytotrophoblast) remain temporally unfused. The proliferating cytotrophoblasts and the syncytiotrophoblasts give rise to a system of trabeculae intermingled with hematic lacunae. From these trabeculae, the primordial villi are generated and distributed over the entire periphery of the chorionic membrane. Villi in contact with the decidua basalis (that part in contact with maternal endometrium) proliferate to form the leafy chorion or chorion frondosum, whereas villi in contact with the decidua capsularis degenerate into the chorion leave. At embryonic days 7 to 14, the ICM differentiates into two layers: the epiblast and the hypoblast. Subsequently, from the epiblast, small cells appear between the trophoblast and the embryonic disc to enclose a space that will ultimately become the amniotic cavity; these cells later constitute the amniotic epithelium. On the opposite side between the hypoblast and cytotrophoblast, the exocoelomic membrane and its cavity modify to form the yolk sac. The extraembryonic mesoderm arranges into a connective tissue that surrounds the yolk sac and amniotic cavity, giving rise to amniotic and chorionic mesoderm. Gastrulation, the process through which the bilaminar disc differentiates into the three germ layers: ectoderm, mesoderm, and endoderm, occurs during embryonic days 14 to 21 [1].

12.3 Placenta Tissue Structure

The full-term placenta is discoid in shape and approximately 20 cm in diameter and 3 cm in thickness. Within the placenta, it is possible to distinguish the trophoblast and fetal membranes (the amnion and chorion). The fetal membranes compose the chorionic plate and enclose the fetus in the amniotic cavity. The chorionic plate consists of a spongy layer and a compact layer of chorionic mesoderm separated from the Langans' fibrinoid stria by an incomplete basement membrane. On the basement membrane are varying numbers of proliferating extravillous cytotrophoblast cells and a Langans' fibrinoid layer composed by cytotrophoblastic cells and fibrinoid. Villi, originated from the chorionic plate, anchor the placenta to the maternal endometrium. From the maternal side, the decidua is invaded by the trophoblast of anchoring villi producing placental septa, which divide the fetal side of the placenta into irregular cotyledons.

Some villi anchor the placenta to the endometrium, while others terminate freely in the intervillous space. A basement membrane separates the stromal core from an uninterrupted multinucleated layer, called the syncytiotrophoblast. Between the syncytiotrophoblast and its basement membrane are single or aggregated Langans' cytotrophoblastic cells, commonly called villous cytotrophoblast cells.

The chorionic plate (great part of the placenta on the fetal side) continue from the edge of the placenta to enclose the amniotic fluid and the fetus. The amnion is a thin, avascular membrane composed of an epithelial layer and an outer layer of connective tissue, which is contiguous, over the umbilical cord, with the fetal skin. The amniotic epithelium (AE) is an uninterrupted, single layer of flat, cuboidal, and columnar epithelial cells in contact with amniotic fluid that is attached to a distinct basal lamina that is connected to the amniotic mesoderm (Figure 12.1). In the amniotic mesoderm closest to the epithelium, an acellular compact layer is distinguishable, composed of fibronectin and collagens I/III. Deeper in the amniotic mesoderm, dispersed fibroblast-like mesenchymal cells and rare macrophages are observed. In 2008 the mesenchymal layer of amnion (Figure 12.1) was described to contain two subfractions: one, with

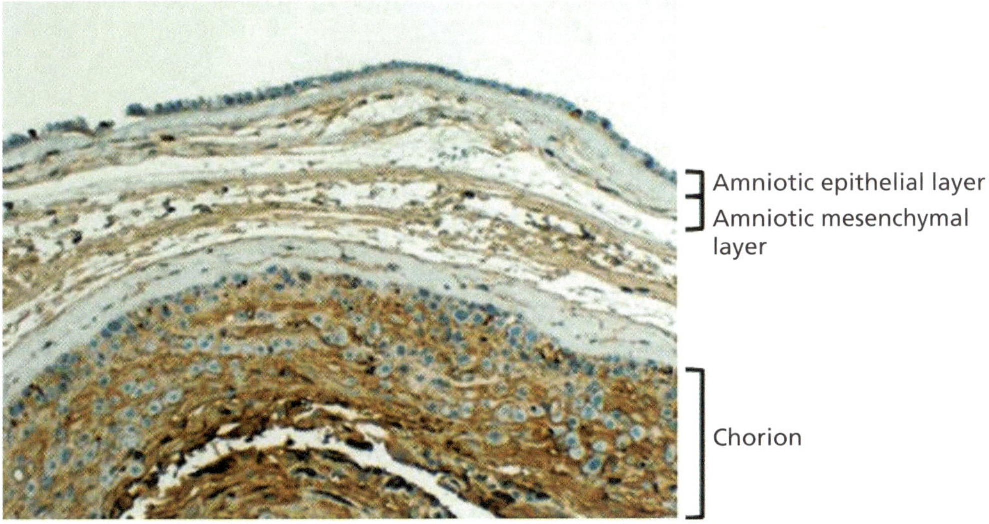

Figure 12.1 Histological section of human term placenta (stained by hematoxylin-eosin; ob. 4x) showing the chorionic plate layers.

mesenchymal phenotype, called amniotic mesenchymal stromal cells, and a second one containing monocyte-like cells [2].

A spongy layer of loosely arranged collagen fibers separates the amniotic and chorionic mesoderm. The chorionic membrane consists of mesodermal and trophoblastic regions. Chorionic and amniotic mesoderm show many similarities in composition. A large and incomplete basal lamina separates the chorionic mesoderm from the extravillous trophoblast cells. The latter, similar to trophoblast cells present in the basal plate, are dispersed within the fibrinoid layer and express immunohistochemical markers of proliferation. The Langhans' fibrinoid layer is composed of matrix type (inner side) and fibrin type (outer side) and usually increases during pregnancy. The trophoblast interdigitates with the deciduas at the edge of the placenta and in the basal plate [2, 3].

12.4 Placenta Immunological Properties

It has been proposed that cells from fetal membranes have immunoregulatory properties. Maternal acceptance of the fetal allograft is not completely understood. The paradox of maternal tolerance of the fetus was first raised more than 50 years ago by Peter Medawar, who explained that the placenta functions as a physical separation between the mother and fetus, which are antigenically immature (the fetus) and immunologically inert (the mother) [4]. Several mechanisms contributing to tolerance induction have been proposed, including the depletion of tryptophan and inhibition of T-cell proliferation by the tryptophan catabolizing enzyme indoleamine 2,3-dioxygenase (IDO) expressed by some cells of the syncytiotrophoblast [5, 6] and the secretion of soluble HLA-G molecules that induce apoptosis in activated CD8-positive and inhibit CD4-positive T-cells [7]. Natural killer cells, also inhibited by HLA-G may help to avoid fetal rejection [8]. Another possible mechanism for contributing to the maintenance of fetomaternal tolerance is through the Fas-ligand pathway: CD95L (Fas ligand)-expressing trophoblast cells have been reported to induce apoptosis of maternal CD95-positive lymphocytes [9]. It is very likely that all pathways described above contribute to the complex phenomenon of fetal–maternal tolerance.

12.5 Human Epithelial Cells from Amnion

AE cells develop very early from the epiblast and prior to gastrulation. Therefore, these cells might maintain some of the plasticity of pregastrulation embryo cells. AE cells have several unique characteristics, such as the presence of many immature or stem cells and low expression of MHC class I antigens [10, 11]. In 2005, Strom's group isolated and characterized AE cells from human-term placenta [12]. We reported that AE cells display stem cell characteristics, such as expression of surface makers normally present on ES and germ cells [12]. AE cells were found to express SSEA-3 (8.7 ± 2.8%), SSEA-4 (43.9 ± 14.8%), TRA1-60 (9.8 ± 4.3%), and TRA1-81 (9.9 ± 4.5%) and do not express SSEA-1 [16, and unpublished report]. In addition, AE cells express the pluripotent stem cell-specific transcription factors Oct-4, Sox-2, and Nanog [12]. These findings were supported by a later study [13, 14]. AE cells do not require other cell-derived feeder layers to maintain

OCT-4 and Nanog expression, do not express telomerase, display a normal karyotype, and are nontumorigenic upon transplantation [12, 14]. Under certain culture conditions, AE cells form spheroid structures that retain stem cell characteristics. AE cells can be easily cultured on regular plastic support in the presence of standard cell culture media. Cultured cells grow to confluence quickly in the presence of serum and growth factor such as epidermal growth factor (EGF) [15]. The morphology of human AE cells in culture is presented in Figure 12.2. It has also been described that AE cells do not form teratocarcinomas or other types of tumors in immunodeficient mice after injection [12, 14]. In addition, no evidence of tumorigenicity was reported in human volunteers when amnion membrane or membrane-derived cells were injected into patients to correct lysosomal storage diseases [16–19]. These characteristics indicate that AE are similar but not identical to ES cells. The hypothesis that AE cells might be pluripotent was supported by the demonstration of the differentiation of the cells. Based on immunohistochemical and genetic analysis, AE cells have the potential to differentiate into all three germ layers: endoderm (liver, pancreas) [12, 20, 21]; mesoderm (cardiomyocyte) [12, 20, 21]; ectoderm (neural cells) [22–25] in vitro. For example, Sakuragawa and coworkers reported that AE cells could be induced to differentiate to cells with neural characteristics [14, 22–28]. Wei and collaborators explored AE differentiation into pancreatic cells and demonstrated the production and release of insulin and a lowering of blood glucose levels following the transplantation of AE cells into diabetic mice [29].

These observations suggest that cells derived from the fetal side of the placenta may retain a multipotent phenotype long after they differentiate from the epiblast.

12.6 Human Mesenchymal Stromal Cells from Amnion

Human mesenchymal stromal cells from amnion (AM) and mesenchymal cells from chorion (CM) are thought to be derived from extraembryonic mesoderm [1]. Both AM and CM cells can be isolated from first-, second-, and third-trimester mesoderm of amnion and chorion, respectively [30, 31]. AM cellular isolation is usually performed

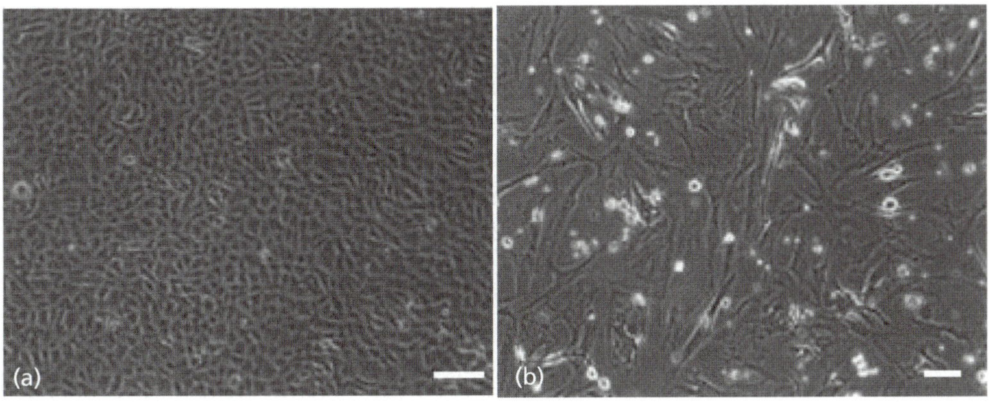

Figure 12.2 Phase-contrast photographs after 2 days of culture of human AE and AM cells. (a) AE cells at confluent status after 3 days of culture in standard culture medium. (b) AM cells after 2 days of culture in standard culture medium. Scale bars = 40 μm.

with term amnion dissected from the deflected part of the fetal membranes to minimize the presence of maternal cells. By a two-step enzymatic procedure, it is possible to obtain a homogenous AM cell population. After removal of the amnion tissue (as described below), the remaining mesenchymal cells are then released by digestion with collagenase [1, 32]. Phase micrographs (Figure 12.2) show the morphology of cultured AM cells. An extensive phenotypical and functional characterization is already available in the literature on AM cells [33–37]. Transmission electron microscopy of AM cells shows mesenchymal and epithelial characteristics. This hybrid phenotype is interpreted as a sign of multipotentiality and is not found in the other mesenchymal cells (CM cells). Expression of CD49d (α4 integrin) on AM cells distinguishes them from AE [37]. As previously described in AE cells, AM cells are also negative for vimentin and cytokeratin 18 expression. The surface marker profile of cultured mesenchymal cells, both from amnion and chorion origin, show a lot of similarities with adult bone marrow mesenchymal cells. All express typical mesenchymal markers (Table 12.1) but are negative for hematopoietic (CD34 and CD45) and monocytic markers (CD14) [29, 30, 35, 37]. Surface expression of SSEA-3 and SSEA-4, as well as mRNA for OCT-4, has been reported [29, 34, 37–39]. There is some evidence for AM differentiation into all three germ layers (i.e., ectoderm, mesoderm and endoderm) as described for AE cells [30, 35, 37]. However questions remain as to the efficiency and proportion of AM cells that exhibit such plasticity [40].

12.7 Human Mesenchymal Stromal Cells from Chorion

Despite the thorough characterization of AE and AM cells, there are few published reports on CM cells [30, 35, 41]. The available details are summarized in Table 12.1. CM cells are isolated after mechanical and enzymatic removal of the trophoblastic layer with dispase. Chorionic mesodermal tissue is then digested with collagenase [30, 35]. Mesenchymal cells have also been isolated from chorionic fetal villi through explant culture, although maternal contamination is more likely [30, 31, 41]. Once isolated, CM cells can be easily plated on regular plastic culture dishes and will proliferate for a few passages. However, after four or five passages, a decreased proliferation rate has been observed. First-trimester-isolated AM cells display a lower rate of proliferation compared to third-trimester CM cells [35]. CM cells are more primitive and metabolically quiescent compared to the AM cells, which was established based on microscopic morphology of the cells. Specifically, compared to AM cells, transmission electron microscopy of CM shows a simpler cytoplasmic organization. The most relevant features include the presence of stacks of rough endoplasmic reticulum cisternae, dispersed mitochondria and glycogen lakes; they lack the presence of assembled contractile filaments, prominence of endocytotic traffic, and junctional communications. Overall, the ultrastructural characteristics of CM cells resemble those found in the hematopoietic progenitors, suggesting their position at the higher levels of the stem cell hierarchy [38]. Also, CM cells have been shown to differentiate into osteoblasts, chondroblasts, and adipocytes (mesodermal lineages) [30, 34, 35, 37]. Human amniotic and chorionic cells successfully and persistently engraft in multiple organs and tissues in vivo. For example, brain, lung, bone marrow, thymus, spleen, kidney, and liver have been found to be repopulated after amnion and chorion cell injections into neonatal swine and rats, indicative of migration consistent with the expression of adhesion and migration molecules (l-selectin, VLA-5, CD29, and P-selectin ligand 1) as well as cellular matrix proteinases (MMP-2 and MMP-9) [42] (Table 12.1).

12.8 Materials

- Hanks' balanced salt solution (HBSS; calcium- and magnesium-free; CMF-HBSS, Lonza, cat. no. 04-315Q).
- 10×HBSS (Invitrogen, cat. no. 14065-056).
- Ethylene glycol tetraacetic acid (EGTA, Sigma, cat. no. E4378).

Table 12.1

Marker	Alternative name	AE	AM	CM
CD3	T3	–	–	
CD10	gp100	+	+	+
CD13	APN	+	+	+
CD14	LPS-R	–	–	–
CD29	b1-integrin	+	+	+
CD31	PECAM-1	–	–	–
CD34		–	–	–
CD44	H-CAM	+	+	+
CD45	LCA	–	–	–
CD49c	a3-integrin	–	+	
CD49d	a4-integrin	–	+	
CD49e	a5-integrin	+	+	+
CD49f	a6-integrin	+		
CD54	ICAM	–	+	
CD90	Thy-1	+	+	+
CD73	SH3-4	+	+	+
CD105	SH2	+	+	+
CD117	c-Kit	+/–	–	–
CD133	AC133/2	–	–	–
CD140b	PDGFRb	+	+	+
CD166	ALCAM	+	+	+
CD271	LNGFR	–	+/–	+/–
CD324	E-cadherin	+	–	–
CD326	EpCAM	+	+	
CD338w	ABCG2	+/–	–	
CD349	Frizzled9	–	+	+
HLA-ABC	MHC I	+	+	+
HLA-DR	MHC II	–	–	–
SSEA1		–	–	
SSEA3		+	–	
SSEA4		+	–	
TRA1-60		+	–	
TRA1-81		+	–	

- Antibiotic-antimycotic solution (Cellgro, cat. no.30-004-CI).
- 0.25% (w/v) trypsin/EDTA (Invitrogen, cat. no. 25200-056).
- Phosphate buffer saline (PBS; EMD, cat. No. 6505).
- Eagle's minimum essential medium (EMEM, Lonza, cat. no. 12-136F).
- Collagenase type IV (Sigma, cat. no. C5138).
- DNase I (Sigma, cat. no. DN25).
- DMEM (with 4.5g of glucose/liter, Lonza, cat. no. 12-604F).
- 100 mM sodium pyruvate (Invitrogen, cat. no. 11360-070).
- Heat-inactivated fetal bovine serum (Invitrogen, cat. no. 16141).
- Nonessential amino acid (Invitrogen, cat. no. 11140-050).
- L-glutamine (Cellgro, cat. no. 25-005-CI).
- 2-mercaptoethanol (Invitrogen, cat. no. 21985-023).
- Antibiotics solution (Cellgro, cat. no. 30-004-CI).
- Human recombinant epidermal growth factor (EGF, Sigma, cat. no. E9644).
- Laminar flow cabinet (BSL-2).
- Absorbent bench paper.
- Sterile field, 16×29 in. (n = 2).
- Sterile scalpel (n = 2).
- 500-ml glass beakers (n = 6).
- Sterile scissors (n = 4) and forceps (n = 4).
- Sterile gloves and sleeves.
- 100-μm nylon cell strainers (n = 4).
- 50-ml conical polypropylene centrifuge tubes (n = 16).
- Refrigerated centrifuge.
- 37°C incubator.
- 37°C water bath.

Additional reagents and equipment for cell counting:

- Trypan blue solution (0.4%, Sigma, cat. no. T8164).
- Hemocytometer.

12.9 Methods

In this section, we will describe in detail the method to isolate human AE and AM cells from full-term placentas. A detailed protocol describing the separation of the amnion membrane from placenta and the isolation of epithelial cells were published previously [15, 43]. All placentas are useful for AE or AM cell isolation, albeit, for sterility purposes, placentas are normally obtained from caesarean section.

First-, second-, and third-trimester placentas can also be used for cell isolation. The procedure is similar to the one described, however volumes of solutions used and times of digestion might be different.

12.9.1 Preparation of Solutions for Isolation and Culture

All the solutions required for tissue transportation and cell isolation, once prepared, are stable up to 4 weeks (if not differently specified) if stored at 4°C in sterile conditions.

- *Transportation medium* (for placenta transportation)*:* CMF-HBSS supplemented with antibiotic–antimycotic solution (100 U/ml penicillin G, 100 μg/ml streptomycin, and 0.25 μg/ml amphotericin B).
- *Predigestion buffer:* 1 L CMF-HBSS supplemented with 0.5 mM EGTA (Once prepared, it is stable up to 4 weeks if stored at 4°C in sterile conditions).
- *Digestion solution:* EMEM supplemented with 1 mg/ml collagenase type IV and 25 μg/ml DNase I (to be prepared fresh within 2h before using in a volume proportional to the amount of tissue).
- *Standard culture medium:*
  - DMEM = 430 mL
  - 100 mM sodium pyruvate = 5 mL
  - 10% (v/v) heat-inactivated fetal bovine serum = 50 mL
  - Nonessential amino acid = 5 mL
  - L-glutamine = 5 mL
  - 2-mercaptoethanol = 0.5 mL
  - Antibiotics solution = 5 mL
  - EGF (added freshly every time medium is changed) = 10 ng/ml

12.9.2 Preparation for Cell Isolation

Before starting the cell isolation technique and before carrying the placenta out of its transportation container, prepare hood for isolation as described below:

- Equip a BSL-2 laminar-flow cabinet with the following: absorbent bench paper, sterile field, sterile scalpel, sterile 500-ml beakers, sterile scissors and forceps, sterile gloves and sleeves, 100-μm nylon cell strainers, and 50-ml Falcon tubes.
- Fill all the 500-ml beakers with 200 ml HBSS.
- Warm the predigestion buffer to 37°C in a water bath.
- Thaw and prewarm the trypsin/EDTA solution to 37°C in a water bath.

From this point on all solutions and equipment must be sterile, and used according to proper aseptic technique.

12.9.3 Release Amniotic Epithelial Cells

The placenta is immersed in transportation solution and must be kept at 4°C during transportation until the isolation procedure starts.

1. The operator, wearing appropriate personal protection equipment, introduces the placenta container in the laminar-flow cabinet and opens the container.
2. The operator, wearing sterile gloves, removes the tissue from the transportation container and places the whole placenta on the sterile field with the maternal surface (rough surface) face down on the paper. In this position, the amnion membrane (smooth surface) faces up and the umbilical cord (if present) lies across the upper

surface (Figure 12.3). The umbilical cord can be removed (by scalpel) and discarded (or utilized for alternative purposes, not discussed in this chapter).

3. A superficial X-shaped incision should be made with a scalpel, through the upper placental surface (amnion), without damaging the lower layer (chorion) (Figure 12.3).
4. The operator can now easily peel the amnion membrane from the underlying chorion layer, starting from the cut edge (Figure 12.3).
5. The amnion can now be placed in the sterile 500-ml glass beaker containing 200 ml HBSS, and gently washed (Figure 12.3). This step must be repeated three (3) times to remove all the blood clots. Each time the amnion needs to be moved to a clean HBSS-containing beaker.
6. Discard the remaining part of the placenta.
7. Remove all the dissecting supplies from the work area.
8. Move the amnion membrane in one or two 50-ml centrifuge tubes (according to the size of the tissue). Usually the volume of membrane tissue from an average placenta is around 20–30 ml.
9. Add 20 ml warm predigestion buffer and gently rock the tissue in the enzymatic solution for 30 sec.
10. Transfer the membrane with sterile forceps to two new 50-ml tubes and add 20 ml of predigestion buffer in the tubes and incubate 10 min at 37°C in rotation.
11. Transfer the membrane into new tubes (discarding the predigestion buffer) and add 30 ml of 0.05% (w/v) trypsin/EDTA solution. The volume of membrane tissue from an average placenta is approximately 20 ml.

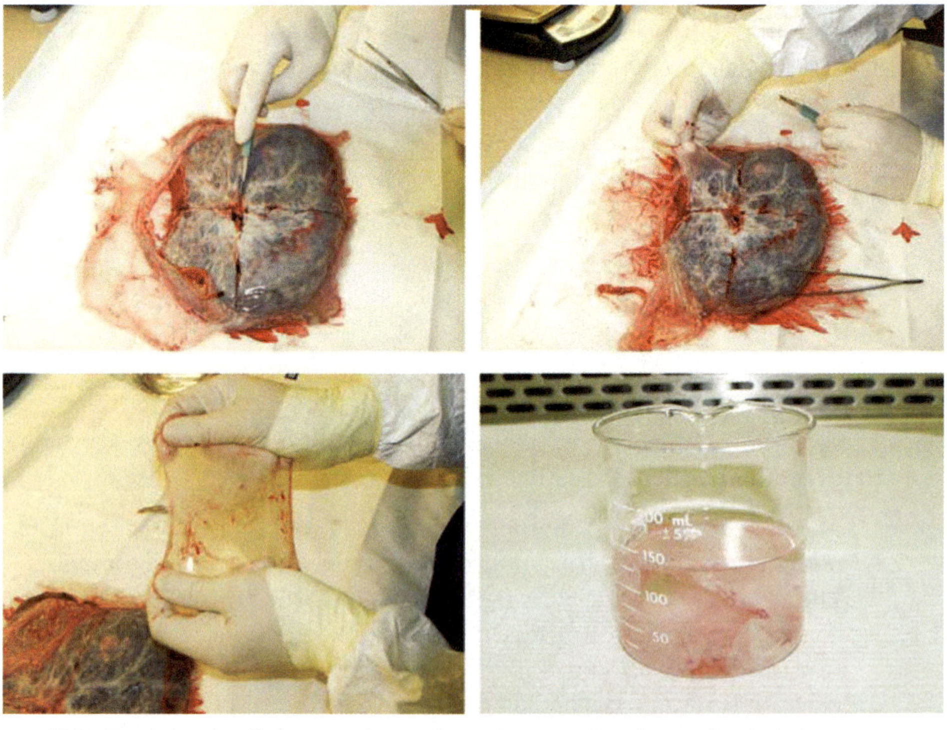

Figure 12.3 Panel showing X-shape cutting and stripping procedure for amnion isolation.

12. Incubate for 40 min at 37°C, in rotation.
13. After 40 min, vigorously shake the tubes in order to release all the AE detached cells and transfer the membrane into new 50-ml tubes containing 30 ml fresh warm trypsin/EDTA, and incubate for an additional 40 min.
14. Save the trypsin digest from step 12.
15. Immediately centrifuge the first digestion at 200×g for 10 min at 4°C to pellet the cell already released from the tissue.
16. Decant the supernatant and resuspend the pellet in 10 ml of standard culture medium. These cells will be combined with the second digestion product.
17. Save the remaining tissue for the AM cell isolation (see below).
18. Perform steps 12 through 15 for the second 40-min digestion.
19. Combine the two cell pellet in one 50-ml tube.
20. Count the cell number and record the cell viability using a hemocytometer and by the Trypan blue method.

At this point, cells can be either frozen using a cryopreservation protocol, used for flow cytometric analysis, or plated for in vitro experimentation.

If cells are to be cryopreserved, proceed with the cryopreservation procedure commonly used for hematopoietic cells.

If the cells are to be characterized by flow cytometry, resuspend the cells in sterile PBS at a concentration of 1×10^7 cells per mL. Proceed immediately to the cell staining according to the common flow cytometry lab procedure.

If cells are to be cultured, plate up to 1×10^5 cells per cm^2 in standard culture medium in regular petri dishes or multi-well plates. Change the medium twice a week. Subculture (1:4) the cells, as needed, when the cells exceed 70% confluence.

12.9.4 Anticipated Results

The yield of AE cells from one full-term placenta is typically about 10 million cells per gram of tissue (for a hypothetically total number of $8–30\times10^7$) cells [15], which will continue to proliferate up to passage 8–10. When cultured in DMEM supplemented, as described in Reagents and Solutions, with 10% serum and 10 ng/ml EGF, the cells grow for 4 to 10 passages. A passage is approximately a 1:4 split when the cells approach confluence.

12.9.5 Release Amniotic Mesenchymal Cells

1. Once the AE isolation procedure is completed, wash the remaining amnion two to three times with 200 ml HBSS to remove all the excess trypsin.
2. Transfer the membrane into two (or more) 50-ml centrifuge tubes.
3. Add the digestion solution at the same volume of tissue in the tube.
4. Incubate on a rotator 1h at 37°C.
5. Equally divide the contents of each tube into two new 50-ml tubes and fill with cold HBSS.
6. Centrifuge at 200×*g* for 5 min at 4°C.
7. Discard the supernatant and resuspend the pellet in HBSS.

8. Centrifuge again at 200×*g* for 5 min at 4°C.
9. Resuspend all the pellets in a small volume of AM standard culture medium and combine them in a single solution.
10. Count the cell number and record the cell viability using a hemocytometer and by trypan blue method.

At this point, cells can be either frozen using a cryopreservation protocol, used for flow cytometric analysis, or plated for in vitro experimentation.

If cells are to be cryopreserved, proceed with the cryopreservation procedure commonly used for hematopoietic cells.

If the cells are to be characterized by flow cytometry, resuspend the cells in sterile PBS at a concentration of 1×10^7 cells per mL. Proceed immediately to the cell staining according to the common flow cytometry lab procedure.

If cells are to be cultured, plate up to 1×10^5 cells per cm^2 in standard culture medium in regular petri dishes or multi-well plates. Change the medium twice a week. Subculture (1:4) the cells, as needed, when the cells exceed 70% confluence.

12.9.6 Anticipated Results

The yield of AM cells from one full term placenta is $2–50\times10^7$ cells, however, in practice, yields are typically 40,000 AM cell per cm^2 of starting material, with a typically fourfold expansion [32, 44].

Overdigestion by an extensive trypsinization of the amnion membrane, in order to release a higher number of cells, can also lead to poor viability and low plating efficiency.

Although the isolation of AE cells as described above yields a virtually pure cell suspension, this is not the case for AM cell preparation. A considerable number of AE cells may still be attached to the membrane before collagenase digestion, and this may result in the presence of AE cell "contamination" in AM cell preparations.

In a recent pubblication, a density separation method to purify AM from AE cells is described [15].

12.10 Placenta-Isolated Cell Characterization

In order to evaluate the homogeneity of the cell population after isolation, characterization by flow cytometry can be useful to determine the quality of the cell preparation. ES cells express a wide variety of antigens; the function of many of these remains unclear. These include stage-specific embryonic antigens (SSEA) 1, 3, and 4, and keratan sulfate associated antigens TRA1-60 and TRA1-81. SSEA-1 is a lactoseries oligosaccharide; SSEA-3 and SSEA-4 are globoseries glycolipids. Undifferentiated mouse ES cells express SSEA-1 and increase the expression of SSEA-3 and SSEA-4 when differentiating. However, the opposite is seen with human ES cells. Undifferentiated human ES cells express SSEA-3, SSEA-4, TRA1-60, TRA1-81, and Thy-1 and differentiation is characterized by increased expression of SSEA-1 and downregulation of SSEA-3 and SSEA-4 [45–48]. Like human ES cells, human placenta-derived cells have been characterized for the expression of many cell surface markers. Both first- and third-trimester placenta-derived cells express low levels of HLA-ABC but not HLA-DR [35, 37], indicating an immunoprivileged status. Table 12.1 summarizes all the markers identified in the three main regions (amnion epithelial, mesenchymal epithelial, and mesenchymal chorion) and their relative expression of each.

12.11 Cell Banking

Taken together, the data clearly indicate that cells with stem or progenitor properties are located in the three different layers: amnion epithelial, amnion mesenchymal, and chorion membrane. Placenta tissue is plentiful; with over 4 million live births/year in the United States, placenta-derived stem cells could be immediately available. Since relatively low technological procedures are needed for isolation and banking, this stem cell source could potentially be available worldwide at a modest cost. So far, the most experience in preservation of placental tissue-derived cells has been gained with cord blood, which contains both hematopoietic and mesenchymal stem cells. When cord blood transplantation proved effective, many cord blood banks were established, offering collection and banking for public (allogeneic) or private (autologous or allogeneic) use. Cord blood can be procured from natural births or caesarean sections as well [49]. Different methods describing the reduction of red blood cells, plasma volume, and cryopreservation are available in scientific literature [50]. Cord blood products containing cryoprotectants (e.g., dimethyl sulfoxide) are frozen at a controlled rate and can be stored in liquid nitrogen for at least 15 years without loss of their engraftment potential in vivo [51]. In contrast to cord blood [52, 53], fetal membrane stem cells are presently preserved mainly for research. However, as these cells gain interest for their regenerative and immunomodulatory properties [37], future medical needs may require concomitant application of cord blood and placental cells from the same donor. Currently, freeze-dried, γ-sterilized, decellularized, glycerol-preserved, and cryopreserved amniotic membranes are used in ophthalmology and wound care [54–57].

12.12 Discussion and Commentary

As recently described by Parolini and collegues in two comprehensive reviews [44, 58], placenta represents a reserve of different types of progenitor/stem cells. These cells can be easily isolated from different compartments and used for cell-based therapy or cryobanked. This stem cell source is a noncontroversial, extensive, and unlimited source. Amnion membrane, like all other placental tissues, is normally discarded following the birth of a baby. However a collection of these tissues does not seem problematic, and in many countries (i.e., the United States, Canada, Brazil, Argentina, Israel, Australia, and most European countries such as the United Kingdom, Spain, Sweden, Italy, and France) is a reality (umbilical cord blood banking). Since the life and development of the fetus is not affected, stem cells derived from placental tissues could avoid most ethical or religious concerns associated with many other stem cell sources.

This chapter delineated a simple laboratory procedure to isolate cells from the amnion layer of human full-term placenta. Moreover, data presented here clearly indicates that cells contained in human full-term placenta have characteristics commonly found in pluripotent stem cells such as ES cells, including the expression of surface markers and genes that maintain pluripotency. Finally, placenta-derived cells, in particular the AE cells, display the ability to differentiate into cell types from all three germ layers, in vitro. In addition, placenta-derived cells display unique properties including their ability to secrete anti-inflammatory factors that might optimize the engraftment of cells following transplantation. The many useful properties of placenta-derived stem cells suggest that these cells may become a useful and noncontroversial source for cells for regenerative medicine.

For these reasons, in 2009 The International Placenta Stem Cell Society (IPLASS; www.iplass.net) was founded, as an independent and nonprofit organization which promotes advances in placenta-derived stem cell research and development, with the final aim of applying these cells for cell therapy-based clinical applications.

12.13 Application Notes

The total time for the whole procedure (from the time the tissue is received to the time AE and AM cells are resuspended) is estimated at around 6h.

The AE cell isolation procedure is approximately 4h, while the AM isolation procedure (from the end of the AE isolation procedure to preparation of the AM cell suspension) is around 2h.

Some variability has been observed in the cell yield between individuals and it has been ascribed to the quality of tissue received. There is no direct evidence of this, but it is reasonable to assume that the time between the delivery of the placenta and the start of the cell isolation is a critical parameter. To minimize the warm ischemia time, the placenta should be transferred to a refrigerator as soon as possible following delivery. Conversely, in order to minimize the cold ischemic time, high viability and yield is achieved for placenta cells isolated within 3h from the tissue collection. Both cell yield and viability are adversely impacted by extended cold ischemic storage of the membrane prior to cell isolation.

The initial viability is not directly indicative of the long-term quality of the cells. For example, in our hands an initial viability over 75% did not always reflect a high efficiency in attachment to the culture dishes or high proliferating ratio. The best estimate of long-term viability of the cells seems to be the efficiency with which the cells are able to attach to the culture dishes, followed by the growth of the cells in the presence of EGF.

During the AE culture, removal of the EGF results in a rapid decrease in proliferation and what appears to be terminal differentiation. However, even in the continued presence of serum and EGF, the cells do not normally proliferate past passages 8–10 under the culture conditions described. Although not optimized at this time, like ES cells, AE cells can be induced to differentiate to specific cell types as previously described. We suggest reference to the original publications for culture conditions used to induce differentiation along these pathways.

In our experience, AM cells have been observed to adhere on tissue culture plastic and proliferate until passages 5–10, albeit with a decline in the AM proliferation beyond second passage. In general, improved proliferation has been observed for third-trimester-derived cells comparing to first-trimester-derived cells [35].

Troubleshooting Table

Problem	Explanation	Potential Solutions
It is difficult to strip the amnion.	It is not possible to find the starting point to strip the amnion.	It may be easier to start the X incision at the point of contact between the placenta and the umbilical cord.
It is not feasible to find the region of the umbilical cord.	An excess of blood on top of the amnion might be present.	It might be useful to pour some HBSS on top of the membrane and gently massage to remove blood clots.
The surface looks rough during the peeling procedure.	The operator is peeling off some extra connective tissue at the interface between chorion and amnion.	The presence of extra connective tissue will likely interfere with the AE cell isolation, it is suggested to pinch the rough part with a second pair of forceps and only collect the amnion.
No viable cells can be found after the predigestion step.	The predigestion step acts only to isolate all the superficial clots and prepare the amnion epithelial layer to the digestion steps.	Discard the solution from the predigestion step.
The viability of the AE cells after the first incubation is low.	The trypsin activity decreases the cell viability.	It is necessary to add two volumes of standard culture medium to the trypsin digest after the first incubation to inactivate the trypsin action.
The viability of the AE cells after the first incubation is low.	Blood clots reduce the efficiency of the trypsin action.	It is necessary perform the washing step (5). It is crucial for the enzymatic digestion to work properly.
The cell pellet after the first digestion appears compact.	The AE cells from the first incubation are in clumps.	Pipette up and down several times to release the cells. Discard any remaining small cell aggregates or filter the cell preparation through a 100-μm filter.
The remaining amnion after the first digestion looks rough.	A lot of cells and debris from the AE isolation are entrapped in the remaining amnion membrane.	It is necessary to move the amnion to a clean beaker each time.
More than two tubes are necessary to procede with the AM isolation.	The amount of wet tissue might be increased after trypsin digestion. The number of tubes depends on the volume of amnion remaining.	In order not to dilute the collagenase solution excessively, divide the tissue as needed so as not to have more than 20–25 ml of wet tissue in 50 ml final volume.
After 45 min the tissue looks undigested.	The incubation time varies according to the membrane thickness.	The incubation time might range between 45 min to 1.5h. It is wise to periodically check the status of digestion starting after the first 30 min, and stop the incubation as soon as the tissue is completely dissolved.

In order to get rid of the presence of AE cell contamination in AM cell preparations it is good practice to change the culture medium 1 to 2h after plating. Considering the slow adhesion of AE cells to culture substrates, as opposed to the fast attachment of AMs, this passage will likely remove many of the AE cells that might be present in the preparation. However, significant contamination of AM with AE cells frequently occurs. The morphological differences between the two cell types might allow you to easily discriminate between AE and AM cells after 24h in culture.

12.14 Summary Points

1. The placenta is an important source of stem/progenitor cells and the placenta-derived stem cells can be easily isolated from different compartments.

2. We have developed a simple procedure to isolate amniotic epithelial cells by enzymatic digestion. The isolation procedure with common laboratory equipment can be performed on human term placenta and the cells can be expanded in culture for 8–10 passages.
3. Amniotic mesenchymal cells can be isolated from the remaining part of the amnion tissue used to release AE cells by collagenase-mediated digestion. AM cells can be easily cultured in the same conditions as AE cells.
4. Like human ES cells, human placenta-derived cells have been characterized for the expression of many cell surface markers, including embryonic markers (SSEA-4, SSEA-3, TRA1-60, TRA1-81), stemness markers (ABCG2, CD34, CD133/2), adhesion molecules (CD29, CD49c-f, CD54, CD326), or other lineage markers.

Acknowledgments

This work was supported by National PKU Alliance and Associazione per la Prevenzione e Cura dell'Epatite Virale "Beatrice Vitiello" ONLUS.

References

[1] Moore, R. M., et al., "Physiological Apoptotic Agents Have Different Effects upon Human Amnion Epithelial and Mesenchymal Cells," *Placenta,* Vol. 24, No. 2–3, 2003, pp. 173–80.

[2] Magatti, M., et al., "Human Amnion Mesenchyme Harbors Cells with Allogeneic T-Cell Suppression and Stimulation Capabilities, *Stem Cells,* Vol. 26, No. 1, 2008, pp. 182–92.

[3] Choi, D., et al., "In Vivo Differentiation of Mouse Embryonic Stem Cells into Hepatocytes," *Cell Transplant.,* Vol. 11, No. 4, 2002, pp. 359–68.

[4] Mellor, A. L., and D. H. Munn, "Immunology at the Maternal-Fetal Interface: Lessons for T Cell Tolerance and Suppression," *Annu. Rev. Immunol.,* Vol. 18, 2000, pp. 367–91.

[5] Munn, D. H., et al., "Inhibition Of T Cell Proliferation by Macrophage Tryptophan Catabolism," *J. Exp. Med.,* Vol. 189, No. 9, 1999, pp. 1363–72.

[6] Munn, D. H., et al., "Prevention of Allogeneic Fetal Rejection by Tryptophan Catabolism," *Science,* Vol. 281, No. 5380, 1998, pp. 1191–93.

[7] Fournel, S., et al., "Cutting Edge: Soluble HLA-G1 Triggers CD95/CD95 Ligand-Mediated Apoptosis in Activated CD8+ Cells by Interacting with CD8," *J. Immunol.,* Vol. 164, No. 12, 2000, pp. 6100–104.

[8] Thellin, O., et al., "Tolerance to the Foeto-Placental 'Graft': Ten Ways to Support a Child for Nine Months," *Curr. Opin. Immunol.,* Vol. 12, No. 6, 2000, pp. 731–37.

[9] Makrigiannakis, A., et al., "Corticotropin-Releasing Hormone Promotes Blastocyst Implantation and Early Maternal Tolerance," *Nat. Immunol.,* Vol. 2, No.11, 2001, pp. 1018–24.

[10] Draper, J. S., et al., "Surface Antigens of Human Embryonic Stem Cells: Changes upon Differentiation in Culture," *J. Anat.,* Vol. 200, Pt. 3, 2002, pp. 249–58.

[11] Reubinoff, B. E., et al., "Embryonic Stem Cell Lines from Human Blastocysts: Somatic Differentiation In Vitro," *Nat. Biotechnol.,* Vol. 18, No. 4, 2000, pp. 399–404.

[12] Miki, T., et al., "Stem Cell Characteristics of Amniotic Epithelial Cells," *Stem Cells,* Vol. 23, No. 10, 2005, pp. 1549–59.

[13] Elwan, M. A., and N. Sakuragawa, "Evidence for Synthesis and Release of Catecholamines by Human Amniotic Epithelial Cells," *Neuroreport,* Vol. 8, No. 16, 1997, pp. 3435–38.

[14] Ilancheran, S., et al., "Stem Cells Derived from Human Fetal Membranes Display Multilineage Differentiation Potential," *Biol. Reprod.,* Vol. 77, No. 3, 2007, pp. 577–88.

[15] Marongiu, F., et al., "Isolation of Amniotic Mesenchymal Stem Cells," *Curr. Protoc. Stem Cell Biol.,* p. Unit 1E 5.

[16] Libbrecht, L., et al., "Hepatic Progenitor Cells in Hepatocellular Adenomas," *Am. J. Surg. Pathol.,* Vol. 25, No. 1, 2001, pp. 1388–96.

[17] Roskams, T., R. De Vos, and V. Desmet, "'Undifferentiated Progenitor Cells' in Focal Nodular Hyperplasia of the Liver," *Histopathology,* Vol. 28, No. 4, 1996, pp. 291–99.

[18] Theise, N. D., et al., "Hepatic 'Stem Cell' Malignancies in Adults: Four Cases," *Histopathology,* Vol. 43, No. 3, 2003, pp. 263–71.

[19] Xiao, J. C., et al., "Small Epithelial Cells in Human Liver Cirrhosis Exhibit Features of Hepatic Stem-Like Cells: Immunohistochemical, Electron Microscopic and Immunoelectron Microscopic Findings," *Histopathology,* Vol. 42, No. 2, 2003, pp. 141–49.

[20] Davila, J. C., et al., "Use and Application of Stem Cells in Toxicology," *Toxicol. Sci.,* Vol. 79, No. 2, 2004, pp. 214–23.

[21] Miki, T., and S .C. Strom, "Amnion-Derived Pluripotent/Multipotent Stem Cells," *Stem Cell Rev.,* Vol. 2, No. 2, 2006, pp. 133–42.

[22] Elwan, M. A., et al., "Evidence for the Presence of dopamine D1 Receptor mRNA and Binding Sites in Monkey Amniotic Epithelial Cells," *Neurosci. Lett.,* Vol. 262, No. 1, 1999, pp. 9–12.

[23] Elwan, M. A., T. Ishii, and N. Sakuragawa, "Characterization of the Dopamine Transporter Gene Expression and Binding Sites in Cultured Human Amniotic Epithelial Cells," *Neurosci. Lett.,*Vol. 342, No. 1–2, 2003, pp. 61–44.

[24] Sakuragawa, N., et al., "Evidence for Active Acetylcholine Metabolism in Human Amniotic Epithelial Cells: Applicable to Intracerebral Allografting for Neurologic Disease," *Neurosci. Lett.,* Vol. 232, No. 1, 1997, pp. 53–56.

[25] Sakuragawa, N., et al., "Expression of Markers for Both Neuronal and Glial Cells in Human Amniotic Epithelial Cells," *Neurosci. Lett.,* Vol. 209, No. 1, 1996, pp. 9–12.

[26] Henderson, J.K., et al., "Preimplantation Human Embryos and Embryonic Stem Cells Show Comparable Expression of Stage-Specific Embryonic Antigens," *Stem Cells,* Vol. 20, No. 4, 2002, pp. 329–37.

[27] Nichols, J., et al., "Formation of Pluripotent Stem Cells in the Mammalian Embryo Depends on the POU Transcription Factor Oct4," *Cell,* Vol. 95, No. 3, pp. 379–91.

[28] Tamagawa, T., I. Ishiwata, and S. Saito, "Establishment and Characterization of a Pluripotent Stem Cell Line Derived from Human Amniotic Membranes and Initiation of Germ Layers In Vitro," *Hum. Cell,*Vol. 17, No. 3, 2004, pp. 125–30.

[29] Wei, J.PP., et al., "Human Amnion-Isolated Cells Normalize Blood Glucose in Streptozotocin-Induced Diabetic Mice," *Cell Transplant.,* Vol. 12, No. 5, 2003, pp. 545–52.

[30] Soncini, M., et al., "Isolation and Characterization of Mesenchymal Cells from Human Fetal Membranes," *J. Tissue Eng. Regen. Med.,* Vol. 1, No. 4, 2007, pp. 296–305.

[31] Zhang, X., et al., "Successful Immortalization of Mesenchymal Progenitor Cells Derived from Human Placenta and The Differentiation Abilities of Immortalized Cells," *Biochem. Biophys. Res. Commun.,* Vol. 351, No. 4, 2006, pp. 853-59.

[32] Casey, M. L,. and P. P. C. MacDonald, "Interstitial Collagen Synthesis and Processing in Human Amnion: A Property of the Mesenchymal Cells," *Biol. Reprod.,* Vol. 55, No. 6, 1996, pp. 1253-60.

[33] Bilic, G., et al., "In Vitro Lesion Repair by Human Amnion Epithelial and Mesenchymal Cells," *Am. J. Obstet. Gynecol.,* Vol. 190, No. 1, 2004, pp. 87–92.

[34] In 't Anker, P. P. S., et al., "Isolation of Mesenchymal Stem Cells of Fetal or Maternal Origin from Human Placenta," *Stem Cells,* Vol. 22, No. 7, 2004, pp. 1338–45.

[35] Portmann-Lanz, C. B., et al., "Placental Mesenchymal Stem Cells as Potential Autologous Graft for Pre- and Perinatal Neuroregeneration," *Am. J. Obstet. Gynecol.,* Vol. 194, No. 3, 2006, pp. 664–73.

[36] Sakuragawa, N., et al., "Human Amnion Mesenchyme Cells Express Phenotypes of Neuroglial Progenitor Cells," *J. Neurosci. Res.,* Vol. 78, No. 2, 2004, pp. 208–14.

[37] Wolbank, S., et al., "Dose-Dependent Immunomodulatory Effect of Human Stem Cells from Amniotic Membrane: A Comparison with Human Mesenchymal Stem Cells from Adipose Tissue," *Tissue Eng.,* Vol. 13, No. 6, 2007, pp. 1173–83.

[38] Pasquinelli, G., et al., "Ultrastructural Characteristics of Human Mesenchymal Stromal (Stem) Cells Derived from Bone Marrow and Term Placenta," *Ultrastruct. Pathol.,* Vol. 31, No. 1, 2007, pp. 23–31.

[39] Zhao, P. P., et al., "Human Amniotic Mesenchymal Cells Have Some Characteristics of Cardiomyocytes," *Transplantation,* Vol. 79, No. 5, 2005, pp. 528–35.

[40] Miki, T., et al., "Identification of Stem Cell Marker-Positive Cells by Immunofluorescence in Term Human Amnion," *J. Reprod. Immunol.,* Vol. 75, No. 2, 2007, pp. 91–96.

[41] Zhang, X., et al., "Mesenchymal Progenitor Cells Derived from Chorionic Villi of Human Placenta for Cartilage Tissue Engineering," *Biochem. Biophys. Res. Commun.,* Vol. 340, No. 3, 2006, pp. 944–52.

[42] Bailo, M., et al., "Engraftment Potential of Human Amnion and Chorion Cells Derived from Term Placenta," *Transplantation,* Vol. 78, No. 10, 2004, pp. 1439–48.

[43] Miki, T., et al., "Isolation of Amniotic Epithelial Stem Cells," *Curr. Protoc. Stem Cell Biol.,* p. Unit 1E 3.

[44] Parolini, O., et al., "Concise Review: Isolation and Characterization of Cells From Human Term Placenta: Outcome of the First International Workshop on Placenta Derived Stem Cells," *Stem Cells,* Vol. 26, No. 2, 2008, pp. 300–11.

[45] Couvelard, A., et al., "Expression of Integrins During Liver Organogenesis in Humans," *Hepatology,* Vol. 27, No. 3, 1998, pp. 839–47.

[46] Gilles, J. M., et al., "Immunophenotypic Characterization of Human Fetal Liver Hematopoietic Stem Cells During the Midtrimester of Gestation," *Am. J. Obstet. Gynecol.,* Vol. 177, No. 3, 1997, pp. 619–25.

[47] Haruna, Y., et al., "Identification of Bipotential Progenitor Cells in Human Liver Development," *Hepatology,* Vol. 23, No. 3, 1996, pp. 476–81.

[48] Zhao, Y., D. Glesne, and E. Huberman, "A Human Peripheral Blood Monocyte-Derived Subset Acts as Pluripotent Stem Cells," *Proc. Natl. Acad. Sci. USA,* Vol. 100, No. 5, 2003, pp. 2426–31.

[49] Rogers Chrysler, G. et al., "Umbilical Cord Blood Banking," in *Cord Blood: Biology, Immunology, Banking and Clinical Transplantation,* H. E. Broxmeyer (ed.), Bethesda, MD: AABB Press, 2004, pp. 219–57.

[50] Lapierre, V., et al., "Cord Blood Volume Reduction Using an Automated System (Sepax) Vs. a Semi-Automated System (Optipress II) and a Manual Method (Hydroxyethyl Starch Sedimentation) for Routine Cord Blood Banking: A Comparative Study," *Cytotherapy,* Vol. 9, No.. 2, 2007, pp. 165–9.

[51] Skoric, D., et al., "Collection Strategies and Cryopreservation of Umbilical Cord Blood," *Transfus. Med.,* Vol. 17, No.2, 2007, pp. 107–13.

[52] Broxmeyer, H.E., et al., "Human Umbilical Cord Blood as a Potential Source of Transplantable Hematopoietic Stem/Progenitor Cells," *Proc. Natl. Acad. Sci. USA,* Vol. 86, No.10, 1989, pp. 3828–32.

[53] Gluckman, E., et al., "Hematopoietic Reconstitution in a Patient with Fanconi's Anemia by means of Umbilical-Cord Blood from an HLA-Identical Sibling," *N. Engl. J. Med.,* Vol. 321, No. 17, 1989, pp. 1174–8.

[54] Burman, S., et al., "Ophthalmic Applications of Preserved Human Amniotic Membrane: A Review of Current Indications," *Cell Tissue Bank,* Vol. 5, No. 3, 2004, pp. 161–75.

[55] Dua, H. S., et al., "The Amniotic Membrane in Ophthalmology," *Surv. Ophthalmol.,* Vol. 49, No. 1, 2004, pp. 51–77.

[56] Ravishanker, R., A. S. Bath, and R. Roy, "'Amnion Bank'—The Use of Long Term Glycerol Preserved Amniotic Membranes in the Management of Superficial and Superficial Partial Thickness Burns," *Burns,* Vol. 29, No. 4, 2003, pp. 369–74.

[57] Tseng, S. C., et al., "Amniotic Membrane Transplantation for Conjunctival Surface Reconstruction," *Am. J. Ophthalmol.,* Vol. 124, No. 6., 1997, pp. 765–74.

[58] Parolini, O., et al., "Toward Cell Therapy Using Placenta-Derived Cells: Disease Mechanisms, Cell Biology, Preclinical Studies, and Regulatory Aspects at the Round Table," *Stem Cells Dev.,* Vol. 19, No. 2, ppp. 143–54.

13

Liver Repopulation by Fetal Liver Stem/Progenitor Cells

Michael Oertel* and David A. Shafritz

Marion Bessin Liver Research Center and Division of Gastroenterology and Liver Diseases, Dept. of Medicine, Albert Einstein College of Medicine of Yeshiva University, Bronx, NY

*Correspondence: Marion Bessin Liver Research Center, Albert Einstein College of Medicine of Yeshiva University, 1300 Morris Park Avenue, Ullmann 603B, Bronx, NY 10461; Phone: (718) 430-4227; Fax: (718) 430-8975; E-mail: michael.oertel@einstein.yu.edu

Abstract

Substantial progress has been made during the past 10 to 15 years concerning the possibility of liver repopulation by transplanted cells. Protocols were evolved to isolate hepatocytes, as well as stem/progenitor cells, and several models were established to demonstrate their capability to replace liver mass under specialized, highly selective circumstances. However, our laboratory has shown that rat fetal liver stem/progenitor cells (FLSPC) have the capacity to proliferate and differentiate in the normal host liver and replace hepatic mass under more physiologic, nonselective conditions. This requires only a liver proliferative stimulus at the time the cells are transplanted. Since there is no need for extensive modification of the host liver, this model is very useful to study the mechanism of tissue replacement and to develop new cell transplantation strategies that ultimately might be useful clinically. This chapter will describe the isolation of FLSPC from wt DPPIV$^+$ F344 rats and their transplantation into DPPIV$^-$ mutant rats and staining techniques for DPPIV in the host liver to follow the fate of transplanted cells.

Key terms ED14, FLSPC, fetal liver stem/progenitor cells, cell isolation, cell transplantation, liver repopulation, DPPIV, enzymehistochemistry, immunohistochemistry

13.1 Introduction

13.1.1 Background

Transplantation of liver tissue was first reported in the early 1900s, including studies on the transplantation of liver fragments into the anterior chamber of the eye [1]. The transplanted liver tissue degenerated rapidly and disappeared within days or weeks. The first major advance in this field came in the 1960s, when methods were developed to isolate single hepatocytes that were viable in culture [2]. These cells were transplanted through the portal circulation in the 1970s, leading to transient reduction in serum bilirubin levels in the Gunn rat hyperbilirubinemia model for the human disorder, Crigler-Najjar Syndrome, Type 1 [3]. In the 1980s, primary isolated hepatocytes were transplanted into various ectopic sites, including the dorsal fat pad [4], spleen [5], peritoneal cavity [6], and under the renal capsule [7]. The cells survived to varying degrees in these sites and demonstrated hepatic morphology and function, including modest proliferation after inducing liver regeneration [8, 9]. These studies formed the basis for more recent work during the past 15 years that has clearly established that the liver can be repopulated by transplanted hepatic cells with restoration of normal structure and function.

13.1.2 Cell Transplantation Strategies in Rodent Model Systems

To determine whether hepatic cells can be effectively transplanted into the liver requires a method to distinguish transplanted cells from host hepatocytes. A model to follow the fate of transplanted hepatic cells is the dipeptidyl peptidase (DPP)IV deficient Fischer (F)344 rat [10], an inbred strain in which wild type (wt) $DPPIV^+$ hepatocytes can be transplanted into $DPPIV^-$ recipients without rejection [11]. However, the major problem is that after transplantation of hepatocytes into a relatively normal liver, it is currently not possible to selectively expand the transplanted cells to obtain significant replacement of hepatic mass. In contrast, under very specialized, highly selective conditions, transplanted adult hepatocytes can repopulate the liver.

Initial studies used transgenic (uPA) or knockout ($FAH^{-/-}$) mice, in which genetic modification of the host liver provided an environment in which there is both extensive and continuous liver injury and strong selection for survival of transplanted wt cells [12, 13]. Other models for liver repopulation by transplanted cells have rendered host hepatocytes incapable of proliferating by introducing DNA damage through treatment with DNA alkylating agents, retrorsine [14], monocrotaline [15,] or x-irradiation [16]. In these various animal models, transplanted hepatocytes effectively replace the recipient liver, because host hepatocytes have been irreversibly damaged and cannot respond to a liver proliferative stimulus. However, in a normal or near-normal liver environment, transplanted wt hepatocytes do not have a selective advantage over host hepatocytes and there is no effective repopulation. Therefore, considerable attention has recently been focused on isolating, characterizing, and transplanting liver stem and progenitor cells that have higher proliferative potential than adult hepatocytes, as well as hematopoietic, mesenchymal, and embryonic stem cells, which also might have the ability to proliferate more efficiently than host hepatocytes and replace liver mass without requiring major manipulation of the host liver (for a detailed review, see [17]).

13.1.3 Fetal Liver Stem/Progenitor Cells as a Source for Liver Repopulation

Fetal liver stem/progenitor cells are the only cells identified to date that effectively repopulate the normal liver. During rodent liver development, foregut endodermal stem cells begin to proliferate and differentiate on embryonic day (ED) 8.0 [18], migrate into the septum transversum at ~ED 8.5, become tissue-specified and begin to express genes in the hepatic lineages at ~ED9.0-9.5, initially α-fetoprotein (AFP) and albumin (Alb) and subsequently cytokeratin (CK)-19 [19]. These hepatic-specified cells proliferate massively, form the liver bud and then diverge along the hepatocytic and cholangiocytic lineages (at ~ED16).

After microdissection of ED14 fetal livers derived from DPPIV$^+$ F344 rats, isolated fetal liver stem/progenitor cells (FLSPC; containing a subpopulation of bipotent AFP$^+$/CK-19$^+$ cells) are injected into the portal vein of the normal mutant DPPIV$^-$ F344 recipient liver in conjunction with 2/3 partial hepatectomy (PH) [20]. Liver repopulation increases progressively over 6 months after transplanting ED14 fetal liver cells, and the bulk of repopulating clusters contain both hepatocytes and mature bile ducts (Figure 13.1). The transplanted cells are morphologically and functionally integrated into the host liver parenchyma and remain viable long-term in the normal host liver [20, 21]. Six months after transplantation of ~50×10^6 unfractionated fetal liver cells into young adult recipients (2–3 months of age), 23.5 ± 2.1% of the hepatic mass is replaced [21]. Liver repopulation occurs by cell–cell competition in which more highly proliferative trans-

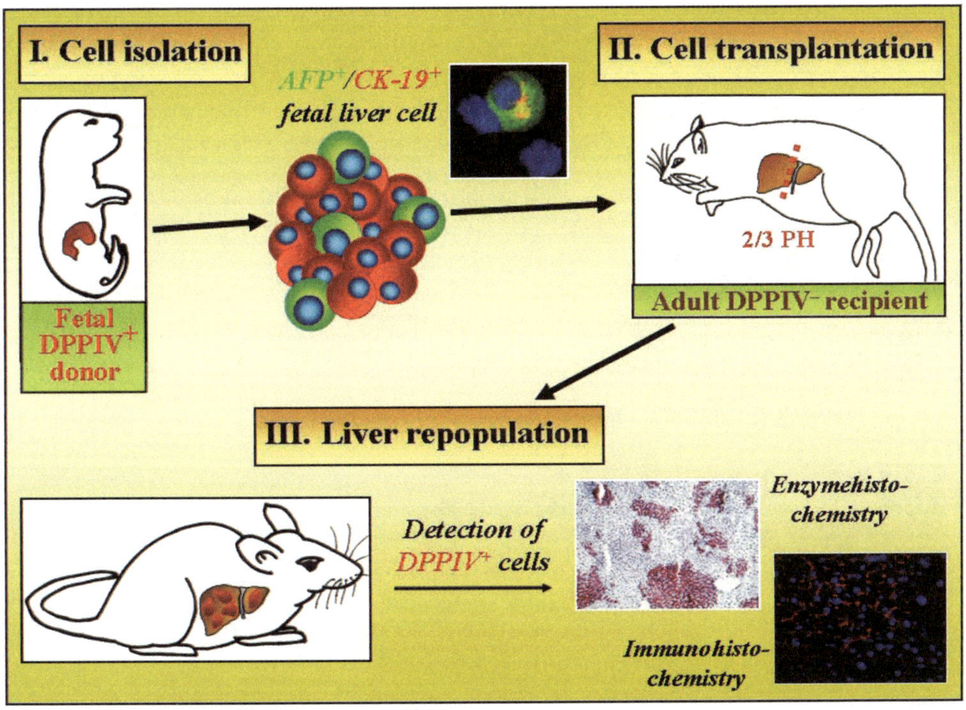

Figure 13.1 Pictorial representation of the DPPIV$^-$ mutant rat model system and experimental steps employed to achieve and follow liver repopulation by transplanted fetal liver cells from DPPIV$^+$ rats. Cells isolated from wt (DPPIV$^+$) ED14 F344 rat liver are isolated and transplanted into the liver of adult DPPIV$^-$ mutant F344 rats immediately after two-thirds PH and repopulation is followed over time by enzymehistochemistry or immunohistochemistry for DPPIV.

planted FLSPC progressively replace less proliferative neighboring host hepatocytes by inducing their apoptosis [21].

13.2 Experimental Design

ED14 fetal liver cells can be easily isolated with reproducible yields using a two-step digestion method; dissected fetal liver tissue is first gently minced and incubated with EGTA, which is then followed by incubation with collagenase [22, 23]. The obtained cell preparations are characterized by high cell viabilities, and therefore serve as a valuable source for cell transplantation studies. For this purpose, an excellent cell transplantation model is the DPPIV$^-$ Fischer rat [10], using DPPIV as marker enzyme to detect transplanted cells derived from syngeneic wt DPPIV$^+$ F344 donors. Importantly, transplantation of FLSPC is performed in conjunction with two-thirds PH [24] to obtain effective liver repopulation [21]. Finally, several detection methods can be used to follow the fate of DPPIV$^+$ FLSPC, as well as their proliferation and differentiation into hepatocytes and bile duct cells after transplantation into mutant DPPIV$^-$ recipients. The most sensitive technique for detection of DPPIV is an enzymehistochemical staining procedure, using Glycyl-proline-4-methoxy-β-naphthylamide as substrate [25], and is, therefore, particularly useful for determining the repopulation capacity of transplanted FLSPC. To further characterize transplanted DPPIV$^+$ cells compared to surrounding DPPIV$^-$ host cells, immunohistochemical methods using specific antibodies for the detection of DPPIV/CD26 in frozen and/or paraffin-embedded liver tissue can be performed. Spectrophotometric detection of DPPIV enzyme activity in liver homogenates can also be used to determine the level of repopulation of DPPIV$^-$ host liver [20], but this assay is not as reliable as enzymehistochemistry with liver tissue when liver repopulation is low (less than 5%).

13.3 Animals

- ED14 time-dated pregnant wt F344 rat (Taconic).
- Mutant DPPIV$^-$ F344 rats (2–4 months of age).

13.4 Materials

13.4.1 Cell Isolation

- Ether (Fisher Scientific).
- 10x Hanks' balanced salt solution (HBSS without Ca^{2+}; phenol red) (GIBCO).
- Magnesium chloride ($MgCl_2$) (SIGMA).
- HEPES (SIGMA).
- Penicillin / streptomycin (GIBCO).
- EGTA ($C_{14}H_{24}N_2O_{10}$) (SIGMA).
- Collagenase (Type I) (Worthington Biochemical).

- Bovine serum albumin (BSA) (SIGMA).
- Calcium chloride ($CaCl_2$) (SIGMA).
- DNase I (Boehringer Mannheim).
- Ethanol (Pharmco-aaper).
- Surgical instruments (ROBOZ).
- 20G1 gauge needle (BD).
- Glass beads (Fisher Scientific).
- Fetal bovine serum (FBS) (Biowest).
- 40-μm nylon-mesh cell strainer (BD Falcon™).
- Dulbecco's modified Eagle medium (DMEM) (GIBCO).
- Trypan blue (SIGMA).

13.4.2 Cell Transplantation

- Isoflurane.
- Ethanol (Pharmco-aaper).
- Surgical instruments (ROBOZ).
- Sutures (Johnson-Johnson).
- Cotton-tipped applicators (Puritan).
- Saline solution (0.9% NaCl) (Fisher Scientific).
- Gauze pads (Henry Schein).
- 30G½, 23G1 gauge needles (BD).
- 1- and 5-mL syringes (BD).
- Surgifoam™ (Johnson-Johnson).

13.4.3 Detection of Transplanted Cells

- Tissue-Tek® OCT Compound (Fisher Scientific).
- 2-methylbutane (Fisher Scientific).
- 10% buffered formalin phosphate (Fisher Scientific).
- Trisma® maleate (SIGMA).
- Sodium hydroxide (NaOH) (Fisher Scientific).
- Sodium chloride (NaCl) (Fisher Scientific).
- Fast blue BB salt (SIGMA).
- Glycyl-proline-4-methoxy-β-naphtylamide (GPMN) (SIGMA).
- Dimethyl-formamide (DMF) (SIGMA).
- Magnesium sulfate, hepatahydrate ($MgSO_4$ x 7 H_2O) (SIGMA).
- Sodium bicarbonate ($NaHCO_3$) (SIGMA).
- *Colorfrost®/Plus* microscope slides and glass coverslips (Fisher Scientific).
- Ethanol (Pharmco-aaper).
- Glacial acetic acid (Fisher Scientific).

- Acetone (Fisher Scientific).
- Cupric sulfate, pentahydrate ($CuSO_4$ x 5 H_2O) (SIGMA).
- Paraformaldehyde (PFA) (SIGMA).
- Harris hematoxylin (Poly Scientific).
- Glycerol (Fisher Scientific).
- Propyl gallate (SIGMA).
- Sodium carbonate (Na_2CO_3) (SIGMA).
- PAP PEN (Fisher Scientific).
- Sodium borohydride ($NaBH_4$) (SIGMA).
- Donkey serum (SIGMA).
- Blocker™ BSA (Thermo Scientific).
- Tween® 20 (SIGMA).
- 4′,6-diamidino-2-phenylindole dichloride (DAPI) (SIGMA).
- Xylene (Fisher Scientific).
- Citric acid ($C_6H_8O_7$) (SIGMA).

13.5 Methods

13.5.1 Cell Isolation

13.5.1.1 Preparation of Reagents

Modified HBSS

- Prepare 500 mL of HBSS containing 0.8 mM $MgCl_2$, and 20 mM HEPES, (pH 7.4).
- Dilute 50 mL of a 10x HBSS solution by adding 435 mL deionized, distilled (dd)H_2O.
- Add 400 μL of 1M $MgCl_2$ and 10 mL of 1 M HEPES (pH 7.4) stock solutions.
- Finally, add 5 mL of a penicillin/streptomycin solution (final concentrations of 100 U/mL and 100 μg/mL, respectively).
- Buffer can be stored at 4°C.

Modified HBSS/EGTA

- Add 500 μL of a 0.2 M EGTA stock solution to 19.5 mL modified HBSS (final concentration of 5 mM EGTA).
- Aliquot solution and store at –20°C.

Modified HBSS/Collagenase

- Dissolve 160 mg collagenase (Type I) and 240 mg BSA in 60 ml modified HBSS.
- Add 80 μL of a 1M $CaCl_2$ stock solution (final concentration of 1 mM) and 80 μL of a 1% DNase I solution (final concentration of 0.001%).
- Bring to a final volume of 80 ml with additional modified HBSS.
- Aliquot solution and store at –20°C.

Note: All solutions must be sterile filtered.

13.5.1.2 Procedure

1. Euthanize an ED14 time-dated pregnant wt F344 rat by ether (or carbon dioxide) inhalation, shave the abdominal region and disinfect with 70% ethanol.
2. Make a 3-to 4-cm midline longitudinal incision through the skin, followed by incision through the abdominal muscles.
3. Dissect and place the uterus in a sterile 10-cm Petri dish containing 15 mL chilled, modified HBSS. Note that the following steps have to be performed on ice and under sterile conditions.
4. Collect the embryos and dissect the livers from the fetuses, using 20G1 gauge needles.
5. Put all fetal livers in 2 mL modified HBSS and add 0.5 mL modified HBSS/EGTA. Triturate gently 5 to 10 times, using a 5-mL serological pipette, to partially disaggregate fetal livers.
6. Centrifuge at 400×g for 5 min at 4°C.
7. Resuspend cell pellet in 4 mL modified HBSS/collagenase and transfer cell solution to a scintillation vial containing 10 glass beads. Place cell solution in a water bath and incubate for 15 min at 37°C under gentle agitation. Incubate no longer than 15 min.
8. Stop the reaction with an equal volume of modified HBSS, containing 20% FBS (final concentration 10% FBS), and filter cell suspension through a 40-μm cell strainer. Use heat-inactivated (56°C for 30 min) FBS.
9. Centrifuge at 400×g for 5 min at 4°C.
10. Wash cell pellet in 45 mL DMEM, containing 10% FBS, and centrifuge at 400×g for 5 min at 4°C. Repeat step once.
11. Resuspend cell pellet in 1 mL DMEM and count for viable cells by trypan blue exclusion. Keep cells on ice until cell transplantation.

13.5.2 Cell Transplantation

13.5.2.1 Procedures

1. Rats are anesthetized by inhalation of 1% to 5% isoflurane.
2. Shave abdomen, disinfect with 70% ethanol. Asepsis must be maintained by appropriate preparation of the surgical site and use of sterile supplies and instruments.
3. Place rat in a supine position and make a 4-to 5-cm midline incision through the skin and the peritoneal wall up to the xiphoid process (Figure 13.2(a)).
4. Lift the xiphoid process and cut the falciform ligament connecting the undersurface of the diaphragm to the liver (Figure 13.2(b)). Ensure not to puncture the thoracic cavity when severing the falciform ligament.
5. Push out the median and left lateral liver lobes, using the forefingers and thumbs (Figure 13.2(c)). After performing a ligation with 2-0 surgical suture at the base of the exposed lobes, lift and cut the liver lobes near the ligation, using curved scissors (Figure 13.2(d)). Use a folded gauze pad to hold and lift the liver lobes.
6. After finishing PH, gently displace the intestine, using cotton-tipped applicators soaked in sterile saline (Figure 13.2(e)). Keep the intestine constantly moist during the remaining procedure.

7. Externalize the portal vein and inject isolated fetal liver cells that have been resuspended in 0.75 mL DMEM, using a 30G 1/2 gauge needle connected to a 1-mL syringe (see Figure 13.2(f)).
8. Remove the needle carefully and stop bleeding immediately by pressing on the puncture site (covered with a small piece of Surgifoam™) with a cotton-tipped applicator soaked in saline.
9. Reinsert the intestine into the peritoneal cavity.
10. After adding a few drops of sterile saline into the peritoneal cavity, close the peritoneal wall, using absorbable suture and swab the wound with a cotton-tipped applicator. Suture the abdominal skin and clean the wound with saline.
11. Inject 5 mL saline subcutaneously, using a 23G1 gauge needle connected to a 5-mL syringe, into the dorsal surface of neck to support recovery of the rat.

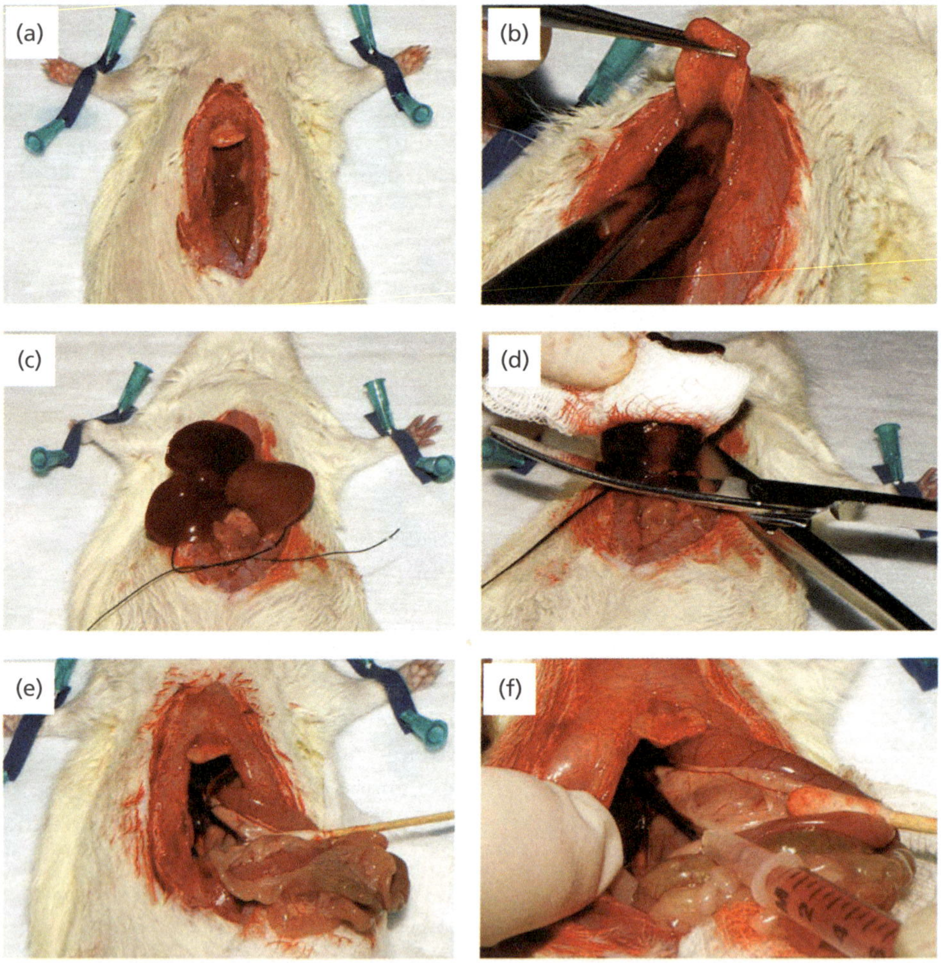

Figure 13.2 Essential microsurgical steps during partial hepatectomy and fetal liver cell transplantation. (a) F344 rat after midline incision through the skin and peritoneal wall. (b) Dissection of the falciform ligament. (c) Ligation of the median and left lateral lobes. (d) Resection of the exposed liver lobes. (e) Exposure of the portal vein. (f) Injection of fetal liver cells into the portal vein.

13.5.3 Detection of Transplanted Cells

13.5.3.1 Tissue Preservation/Fixation

After sacrificing the rats and removing the livers, cut the liver lobes into several pieces and freeze them immediately in 2-methylbutane or fix them in 10% formalin solution.

Snap-Freezing

- To avoid crystal formation during the freezing process, dab the liver tissue sample with a paper towel to remove excessive liquid.
- Add Tissue-Tek® OCT Compound on a piece of cork, press specimen flat into the OCT and drop the sample immediately into chilled 2-methylbutane (prechilled in a beaker on dry-ice).
- Let tissue freeze completely and store at –80°C until use.

Formalin Fixation

- Fix tissue samples in 10% neutral buffered formalin solution for 12 to 24 h (depending on the thickness of liver slices) at 4°C.
- Tissue processing through graded alcohol solutions, and xylene and paraffin embedding should be performed using standard methods.

13.5.3.2 Enzymehistochemistry for DPPIV in Frozen Tissue

Preparation of Reagents

TMS Buffer (0.1M Trisma® Maleate/0.1M NaCl; pH 6.5)

Prepare 200 ml of 1M Trisma® Maleate Stock Solution

- Dissolve 47.4g Trisma® maleate in 140 ml ddH_2O.
- Using a pH meter, adjust the pH to 6.5 with 10 N NaOH. While monitoring the pH, precipitates will form at ~ pH 4.3. Solution will clear again after reaching ~ pH 6.3.
- Bring to a final volume of 200 ml with additional ddH_2O.
- Stock solution can be stored at RT.

Prepare 500 ml of 0.1M Trisma® Maleate/0.1 MNaCl (TMS) Buffer Solution

- Add 10 ml of a 5M NaCl stock solution to 50 ml of 1M Trisma® maleate solution.
- Bring to a final volume of 500 ml with additional ddH_2O.

Substrate Solution

- Dissolve 100 mg Fast blue BB salt in 100 mL TMS buffer.
- Dissolve 50 mg GPMN in 3 mL DMF.
- Mix the two solutions immediately before use.

Scott's Water (10x Stock Solution)

- Dissolve 100 mg $MgSO_4$ x 7 H_2O in 250 ml ddH_2O.
- Dissolve 10g $NaHCO_3$ in 250 ml ddH_2O.
- Mix the two solutions and store at RT; dilute 1:10 in ddH_2O before use.

Procedure

1. Cut fresh sections of 5 μm from snap-frozen block of liver tissue and pick them up on *Colorfrost®/Plus* microscope slides. Cryostat sections can be stored for several months at –20°C before staining. The quality of DPPIV enzyme histochemistry will not be affected.
2. Air-dry cryostat sections for 10 to 20 min at RT.
3. Fix liver sections for 5 min in cold 95% ethanol/5% glacial acetic acid (99:1 vol/vol) (prechilled in a beaker on ice for 20 min). Acetone can also be used as a fixative. Fix the liver sections in 100% acetone on ice for 5 min and proceed with step 5.
4. Wash for 5 min in cold 95% ethanol.
5. Air-dry the slides for 20 to 30 min at RT.
6. Incubate liver sections in the substrate solution for 30 min at 37°C.
7. Wash the slides in TMS buffer for 2 min at RT. Repeat wash step two times. In all wash steps, 0.14 M NaCl solution can be used instead of TMS buffer.
8. To stabilize the red color, incubate the slides for 5 min in 0.1 M $CuSO_4$ solution.
9. Rinse the liver sections in TMS buffer with two changes over 2 min.
10. For long-term preservation of the stained cryosections, perform an additional fixation step in 4% PFA solution (made in PBS) for 10 min at 4°C.
11. Wash three times for 10 min in TMS buffer at RT and rinse briefly with water.
12. Counterstain liver sections with Harris hematoxylin for 1 to 3 min. Rinse in water for 5 min.
13. As a "blueing" step for a crisp blue-violet nuclear counterstaining, incubate the slides for 20 sec in Scott's water, followed by a wash step with water.
14. Mount the slides in 100% glycerol and cover with a glass cover slip. After evaluation under a microscope, remove the glass cover slip and rinse slides with water.
15. Finally, air-dry stained liver sections for 30 min. Slides can be permanently stored in the dark at RT.

13.5.3.3 Indirect Immunohistochemistry for DPPIV/CD26

Preparation of Reagents

Anti-Fading Mounting Medium

- Add 0.4 g propyl gallate to 10 ml glycerol/0.5M Na_2CO_3, pH 8.0, (9:1 vol/vol) in a glass vial.
- Place glass vial on a magnetic stirrer and mix overnight. Medium can be stored at 4°C.

Procedure (Frozen Tissue)

1. Cut fresh sections of 5 μm from snap-frozen block of liver tissue and pick them up on *Colorfrost®/Plus* microscope slides. Cryostat sections can be stored for several months at –20°C before staining.
2. Air-dry cryosections for 10 to 20 min at RT.
3. Encircle the liver sections on the glass sides with a PAP PEN and dry for 2 min at RT.

4. Fix liver sections for 10 min in cold 4% PFA solution (made in PBS). Acetone can also be used as a fixative. Fix liver sections in 100% acetone on ice for 5 min, air-dry for 20 to 30 min at RT, soak in PBS for 2 min and proceed with step 8.
5. Wash the slides with PBS for 10 min. Repeat wash step two times.
6. To reduce free aldehyde groups, incubate liver sections in freshly prepared 0.1% $NaBH_4$ (dissolved in PBS) for 30 min at RT.
7. Wash the slides three times with PBS, 10 min each step.
8. To block nonspecific binding of the fluorescently labeled secondary antibody, incubate liver sections in PBS (containing 2% donkey serum, 1% BSA and 0.05% Tween 20) for 1h at RT in a humid chamber. Depending on the source of secondary antibody, use heat-inactivated (56°C for 30 min) serum from the same species.
9. Dilute primary antibody at its optimal concentration in PBS (containing 1% donkey serum, 1% BSA and 0.05% Tween), apply antibody to the sections and incubate for 2 h at RT *or* overnight at 4°C in a humid chamber. Several primary antibodies against DPPIV/CD26 (mouse anti-DPPIV/CD26) are commercially available. In our hands, optimal dilutions were 1:50 or 1:100. Nevertheless, check the working dilutions suggested by the supplier.
10. Wash the slides three times with PBS, 10 min each step.
11. Dilute the Cy™3-conjugated donkey anti-mouse IgG (Jackson) in PBS (containing 1% donkey serum, 1% BSA and 0.05% Tween 20) at 1:200. Incubate tissue sections with the secondary antibody for 60 min at RT in a humid chamber. Different or the same secondary antisera conjugated with different fluorescent dyes (Cy™2, FITC, PE, TRITC, etc.) are commercially available. The optimal dilution should be determined by the user. DAPI, which is selective for DNA, can be used to visualize cell nuclei. Steps 11–14 should be performed in the dark.
12. Wash the slides three times with PBS for 10 min each step.
13. Mount tissue sections with anti-fading mounting medium and cover with a glass cover slip.
14. Examine liver sections using a fluorescence microscope with appropriate filter sets.

Procedure (Paraffin-Wax Embedded Tissue)

1. Cut sections of 5 μm from liver tissue embedded in paraffin.
2. To remove the paraffin, soak slides two times in xylene for 7 and 5 min at RT.
3. To hydrate tissue sections, place the slides in a staining jar with 100% ethanol for 2 min at RT. Transfer slides to 93% ethanol and then to 70% ethanol for 2 min each step.
4. Drain and immerse liver sections in PBS for about 5 min.
5. Perform heat-mediated antigen retrieval by placing the slides in a slide rack in a microwave-safe container filled with 10 mM citrate buffer (pH 6) and boil for 17 min in a microwave.
6. Let the slides cool to RT for 30 min.
7. Place tissue sections in PBS for 5 min.
8. Remove excess liquid around the section and encircle liver sections on the glass slides with a PAP PEN and dry the ring for 2 min at RT. During this time, do not allow tissue sections to desiccate. If necessary, add a drop of PBS on the liver section.
9. Proceed with immunohistochemical staining as described above (Steps 6–14).

13.6 Anticipated Results and Discussion

Currently, liver transplantation is the only therapeutic option for patients with end-stage chronic liver disease and for severe acute liver failure. Since the major problem is donor organ scarcity, attention has been focused on the possibility to restore liver function through cell transplantation.

Several techniques are available describing the isolation of candidate cells from different sources and their transplantation. However, in these experimental models, the recipient liver must be in a continuous state of massive injury to severely impair host hepatocyte proliferation and a significant selection pressure must be present to favor proliferation/survival of transplanted cells compared to host hepatocytes. The only experimental model currently available that demonstrates effective repopulation in a normal or near-normal liver environment is the FLSPC transplantation model developed in our laboratory.

13.6.1 Isolation of ED14 Fetal Liver Cells

Putative stem/progenitor cells have been considered as a potential source for future clinical cell transplantation. Fetal liver stem/progenitor cells exhibit major properties of stem cells: extensive proliferative activity, multipotency (differentiation into hepatocytes and bile duct cells), and functional long-term tissue reconstitution [17]. On ED14, fetal liver cells are at an uncommitted stage of liver development and, therefore, are suitable for cell transplantation studies. After ED16, fetal liver epithelial cells have already begun to diverge into the hepatocytic or cholangiocytic lineages, and the repopulation capacity of these cells is markedly reduced after transplantation into a normal liver environment [20]. However, they are still capable of fully repopulating the liver under highly selective conditions (e.g., pretreatment of the liver with retrorsine and performing PH at the time of cell transplantation (Sandhu et al., unpublished data).

Depending on the number of embryos dissected (up to 13 embryos per rat), the method described in this chapter allows the isolation of unfractionated ED14 fetal liver cells with high cell yield. It should be noted that unfractionated fetal liver cells contain only 2% to 5% FLSPC that are capable of repopulating the normal adult liver. The isolation of stem/progenitor cells from the adult liver is problematic, because there are no specific surface markers identified to date for these cells. However, enrichment of FLSPC from ED14 fetal liver up to 95% is also possible, but requires additional steps [26].

13.6.2 Transplantation of ED14 Fetal Liver Cells

The exoprotease DPPIV, first discovered by Hopsu-Havu and Glenner [27], is highly expressed in the liver and other organs of wt F344 rats [28]. Mutant DPPIV$^-$ Fischer rats are lacking this enzyme and, therefore, can be used as recipients for wt DPPIV$^+$ hepatic cells. The advantage of this model is that the transplanted cells are readily detectable through their expression of DPPIV in the host liver and no prior cell marking procedures are necessary. Furthermore, this model represents a syngeneic cell transplantation model and, therefore, immune rejection of transplanted cells does not occur [10].

Since DPPIV is stably expressed in the progeny of FLSPC, this model is also useful for long-term repopulation studies. It should be noted that DPPIV is not expressed in ED14 fetal liver cells; however, its expression begins later during fetal development (ED17 and beyond) [29]. Thus, engraftment studies using this model are not recommended because transplanted fetal liver cells are untraceable through DPPIV detection in the first 3 days after cell transplantation.

Importantly, since the size of fetal liver cells is much smaller compared to hepatocytes, this model allows transplantation of significantly higher numbers of fetal liver cells (up to 6×10^7 cells) than with hepatocytes, without blocking liver sinusoids causing tissue infarction [21]).

Finally, ED14 fetal liver cell transplantation has to be performed in conjunction with two-thirds PH, as a proliferative stimulus is necessary to initiate liver repopulation by FLSPC. In its absence, only rare, very small clusters of $DPPIV^+$ hepatocytes, bile ducts or hematopoietic cells are observed [21].

13.6.3 Detection of Transplanted $DPPIV^+$ Cells

The enzyme histochemical method for detection of DPPIV described above is a simple, fast, and efficient procedure to determine the repopulation capacity of wt liver cells after their transplantation into mutant $DPPIV^-$ F344 rats. Unfractionated fetal liver cells contain approximately 2% to 5% FLSPC that are capable to proliferate massively in the recipient liver and differentiate into hepatocytes and bile duct cells. These cells can be readily detected and distinguished from each other by enzyme histochemistry for DPPIV that is expressed in a membranous bile canalicular pattern unique to hepatocytes (Figure 13.3(a)) and in a characteristic diffuse cytoplasmic pattern in bile duct epithelial cells (Figure 13.3(b)) [20]. After transplantation of ~ 5×10^7 fetal liver cells, FLSPC form cell clusters that increase in number and size for up to 6 months (and beyond) and replace up to 24% of host liver tissue (Figure 13.3(d)) [20, 21]. Since ED14 fetal liver cells contain more than 80% hematopoietic cells, some hematopoietic cells will be observed in the recipient liver after cell transplantation. In addition, endothelial cells are present in isolated fetal liver cell preparations. These cells also express DPPIV in a diffuse pattern but are readily distinguished by their unique morphologic appearance (Figure 13.3(c) and [20]). Transplanted endothelial cells can also be identified by double label immunohistochemistry for DPPIV and an endothelial cell marker, RECA-1 in rats [20] or PECAM-1 in mice [30]. After fetal liver cell transplantation, $DPPIV^+$ hematopoietic cells can be detected in other organs (e.g., spleen, bone marrow, and intestine) [26].

To further characterize transplanted and/or host cells, it is necessary to visualize DPPIV and other markers simultaneously. Enzymehistochemistry for the detection of DPPIV can be combined with immunohistochemical methods using specific antibodies for the detection of surface or intracellular markers, as previously reported [26]. However, the use of such combined protocols is technically complicated and, therefore, not recommended for routine immunohistochemistry.

The staining pattern for hepatocytes and bile duct cells by DPPIV immunohistochemistry using formalin-fixed, paraffin-embedded tissue (Figure 13.3(e)) is the same as obtained by enzymehistochemistry with frozen tissue (Figure 13.3(a) and 13.3(b)). Even though the intensity of staining by immunohistochemistry with forma-

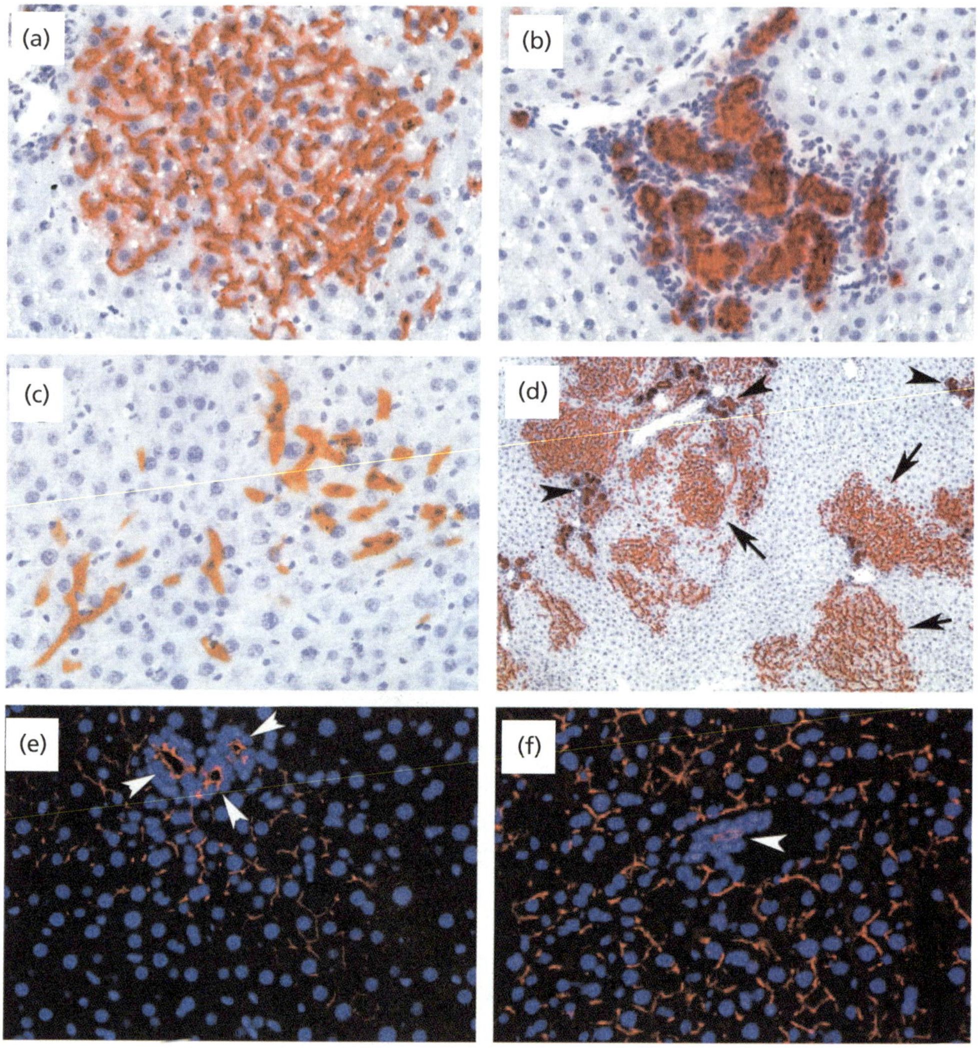

Figure 13.3 Detection of differentiated DPPIV$^+$ cells derived from transplanted ED14 fetal liver stem/progenitor cells in normal rats. (a–d) Six months after transplantation of unfractionated fetal liver cells into mutant DPPIV$^-$ F344 rats, DPPIV$^+$ hepatocytic (a), bile duct epithelial (b), and endothelial cells (c) were detected by enzymehistochemistry in frozen liver tissue. (d) Selected area of high repopulation showing mixed clusters containing hepatocytic (arrows) and bile duct structures (arrowheads). (e) Using immunohistochemistry for CD26, DPPIV$^+$ cells with hepatocytic and cholangiocytic (arrowheads) appearances were detected in paraffin-embedded liver tissue. (f) Immunohistochemical detection of CD26 (DPPIV) in snap-frozen liver tissue derived from a wt DPPIV$^+$ F344 rat ("positive" control).

lin-fixed, paraffin-embedded tissue is lower than that obtained with frozen tissue (Figure 13.3(f)), tissue morphology is better preserved in the former. Importantly, immunohistochemistry for DPPIV can be easily combined with immunohistochemistry for other markers to further characterize transplanted hepatic cells in the host parenchyma.

Troubleshooting Table

Problem	Explanation	Potential Solutions
Low cell yield after isolation.	Cell loss during digestion with EGTA and/or collagenase.	Be sure that the fetal livers are sufficiently "minced" after trituration in HBSS/EGTA solution. Only small tissue pieces should be left. Avoid adhesion of liver pieces to the wall of the serological pipette. Test the activity of collagenase (see below).
	Formation of cell aggregates during cell isolation procedure.	Keep cells always at 4°C (except during digestion with HBSS/collagenase).
Low viability of fetal liver cells.	Cells were triturated too vigorously or for too long of a time in HBSS/EGTA.	The golden rule is to triturate cells with HBSS/EGTA for as short a time as possible to achieve the desired effect.
	Low quality of collagenase.	Before purchasing commercially available collagenase, test a sample of a particular lot by performing an isolation of ED14 fetal liver cells. The viability of the cells should be higher than 95% at the end of the procedure.
Reduced survival rate of the recipients.	Bleeding after PH.	Ensure that the suture around the base of the exposed liver lobes is tightened sufficiently and the liver stump is not bleeding after PH.
	Bleeding after cell infusion.	Avoid post-injection bleeding by pressing on the puncture site in the portal vein for l to 3 min. Double-check the possibility of bleeding before suturing the abdominal incision.
	Liver infarct after cell infusion caused by cell aggregates.	Transplant cells immediately after their isolation. Check cell suspension under the microscope before cell infusion. If there are cell aggregates, filter the cell suspension through a 40-μm cell strainer.
Poor repopulation.	Cell preparation contained a low number of FLSPC.	Prepare cytospins of unfractionated fetal liver cells after isolation. Slides can be stored at –20°C and stained at a later time for AFP and CK-19 to determine the percentage of FLSPC [31].
	Bleeding after cell infusion, followed by cell leakage into the abdominal cavity.	Infuse cell suspension slowly to minimize injection pressure. Avoid postinjection bleeding (see above).
	Low engraftment efficiency of transplanted cells.	Cells were not prepared properly.
No or poor staining (EHC).	No repopulation by transplanted FLSPC.	Always include a "positive" control (liver section from DPPIV$^+$ F344 rat) to make sure that the staining worked.
	Incorrect concentrations of the compounds of the substrate solution.	Check the concentrations of ingredients in the TMS buffer and the substrate solution.
	Inadequate substrate incubation temperature and/or incubation time too short.	Incubation should be at 37°C for 30 to 45 min (but not longer than 1h).
	Losing staining intensity during mounting procedure.	Do not dehydrate, clear and mount liver sections in a permanent non-aqueous mounting medium.
No or poor staining (IHC).	No repopulation by transplanted FLSPC.	Always include a "positive" control to make sure that the staining worked.
	Concentration of primary and/or secondary antibody too low.	Titrate the antibodies to determine the optimal concentrations.
High background (IHC).	Inappropriate wash steps.	Increase washing time and be sure to change wash solution three times between steps.
	Incubation time with antibodies too long.	Reduce the incubation time (especially for primary antibodies).
	Insufficient blocking.	Increase blocking time (2h). Omit BSA in blocking solution.
	Concentration of primary and/or secondary antibody too high.	Titrate the antibodies to determine the optimal concentrations with minimal background. Always include "negative" controls (omit primary antibody) to determine background caused by secondary antibodies.

EHC = enzymehistochemistry
IHC = immunohistochemistry

13.7 Summary Points

1. We have developed a liver cell transplantation model, using ED14 fetal liver cells as donor cells. Using the described two-step digestion method, the obtained cell preparations are characterized by high cell quantity and viability, as well as reduced cell aggregation, that enables the infusion of high numbers of fetal liver cells into the liver through the portal venous system.
2. Fetal liver stem/progenitor cells contain epithelial liver stem/progenitor cells with massive proliferative capacity and differentiation potential in the liver. These cells can replace ~25% of the liver mass in the normal adult liver at 6 months after cell transplantation; the repopulated liver has a totally normal structure and the transplanted cells show normal physiologic and biochemical function.
3. FLSPC significantly repopulate the normal adult liver and produce results that are orders of magnitude better than those obtained using hematopoietic stem cells, mesenchymal stem cells or embryonic stem cells in a normal liver microenvironment. Therefore, this model represents a unique tool to further characterize and describe the properties and behavior of candidate stem cells for liver repopulation studies in vivo, as well as to develop new strategies to replace diseased liver tissue.
4. The distinctive properties of fetal liver stem/progenitor cells in repopulating the normal adult liver leads to potential application of these cells for therapeutic liver repopulation. However, before such studies are initiated, it will be important to understand at the molecular and cellular level precisely how these cells induce and maintain repopulation in the liver.

Acknowlegments

The authors thank their colleagues and postdoctoral fellows for their contributions to studies from our laboratory cited in this review.

References

[1] Bock, I., and H. Popper, "Über Lebertransplantation in die Vorderkammer des Auges," *Virchows Arch. [A]*, Vol. 299, 1937, pp. 219–34.

[2] Berry, M. N., and D. S. Friend, "High Yield Preparation of Rat Liver Parenchyma Cells: A Biochemical and Fine Structural Study," *J. Cell Biol.*, Vol. 43, No. 3, 1969, pp. 506–20.

[3] Matas, A. J., et al., "Hepatocellular Transplantation for Metabolic Deficiencies: Decrease of Plasma Bilirubin in Gunn Rats," *Science*, Vol. 192, No. 4242, 1976, pp. 892–94.

[4] Jirtle, R. L., C. Biles, and G. Michalopoulos, "Morphologic and Histochemical Analysis of Hepatocytes Transplanted into Syngeneic Hosts," *Am. J. Pathol.*, Vol. 101, No. 1,) 1980, pp. 115–26.

[5] Kusano, M., and M. Mito, "Observations on the Fine Structure of Long-Survived Isolated Hepatocytes Inoculated Into Rat Spleen," *Gastroenterology*, Vol. 82, No. 4, 1982, pp. 616–28.

[6] Demetriou, A. A., et al., "Survival, Organization and Function of Microcarrier-Attached Hepatocytes Transplanted in Rats," *Proc. Natl. Acad. Sci. USA*, Vol. 83, No. 19, 1986, pp. 7475–79.

[7] Ricordi, C., et al., "Trophic Factors From Pancreatic Islets In Combined Hepatocyte-Islet Allografts Enhance Hepatocellular Survival," *Surgery*, Vol. 105, No. 2, Pt 1, 1989, pp. 218–23.

[8] Jirtle, R. L. and G. Michalopoulos, "Effects of Partial Hepatectomy on Transplanted Hepatocytes," *Cancer Res.*, Vol. 42, No. 8, 1982, pp. 3000–04.

[9] Gupta, S., et al., "Transplanted Isolated Hepatocytes: Effect of Partial Hepatectomy on Proliferation of Long-Term Syngeneic Implants in Rat Spleen," *Pathology*, Vol. 19, No. 1, 1987, pp. 28–30.

[10] Thompson, N. L., et al., "A Fischer Rat Substrain Deficient in Dipeptidyl Peptidase IV Activity Makes Normal Steady-State RNA Levels and an Altered Protein. Use as a Liver-Cell Transplantation Model," *Biochem. J.*, Vol. 273, Pt 3, 1991, pp. 497–502.

[11] Rajvanshi, P. A., et al., "Studies of Liver Repopulation Using the Dipeptidyl Peptidase IV Deficient Rat and Other Rodent Recipients: Cell Size and Structure Relationships Regulate Capacity for Increased Transplanted Hepatocytes Mass in the Liver Lobule," *Hepatology*, Vol. 23, No. 3, 1996, pp. 482–96.

[12] Rhim, J., et al., "Replacement of Diseased Mouse Liver by Hepatic Cell Transplantation," *Science*, Vol. 263, No. 5150, 1994, pp. 1149–52.

[13] Overturf, K., et al., "Hepatocytes Corrected by Gene Therapy Are Selected In Vivo in a Murine Model of Hereditary Tyrosinaemia Type I," *Nat. Genet.*, Vol. 12, No. 3, 1996, pp. 266–73.

[14] Laconi, E., et al., "Long-Term, Near Total Liver Replacement by Transplantation of Isolated Hepatocytes," *Am. J. Pathol.*, Vol. 153, No. 1, 1998, pp. 319–29.

[15] Witek R. P., S.H. Fisher, and B.E. Petersen, "Monocrotaline, an Alternative to Retrorsine-Based Hepatocyte Transplantation in Rodents," *Cell Transplant.*, Vol. 14, No. 1, 2005, pp. 41–47.

[16] Guha, C., et al., "Amelioration of Radiation-Induced Liver Damage in Partially Hepatectomized Rats by Hepatocytes Transplantation," *Cancer Res.*, Vol. 59, No. 23, 1999, pp. 5871–74.

[17] Oertel, M., and D.A. Shafritz, "Stem Cells, Cell Transplantation and Liver Repopulation," *Biochim. Biophys. Acta.*, Vol. 1782, No. 2, 2008, pp. 61–74.

[18] Zhao, R., and , S.A. Duncan, "Embryonic Development of the Liver," *Hepatology*, Vol. 41, No. 5, 2005, pp. 956–67.

[19] Zaret, K., "Molecular Genetics of Early Liver Development," *Annu. Rev. Physiol.*, Vol. 58, 1996, pp. 231–51.

[20] Sandhu, J. S., et al., "Stem Cell Properties and Repopulation of the Rat Liver by Fetal Liver Epithelial Progenitor Cells," *Am. J. Pathol.*, Vol. 159, No. 4, 2001, pp. 1323–1334.

[21] Oertel, M., et al., "Cell Competition Leads to High Level of Normal Liver Reconstitution by Transplanted Fetal Liver Stem/Progenitor Cells," *Gastroenterology*, Vol. 130, No. 2, 2006, pp. 507–20.

[22] Sigal, S. H., et al., "Characterization and Enrichment of Fetal Hepatoblasts by Immunoadsorption ("Panning") and Fluorescence-Activated Cell Sorting," *Hepatology*, Vol. 19, No. 4, 1994, pp. 999–1006.

[23] Dabeva, M. D., et al., "Proliferation and Differentiation of Fetal Liver Epithelial Progenitor Cells after Transplantation into Adult Rat Liver," *Am. J. Pathol.*, Vol. 156, No. 6, 2000, pp. 2017–31.

[24] Higgins, G. M., and R.M. Anderson, "Experimental Pathology of the Liver. I. Restoration of the Liver of the White Rat Following Partial Surgical Removal," *Arch. Pathol.*, Vol. 12, 1931, pp. 186–202.

[25] Lojda, Z., "Studies on Dipeptidyl(Amino)Peptidase IV (Glycyl-Proline Naphthylamidase). II. Blood Vessels," *Histochemistry*, Vol. 59, No. 3, 1979, pp. 153–66.

[26] Oertel, M., et al., "Purification of Fetal Liver Stem/Progenitor Cells Containing All the Repopulating Potential for the Normal Adult Rat Liver," *Gastroenterology*, Vol. 134, No. 3, 2008, pp. 823–32.

[27] Hopsu-Havu, V. K., and G.G. Glenner, "A New Dipeptide Naphthylamidase Hydrolyzing Glycyl-Prolyl-Beta-Naphthylamide," *Histochemie*, Vol. 7, No. 3, 1966, pp. 197–201.

[28] Hartel, S., et al., "Dipeptidyl Peptidase (DPP) IV in Rat Organs. Comparison of Immunohistochemistry and Activity Histochemistry," *Histochemistry*, Vol. 89, No. 2, 1988, pp. 151–61.

[29] Feracci, H., et al., "The Establishment of Hepatocyte Cell Surface Polarity During Fetal Liver Development," *Dev. Biol.*, Vol. 123, No. 1, 1987, pp. 73–84.

[30] Nierhoff, D., et al., "Purification and Characterization of Mouse Fetal Liver Epithelial Cells with High In Vivo Repopulation Capacity," *Hepatology*, Vol. 42, No. 1, 2005, pp. 130–39.

[31] Oertel, M., et al., "Properties of Cryopreserved Fetal Liver Stem/Progenitor Cells That Exhibit Long-Term Repopulation of the Normal Rat Liver," *Stem Cells*, Vol. 24, No. 10, 2006, pp. 2244–51.

14

Myoblast Transplantation: Techniques in Nonhuman Primates as a Bridge to Clinical Trials

Daniel Skuk*, Marlyne Goulet, Martin Paradis, and Jacques P. Tremblay

Neurosciences Division—Human Genetics, CHUQ Research Center—CHUL, Quebec, Canada
Author for correspondence:
*Correspondence: Neurosciences Division—Human Genetics, CHUQ Research Center—CHUL
2705 boulevard Laurier, Québec, Canada G1V 4G2; Phone: (418) 654-2186; Fax: (418) 654-2207;
E-mail: Daniel.Skuk@fmed.ulaval.ca

Abstract

Myoblast transplantation is an experimental strategy for the potential treatment of some myopathies. To be efficient, it needs to obtain a homogeneous incorporation of the transplanted myoblasts into myofibers throughout the cell-grafted skeletal muscles. So far, the only method acceptable for clinical applications enabling this result is to perform very close cell injections throughout the cell-grafted skeletal muscle. This must be complemented with appropriate immunosuppression to ensure the survival of the graft in allogeneic conditions. Tacrolimus was the only immunosupressant that proved so far to be efficient in this context. The experiments conducted in nonhuman primates enabled us to design better clinical protocols than those used in the initial stages of this approach. For this reason, this chapter describes the protocol of myoblast transplantation routinely done in our laboratory for experiments in nonhuman primates and discusses the adjustments needed for clinical application.

Key terms clinical trial, genetic complementation, intramuscular cell injection, myoblast transplantation, myopathies, nonhuman primate experiments, preclinical model, tacrolimus immunosuppression

14.1 Introduction

Myoblast transplantation is a type of cell transplantation in which the transplanted cells are the immediate precursors of the skeletal muscle parenchyma (i.e., myoblasts) [1, 2]. It is a potential therapeutic strategy that is being experimentally tested as a treatment for diseases of the skeletal muscle and the heart. This chapter resumes the method used so far by our laboratory for myoblast transplantation in skeletal muscles from a clinical perspective. Given our large experience with nonhuman primates as a bridge for human extrapolations, we describe here the protocol used in this animal model, discussing in the final part the adjustments necessary for human applications.

The close phylogenetic relationship between human and nonhuman primates provides the best anatomical, physiological, antigenic, and immunological similarities for preclinical transplantation research [3–5]. In contrast to mice, macaque monkeys share with humans important transplantation immune parameters, such as a well-conserved homology in the genes encoding the major histocompatibility complex molecules and the T-cell receptor [6–9]. Rejection of organ allografts in nonhuman primates is driven by the same immune elements as humans, showing comparable histological features and cadence [5]. In myoblast transplantation, cell isolation, cell culture, cell scale-up, and final cell population for transplantation are similar between monkeys and humans, which is not the case for other animal models such as mice and dogs. In addition, due to the large size of monkey and human muscles, the problems posed with intramuscular cell injections are essentially the same, which is not the case in the tiny muscles of mice. For all these reasons, our experiments in nonhuman primates [10–15] enabled us to plan clinical strategies of myoblast transplantation [16–18] more efficiently than those used in former clinical trials [19–26].

In order to produce a clinical benefit, the two main duties of transplanted myoblasts in the skeletal muscle are: (1) to fuse with the recipient myofibers to produce "genetic complementation" (this would be beneficial in the case of recessive genetic myopathies), and (2) to form new functional skeletal muscle (this would be beneficial in the case of muscle weakness due to loss of contractile parenchyma). The first property is the best documented in different preclinical [10–15, 27–30] and clinical conditions [16–18, 24], while the second property was reported so far only in some experimental conditions in mice [31–34]. For this reason, the focus of the chapter is on myoblast transplantation as a tool for genetic complementation.

Genetic complementation implies that, following fusion of the transplanted myoblasts with the recipient myofibers, the syncytium of these myofibers will express proteins coded by the exogenous and the host nuclei [35]. These myofibers, referred to as hybrid, can thus express a protein that is not expressed by the host genome [36]. Clinical examples of genetic complementation by myoblast transplantation were clearly observed in the last clinical trials performed by our laboratory in patients with Duchenne muscular dystrophy [16–18]. Duchenne muscular dystrophy is a severe degenerative genetic myopathy produced by a severe deficiency of dystrophin, a subsarcolemmal protein. The allotransplantation of myoblasts obtained from normal (i.e., nonmyopathic) donors in patients with Duchenne muscular dystrophy on tacrolimus immunosuppression, leads to 3% to 34.5% of the recipient myofibers expressing the donor-derived dystrophin in cross-sections of muscle biopsies performed in the cell-grafted sites [16–18].

14.2 Experimental Design

The reasons that conditioned the protocol for myoblast transplantation in nonhuman primates as described above are essentially the following:

1. Our techniques of transplantation seek to implant exogenous cultured myoblasts as efficiently as possible into skeletal muscles of nonhuman primates, humans, and potentially other large animals. Intramuscular injection is so far the only appropriate method to deliver exogenous myoblasts to skeletal muscles, but its main limitation is that the injected myoblasts fuse mostly with the myofibers around the injection trajectories [14, 15]. Since the intracellular proteins coded by a nucleus in a myofiber remain localized to a region known as nuclear domain [37], gene complementation following the fusion of transplanted myoblasts in a single site of a myofiber does not ensure the expression of donor-derived proteins throughout the myofiber but, instead, to a short distance around the donor-derived nuclei. Therefore, a homogeneous incorporation of the implanted myoblasts throughout the target skeletal muscle is needed. The only approach that we found so far to reach this objective was to perform cell injections very close to each other throughout the target region, delivering the cell suspension homogeneously during the needle withdrawal along the target muscle depth [13–18].
2. Since there are no colonies of myopathic monkeys, we must work with nonmyopathic animals. In this case, the success of the graft in terms of degree of integration of the transplanted myoblasts into the recipient's myofibers cannot be evaluated by detecting a protein normally present in the donor and absent in the diseased recipient. The use of reporter proteins is thus mandatory, and in our case we routinely introduce into the donor-derived myoblasts the LacZ gene to detect the hybrid myofibers by the expression of β-galactosidase.
3. In order to avoid an open surgery, both to make the transplantation less invasive as possible and to accelerate the procedure, we perform percutaneous cell injections in clinical trials and in most preclinical experiments in monkeys. A percutaneous procedure also facilitates the placement of a transparent adhesive dressing with a grid on the region to be grafted, because adhesive dressings adhere easily to the dry skin but not to the muscle. These dressings are very helpful in guiding the sequence of injections and can be preserved as a testimony of the injection procedure.
4. When small muscle regions are grafted, they need to be labeled to identify them later if a biopsy is required for analysis. In monkeys and other large animals, we place two points of non-absorbable neutral suture, intramuscularly, at each side of the grafted region. Obviously, labels are not mandatory when a whole muscle is grafted.

14.3 Materials

14.3.1 Facilities/Equipment

- Cell culture facilities.
- Flow cytometry facilities.
- Animal room with nonhuman primate facilities.
- Surgical room for large animals with equipment for general anesthesia.

14.3.2 Reagents

- MCDB 120 or MB-1 culture media (Hyclone, custom made according to the recipe of Ham et al. 1988 [38]).
- DMEM culture medium (Gibco).
- Fetal bovine serum (FBS) (Wisent or Hyclone).
- Basic fibroblast growth factor (Feldan).
- Bovine serum albumin (Sigma).
- Dexamethasone (Sigma).
- Insulin (Sigma).
- Antibiotic-antimycotic solution 100X (penicillin G/Amphotericin B/Streptomycin) (Wisent).
- Collagenase 1A (Sigma).
- Trypsin-EDTA (Gibco).
- Hanks' balanced salt solution (HBSS) (Gibco).
- Polybrene (Sigma).
- Geneticin G418 (Gibco).
- Phycoeritrin-labeled CD56 antibody (Beckman Coulter).
- Dimethyl sulfoxide (DMSO) (Sigma).
- Trypan blue solution 0.4% (Sigma).
- Tacrolimus powder for intramuscular injection (Astellas Pharma Inc.).
- Isoflurane (Abbott).
- 0.05% chlorhexidine gluconate aqueous antiseptic solution with 4% isopropylic alcohol.
- 70% isopropyl alcohol.
- 1% povidone-iodine.
- 0.5% chlorhexidine gluconate in 70% isopropyl alcohol.

14.3.3 Cell Culture Material

- Petri culture dish, 100×20 mm (Sarstedt).
- Dissection forceps.
- Fine scissors with curved tips.
- Sterile disposable pipettes (10 ml, 5 ml, and 2 ml).
- Pipette aid filler/dispenser.
- Fine pipetters (2–20 μl, 10–100 μl, and 100–1,000 μl).
- Pipetter tips (1–200 μl and 100–1,000 μl).
- 50-ml centrifuge tubes (Sarstedt).
- 15-ml centrifuge tubes (Sarstedt).
- 25-cm^2 cell culture flasks (Sarstedt).
- 75-cm^2 cell culture flasks (Sarstedt).
- Spinner flask.
- 6-well culture plates (Corning-Costar).

- 2-ml cryovials (Simport).
- 5-ml polystyrene round-bottom tube (Falcon).
- Microtubes (Sarstedt).
- 10-ml injection vials (Omega).

14.3.4 Surgical Material

- Surgical attire: mask, cap, cover-shoes, sterile jacket, and sterile surgical gloves. Due to the risks of zoonosis and especially of simian herpes B virus contamination, use two pairs of gloves as a protection for any accidental injury with the injection or suture needles, and eye protection (face shield or glasses).
- Sterile fenestrated surgical drapes.
- OpSite sterile transparent dressings (Smith & Nephew).
- OpSite sterile transparent dressings with a 5-mm grid (Smith & Nephew).
- Hamilton syringes of 50, 100, or 250 μl. The 50- and 100-μl syringes are preferable for manual injections (50-μl syringes ensure a more precise control delivering small volumes of cell suspension). The 250-μl syringes are the preferable size for repeating dispensers, with which 5 μl of cell suspension can be delivered at each time.
- PB600-1 repeating dispenser (Hamilton).
- Injection needles of 25 to 27 gauge (the length will depend on the depth to be injected).
- Scalpel handle #3.
- Scalpel blade #15.
- Surgical scissors with straight blunt.
- Fine scissors with curved handle and blunt.
- Skin retractor of 2.5 cm spread.
- Colibri retractor of 1.5 spread (for fascia).
- Micro-Mosquito hemostat forceps.
- Adson forceps with finely serrated tips.
- Adson forceps with teeth.
- Fine needle holder.
- Sterile gauzes.
- Vicryl suture 5.0 or 4.0 (Ethicon Inc.).
- Nonabsorbable polypropylene suture (Prolene or similar), 5.0 or 4.0.
- Skin stapler.

14.3.5 Cell Line

A cell line producing a replication-defective retroviral vector encoding a LacZ reporter gene and a neomycin-resistance gene. (The authors use a PA317 packaging cell line producing a retroviral vector LNPOZC7, gift from Dr. Constance Cepko, Harvard University, Boston, MA. See [39] for protocols of Dr. Cepko regarding preparation of a specific retrovirus producer cell line).

14.4 Methods

14.4.1 Cell Culture

14.4.1.1 Production of Supernatant with Retroviruses Encoding LacZ

1. Prepare culture medium: DMEM supplemented with 10% FBS.
2. Thaw one cryovial with ~ 3×10^6 cells from the cell line producing the retroviral vector encoding the LacZ reporter gene.
3. Replace the freezing medium by culture medium, and transfer the cells to one 75-cm^2 flask. Place the cells in a humidified incubator at 37°C and 5% CO_2.
4. After 1 week, detach the cells from the flasks using 0.125% trypsin-EDTA, neutralize trypsin with culture medium and centrifuge 5 min at 400 g to remove medium. Resuspend the final cell pellet in culture medium and seed the cell suspension into 10 culture flasks of 75 cm^2 at 2×10^5 cells per flask. Place the flasks in a humidified incubator at 37°C and 5% CO_2.
5. After 3 days (e.g., from Friday to Monday) change the culture medium placing 6 ml in each flask and incubate at 37°C and 5% CO_2.
6. During the following 4 days (e.g., from Tuesday to Friday), collect the supernatant daily (i.e., the culture medium containing the retroviruses) in 15-ml tubes. During the first 3 days of supernatant collection, replace it each time by 6 ml per flask of fresh culture medium, incubating the cultures at 37°C and 5% CO_2. Store the tubes with the supernatant in a freezer at –80°C.
7. The fourth day of supernatant collection (cells should be confluent) select one culture flask and repeat steps 4 to 6. Discard the other 9 flasks. Depending on the amount of supernatant desired, repeat steps 4 to 6 for another week or stop here and discard all the culture flasks.

14.4.1.2 Myoblast Culture

1. Prepare culture medium: MCDB 120 or MB-1 supplemented with 15% FBS, 10 ng/ml of basic fibroblast growth factor, 0.5 mg/ml of bovine serum albumin, 0.39 μg/ml of dexamethasone, and 5 μg/ml (0.14 U/ml) of insulin.
2. Perform a biopsy under aseptic conditions in a skeletal muscle of a macaque monkey as detailed in Section 14.1.3.
3. Immediately after excision, place the biopsy in a sterile 50-ml tube with sterile culture medium with 1% of the antibiotic-antimycotic solution. Place the tube on ice and bring the biopsy to the culture room.
4. Inside the laminar flow hood, mince the muscle biopsy in a Petri dish containing culture medium using dissection forceps and fine scissors with curved tips. Mince into fragments of less than 1 mm^3 and rinse three times by settling with HBSS with 1% of the antibiotic–antimycotic solution.
5. Place the minced muscle into a 25-cm^2 culture flask containing the same culture medium without antibiotics, in order to ensure that the tissue has no microbial contamination before resuming the cell culture.
6. Leave the minced muscle in an incubator at 37°C for 24 to 48h.

7. If there is no microbial contamination, digest the minced muscle on a spinner flask 1h at 37°C using 10 to 15 ml collagenase 0.2% in HBSS. The volume of collagenase solution will depend on the size of the biopsy.
8. Collect the suspension of the digested muscle and centrifuge 1 minute at 75g to precipitate the myofiber fragments.
9. Harvest the supernatant (which contains cells) and centrifuge 5 min at 400g. Resuspend the cell pellet in 5 ml of culture medium.
10. Resuspend the myofiber fragments remaining from the centrifugation in step 8 with 10 to 15 ml of 0.125% trypsin-EDTA in HBSS. Digest for 45 min in a spinner flask at 37°C.
11. Harvest the digested suspension and centrifuge 5 min at 400g.
12. Resuspend the pellet with the previous 5 ml of medium containing the cells collected in step 9.
13. Transfer the cell suspension to five 25-cm^2 culture flasks and place them in a humidified incubator at 37°C and 5% CO_2.
14. After four days in culture, harvest the culture medium from the flasks with the floating myofiber fragments and transfer it to two 15-ml centrifuge tubes. Rinse the culture flasks twice with HBSS and add 5 ml of culture medium to each flask. Return the culture flasks to the incubator.
15. Centrifuge the medium collected in step 14 at 400g for 5 min. Remove the supernatant and resuspend the myofiber fragments in 6 ml of culture medium. Place the culture medium with the myofiber fragments in a 75-cm^2 flask, to allow growth of any satellite cell still remaining in the myofiber fragments, and place the flask in a humidified incubator at 37°C and 5% CO_2.
16. Four days after step 15, rinse the flask with HBSS to remove the myofiber debris, and add 12 ml of culture medium. Continue incubation at 37°C and 5% CO_2.
17. Change the culture medium every 2 to 3 days until cell confluence, both in cell cultures obtained at steps 14 and 16.
18. At cell confluence, detach the cells from the flasks using 0.125% trypsin-EDTA, neutralize trypsin with culture medium.
19. Use a sample of these cells to perform a flow cytometry analysis after incubation with an anti-CD56 antibody, to verify the percentage of myoblasts in the cell culture.
20. Seed as much as needed of the rest of the cells on 6-well culture plates (150 to 200×10^5 cells per well) with culture medium (3 ml per well) and incubate them for 24h at 37°C and 5% CO_2.
21. Thaw as many tubes of supernatant containing the retroviral vector encoding the LacZ gene as needed to place 3 ml per well in the culture plates. Replace the culture medium with this supernatant, adding 5% more FBS and 8 μl/ml from a solution of 1 mg/ml polybrene. Incubate for 15 min at 37°C and 5% CO_2.
22. Centrifuge the plates 30 min at 400g.
23. Replace the culture medium with the retroviral vector for culture medium and incubate for 24 h at 37°C and 5% CO_2.
24. Repeat steps 21 to 23 and incubate for 48h at 37°C and 5% CO_2.
25. Select the transduced cells with 4 ml per well of culture medium containing 600 μg/ml Geneticin within 4 days.

26. Remove the culture medium with Geneticin, rinse the cultures three times with HBSS to remove the dead cells, and add culture medium. Incubate at 37°C and 5% CO_2.
27. At confluence, detach the cells from the 6-well culture plates using 0.125% trypsin-EDTA and neutralize trypsin with culture medium.
28. Count a sample of the cells to know the total amount of cells and resuspend them in enough culture medium with 10% DMSO to allow for obtaining aliquots of 1 ml containing 5×10^6 cells each.
29. Add aliquots of 1 ml of the cell suspension to 2-ml cryovials. Place the cryovials at –80°C in a box of plastic foam, allow to cool slowly, and store in liquid nitrogen after 24 h.

14.4.1.3 Skeletal Muscle Biopsy for Myoblast Culture

1. With the monkey under general anesthesia in the surgical room, shave the region over the muscle to be biopsied. Our preferred muscles for these biopsies are the gastrocnemius, the quadratus femoris or the biceps brachium. If the animal will be used later as a recipient of myoblast transplantation, choose a muscle that you do not plan to use for grafts.
2. Clean the region with 0.05% chlorhexidine gluconate aqueous antiseptic solution with 4% isopropylic alcohol.
3. Perform surgical antisepsis of the operative region: three passages of 70% isopropyl alcohol and povidone-iodine or three passages of 0.5% chlorhexidine gluconate in 70% isopropyl alcohol.
4. Place a sterile fenestrated surgical drape over the operative region. Adhere an OpSite sterile transparent dressing over the region to be biopsied, covering the opening of the surgical drape.
5. Using the scalpel, make a longitudinal linear incision of 4 to 5 cm in the skin. Dissect the plane between the skin and the muscle using a Mosquito hemostat forceps and then place a skin retractor (2.5 cm spread) to maintain the muscle surface exposed. The quadratus femoris (this is not the case in the biceps brachium and the gastrocnemius) has a thick fascia that needs also a longitudinal linear incision with the scalpel, dissection from the underlying muscle, and eventually placement of a Colibri retractor (1.5 spread) to expose the muscle.
6. Perform the excision of a muscle fragment using scalpel and fine curved scissors. One biopsy of roughly 1 gram is usually enough. If the animal will be maintained alive, perform a thin longitudinal excision of 2 to 3 cm following the muscle fascicles to disturb as little as possible the muscle anatomy. Perform adequate hemostasis and, if needed to avoid a cavity in the tissue, approximate the excision edges with a continuous suture with Vicryl. In the quadratus femoris, close the fascia incision also with a continuous suture with Vicryl.
7. Close the skin incision with surgical staples and clean the skin in the operative region.

14.4.2 Cell Transplantation

14.4.2.1 Pretransplantation Procedures

1. Thaw cells 10 days before the date of transplantation. Generally, one cryovial is enough to obtain more than 200×10^6 cells in this interlude.
2. Replace the freezing medium by culture medium, and transfer the cells to five 75-cm^2 flasks. Place them in a humidified incubator at 37°C and 5% CO_2.
3. At cell confluence (in around 5 days) detach the cells from the flasks using 0.125% trypsin-EDTA, neutralize trypsin with culture medium and centrifuge 5 min at 400 g to remove medium. Resuspend the final cell pellet in culture medium and transfer the cell suspension to as many 75-cm^2 culture flasks as required to obtain the amount of cells required for the experiment. Place them in a humidified incubator at 37°C and 5% CO_2.
4. Begin tacrolimus immunosuppression of the monkey 5 to 7 days before transplantation. Whenever possible, use an intramuscular formulation, administering tacrolimus oncc a day and beginning at 0.5 mg/kg/day.
5. Adjust tacrolimus doses to target blood levels between 50 and 70 ng/ml during the first month post-transplantation. Our present evidence indicates that later the target blood levels can be reduced between 40 and 20 ng/ml.

14.4.2.2 Preparation of Cells for Transplantation

1. Allow at least 2h between the beginning of cell preparation and the transplantation.
2. Detach the cells from the flasks using 0.125% trypsin-EDTA, neutralize trypsin with culture medium and rinse the cells three times in HBSS by centrifugation.
3. Verify the cell viability with a trypan blue exclusion method on a sample of cells.
4. Resuspend the cells in HBSS, in accordance with the desired final concentration and volume of injection. This will depend on experiment variables such as the volume of muscle to be grafted or the interinjection distance to be used. For an optimal transplantation, 10^5 cells per cm of injection trajectory must be calculated. Place the cell suspension either in sterile microtubes or in sterile injection vials.
5. Place the vials with the cell suspension on ice, enveloped in gauze, together with two 5-ml polystyrene round-bottom tubes with sterile HBSS, and carry them to the surgical room.

14.4.2.3 Procedures in the Surgical Room

1. With the monkey under general anesthesia, shave the region to be transplanted.
2. Clean the region with 0.05% chlorhexidine gluconate aqueous antiseptic solution with 4% isopropylic alcohol.
3. Perform surgical antisepsis of the operative region: three passages of 70% isopropyl alcohol and povidone-iodine or three passages of 0.5% chlorhexidine gluconate in 70% isopropyl alcohol.
4. Place a sterile fenestrated surgical drape over the region to be grafted with cells.
5. Adhere an OpSite sterile transparent dressing with a 5-mm grid over the region to be grafted, covering the opening of the surgical drape (Figure 14.1(a)). If necessary,

circumscribe further the operative region on the fenestrated drape with sterile gauzes before placing the Opsite.

6. For the cell injections, use either 50- or 100-μl Hamilton syringes operated manually or a PB600-1 repeating dispenser attached to a 250-μl Hamilton syringe (Figure 14.1(b) and (c)). With the PB600-1 repeating dispenser the procedure may be twice as fast as using a manually operated syringe and is less tiring during long transplantations. Sterilize at least two syringes and at least two repeating dispensers for a transplantation session in order to replace one of them if it does not slide adequately during the cell injection procedure.
7. Before filling the cell-injection syringe with the cell suspension, aspirate sterile HBSS with it from one of the 5-ml tubes and, with the tip of the syringe in the HBSS, make

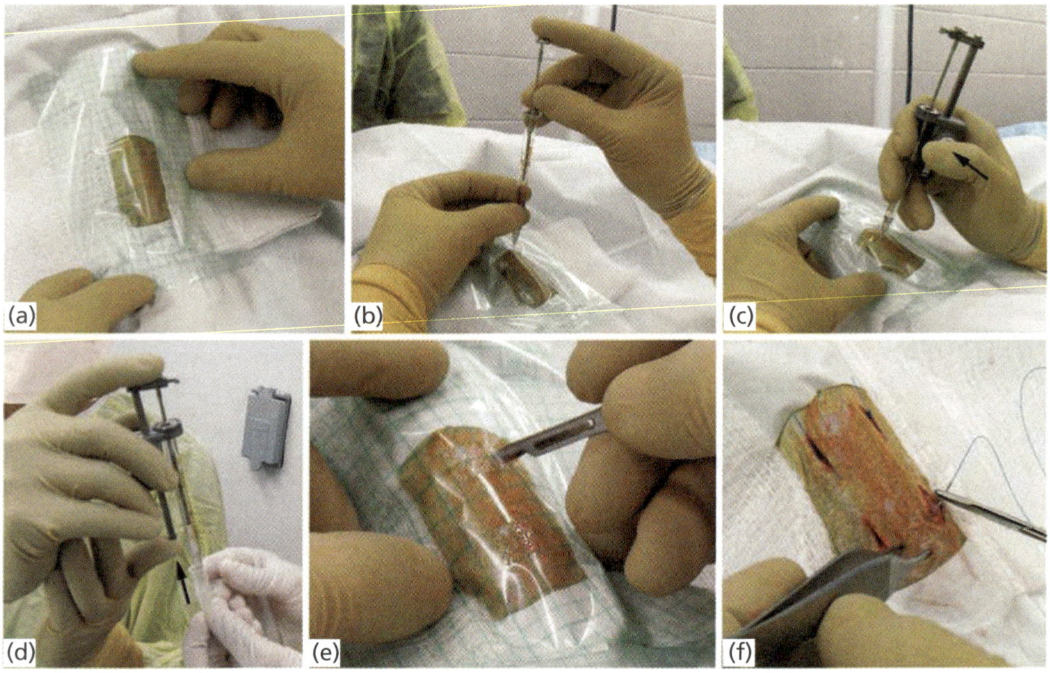

Figure 14.1 Technique of myoblast transplantation in the biceps brachii of a macaque monkey. (a) After performing antisepsis of the operative region and placing a sterile fenestrated surgical drape over it, an OpSite sterile transparent dressing with a 5-mm-grid is adhered. (b) and (c) illustrate the way to perform the cell injections, either using a 50- to 100-μl Hamilton syringe operated manually (b), or a PB600-1 repeating dispenser attached to a 250-μl Hamilton syringe (c). (b) The use of only a Hamilton syringe requires both hands to manipulate it. Needle penetration and withdrawal are done by moving the syringe with both hands while the index finger of the hand on the top pushes the piston homogeneously during the needle withdrawal and delivering few milliliters of cell suspension. (c) The PB600-1 repeating dispenser facilitates the procedure because it can be operated with only one hand, liberating the other hand for tasks such as holding the muscle or cleaning blood with a gauze; the small amount of cell suspension delivered in each single injection is constant and the surgeon needs only to concentrate on pushing the button (arrow) during the needle withdrawal and not in coordinating both hands to control the accurate displacement of the piston in parallel to the syringe withdrawal; and since a larger syringe can be used, pauses for refilling are less frequent. (d) How to fill the syringe attached to the PB600-1 repeating dispenser (in the picture, the cell suspension is contained in a microtube). Pushing the bottom of the jagged rod of the dispenser (arrow) with one finger is easier than pulling the piston. (e) In the figure, two regions of 1 cm^2 were grafted and they can be identified by the punctures left in the plastic sheet of the Opsite dressing. To label these regions for future selective biopsy, a small incision is done with the scalpel at each side of these regions before removing the plastic sheet of the dressing. (f) Thereafter, the plastic sheet of the dressing is removed and single points of a non-absorbable polypropylene suture are placed in the muscle.

some up-and-down movements with the piston to ensure that it slides well. Then, fill the syringe with enough HBSS to fill the injection needle after placement of the latter in the syringe. Afterwards, insert the injection needle in the syringe tip, point the needle up, and evacuate the HBSS of the syringe insuring that no air remains into the syringe and needle.

8. Fill the syringe with the cell suspension, avoiding the formation of air bubbles in the syringe. The first time, make a few up-and-down movements of the piston to dilute the HBSS remaining in the syringe into the cell suspension (Figure 14.1(d)).
9. For the cell injection, perform parallel penetrations of the needle, roughly perpendicular to the grafted surface, and deliver 1 to 10 μl of cell suspension per cm of injection trajectory as homogeneously as possible during the needle withdrawal. Distribute the cell injections following a matrix covering homogeneously the selected region and maintaining a constant inter-injection distance. The punctures left by the injection needle in the Opsite (Figure 14.1(e)) allow controlling the pattern of injections and the interinjection distance (this is very difficult to do in the naked skin, because there is no visible trace remaining in most injection sites).
10. After ending cell injections in a whole muscle, remove the plastic sheet of the Opsite, clean it in a disinfectant solution and preserve it in your files, writing on it the identification of the animal, the date, the cell-grafted muscle and an indication of the proximal and distal position during the cell injection procedure.
11. If the cell injections were done in small regions to be selectively sampled later, perform two small incisions (<1 cm), one at each side of the myoblast-injected region and at ~ 5 mm from it (Figure 14.1(e)). Remove the plastic sheet of the Opsite and preserve it for your files as in the previous step. Without removing the adherent sheet of the Opsite, place two single points of a nonabsorbable polypropylene suture (Prolene or similar) in the muscle through the two small incisions in the skin (Figure 14.1(f)). Leave a redundant suture knot (e.g., three loops of double throws) with long tails (5–8 mm) to facilitate finding them at the moment of the biopsies. Place the suture tails towards the exterior of the biopsy region by creating small subcutaneous pockets with a Mosquito hemostat forceps. Close each skin incision with one or two surgical staples.

14.5 Data Acquisition, Anticipated Results, and Interpretation

Histological analyses of the grafted regions following muscle biopsies in those sites are a crucial source of data, although functional analyses depicting any potential clinical improvement in the case of testing myoblast transplantation on myopathic animals or patients will be the ultimate and relevant data from a clinical point of view. In the case of nonhuman primates, the lack of colonies of myopathic monkeys leads only to the possibility of the histological analysis (i.e., to evaluate the success of the graft in terms of degree of integration of the transplanted myoblasts into the recipient's myofibers). This can be extrapolated to recessive genetic myopathies in which the disease is caused by the severe deficiency of a protein in the myofibers (i.e., dystrophin in Duchenne muscular dystrophy), because the capacity to prevent the progressive muscle degeneration characteristic of these diseases will depend on the ability to make the absent protein being expressed in as much muscle volume as possible. However, differences in the nuclear domain or in the methods of detection between reporter proteins and therapeutic proteins must be considered for the analysis from a clinical perspective.

To evaluate the success of the graft in terms of degree of integration of the transplanted myoblasts into the recipient's myofibers, the total number of myofiber profiles and the number of hybrid myofiber profiles must be counted in cross-sections of the biopsies performed in the myoblast-grafted sites at least 4 weeks after transplantation. The use of reporter proteins such as β-galactosidase is obviously possible only in animal experiments and, again due to the lack of colonies of monkeys with genetic myopathies, it is the only possibility in this animal model (Figure 14.2). In clinical conditions, the percentage of myofiber profiles expressing the therapeutic protein (such as dystrophin in Duchenne muscular dystrophy) must be calculated by performing the immunohistochemical detection of the protein in cross sections of the muscle biopsies. The second histological parameter that is important to analyze is whether there are signs of specific immune rejection of the graft (i.e., infiltration by lymphocytes). Hematoxylin-eosin staining clearly depicts the presence of lymphocyte accumulations around the hybrid myofibers, characteristic of acute rejection in this context, but immunohistochemical techniques to define their phenotype as CD8+ and CD4+ T-lymphocytes is an essential complement. Finally, a third histological parameter to consider in clinical conditions is any potential improvement of the skeletal muscle structure following the transplantation.

With the method described above for cell isolation and scale-up, the final cell population for transplantation in monkeys is usually >90% CD56+. The cell death at the

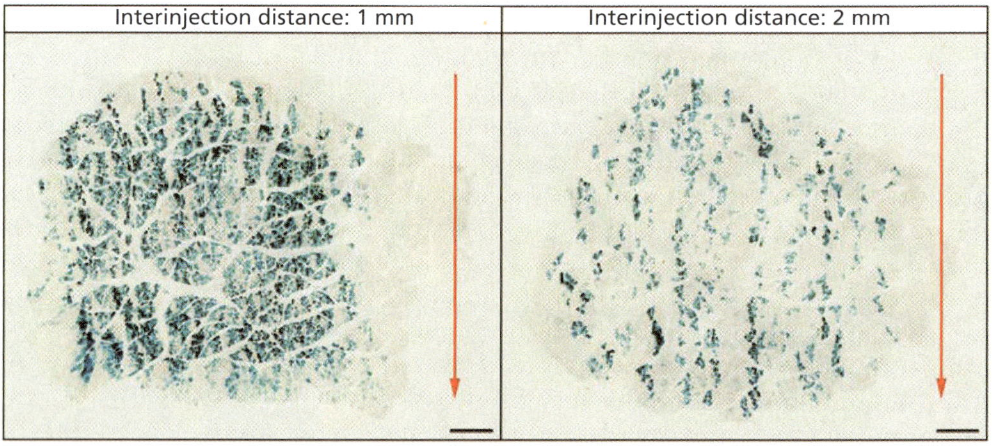

Figure 14.2 Whole cross sections of muscle biopsies obtained 1 month after myoblast transplantation in two sites of ~ 1 cm^2 of surface in the biceps brachii of a cynomolgus monkey. β-galactosidase-labeled myoblasts were injected using the PB600-1 repeating dispenser attached to a 250-μl Hamilton syringe, testing two inter-injection distances (i.e., 1 and 2 mm). The cross sections were histochemically treated to reveal β-galactosidase expressing myofibers (dark precipitate). The density of β-galactosidase-positive myofiber profiles is higher in the site in which myoblasts were delivered by injections placed 1 mm from each other than in the site in which myoblasts were delivered by injections placed 2 mm from each other. The arrows indicate the direction of the cell-injection trajectories done one month before. The topographical distribution of the β-galactosidase-positive myofiber profiles reproduces the pattern of the cell-injection trajectories due to the fact that the injected myoblasts fused essentially with the myofibers damaged by the needle. Bars = 1 mm (both images are at the same magnification).

moment of transplantation according to the trypan blue exclusion test is generally around 5% to 6% and on several occasions undetectable. Intraoperative bleeding is minimal if fine needles are used as recommended (25 to 27 gauge). Fine needles, in addition, ensure excellent muscle regeneration in spite of the repetitive injection damage. In the case of grafting cells throughout whole muscles, local swelling will be present for some days, but this postoperative inflammation usually resolves in less than a week in nonhuman primates, after which the muscle returns to its preoperative size and texture.

With the present technology, the results of the graft in terms of amount of muscle expressing the donor-derived proteins will depend on the amount of cells injected, on the proper delivery of the cell suspension during the needle withdrawal, on the interinjection distance and on appropriate control of acute rejection. In monkeys allotransplanted with β-galactosidase-expressing myoblasts, tacrolimus levels higher than 50 ng/ml will ensure good control of acute rejection in this context. It seems important to maintain this tacrolimus level at least for the first posttransplantation month, although our present evidence indicates that later it could be decreased to 20 ng/ml. According to our results, grafting $>10^5$ cells per cm of injection trajectory and using an interinjection distance of 1 mm, the intramuscular transplantation of β-galactosidase-labeled myoblasts should produce between 25% to 70% of β-galactosidase-positive myofibers in monkey limb muscles [14]. These percentages should decrease to 5% to 15% when the interinjection distance is 2 mm [14]. Figure 15.2 illustrates the results obtained using these two interinjection distances in two different sites of the biceps brachii in a cynomolgus monkey. The reasons by which a large range of results can be observed using a same pattern of cell injection is still not clear.

14.6 Discussion and Commentary

To reach its therapeutic objective, myoblast transplantation needs to induce the expression of therapeutic proteins in a large proportion of the skeletal muscle. To achieve that, myoblast transplantation needs to succeed in three steps:

1. An appropriate amount of living myogenic cells needs to be available for transplantation.
2. These cells must be properly delivered to the muscle.
3. The graft must survive in the host.

Our method of myoblast isolation and scale-up is simple and fulfills the first requirement. Our strategy of intramuscular delivery through very close injections provides the best cell delivery that is attainable with the present technology in acceptable clinical conditions. Tacrolimus-based immunosuppression is the only method successfully tested to control acute rejection after myoblast transplantation in nonhuman primates, insuring long-term survival of the graft. Our last clinical trials [16–18] were in fact based on these premises. The protocol that we tested in humans was quite similar to the protocol described above, although with some adaptations from which the most important are the following:

1. There is no genetic modification of the cells for allogenic transplantation. However, in the future it may be possible to transplant the patient's own genetically corrected myoblasts.
2. Two days before transplantation, the culture medium with FBS and albumin is replaced with culture medium containing 10% of serum obtained from the cell donor.
3. Percutaneous cell injections have challenges that are not present in monkeys. The skin in monkeys is thin and the skeletal muscles that we use for grafts (biceps brachii, quadratus femoris, and gastrocnemius) are easy to target. In humans, the surgeon must reach the region to inject the cells keeping in mind the distance from the skin surface to the muscle surface, which varies from one patient to another. In addition, to target a specific muscle in regions with an intricate muscle anatomy such as the forearm, with the special aggravating factor that in myopathic patients one or more muscles can be degenerated, has particular challenges. Thus, careful image analyses must be done before the transplantation, and, importantly, ultrasound analyses under muscle contraction to properly identify the target muscles. Labeling the skin with a surgical pencil during the ultrasound analysis, to be used as references during transplantation, is recommended. The distance from the skin surface to the muscle surface and the width of the muscle at different places must be measured during the ultrasound analysis and the length of the needles for cell injection must be chosen according to these measurements. During transplantation, the surgeon must be attentive to the course of the needle required to penetrate the skin and traverse the muscle, to properly deliver the cells in the muscle, and to avoid wasting cells in the hypodermis.
4. Local anesthesia by infiltration is preferred in humans if the surface to graft is not very large and if the time predicted for the cell injections does not exceed the time of the effect of the chosen anesthetics. Other forms of anesthesia may be discussed with an anesthetist in other cases.
5. When small muscle regions are grafted in humans and a label is needed to identify them later for a biopsy, we prefer methods less invasive than placing points of nonabsorbable neutral suture in the muscle. Thus, we delimited the grafted region by drawing a square on the skin using a surgical pencil and we asked the patient and his family to preserve the label until the biopsy. It is important in this case to ensure that the label is preserved until the biopsy, and even that it is not erased during the surgical preparation. Again, to find exactly the grafted region by an open surgery through thick skin may be a challenge and grafts in very small regions (e.g., $< 1cm^2$) should be avoided.
6. Ultimately, when grafting cells in entire muscles, strength measurements to confirm the clinical effects of the transplantation become a crucial test, without neglecting the histological findings in a biopsy.
7. Tacrolimus is given in its clinical oral formulation (Prograf) beginning at 0.2 mg/kg/day. It must be noted that in monkeys, we can administer doses of tacrolimus that are as high as needed to ensure good control of rejection in any case. In humans, there is a ceiling for tacrolimus administration, dependent on an acceptable balance between reducing the possibilities of acute rejection and reducing the toxicity of high tacrolimus doses. Thus, the possibility of developing an episode of acute rejection is more present in the clinics than in monkey experiments. In fact, our clinical

Troubleshooting Table

Problem	Explanation	Potential Solutions
During cell culture, the percentage of myogenic (CD56+) cells is under the desired level (i.e., <80%).	Fibroblast overgrowth.	Perform a preplating technique [40, 41] to eliminate as many fibroblasts as possible by differential attachment.
In the muscle biopsy done in the myoblast-grafted site, abundant hybrid myofibers are present only in an edge of the biopsy, without evidence of immune rejection.	The biopsy was eccentric to the cell-grafted region.	Be careful to place the reference non-absorbable sutures at the center of both sides of the grafted region, and at a distance no larger than ~ 5 mm.
In the muscle biopsy done in the myoblast-grafted site, muscle regeneration is incomplete.	The biopsy was performed too early after transplantation or larger needles than those recommended were used for cell injection.	Leave at least 1 month between myoblast transplantation and sampling. Use 25- to 27-gauge needles for cell injection.
Histological or serum evidence of acute rejection of the myoblast graft.	The immunosuppression was insufficient to completely control acute rejection.	Maintain tacrolimus levels over 50 ng/ml during the first month and not lower than 20 ng/ml for the rest of the experiment. If other immunosuppressants were used instead of tacrolimus, test higher dosages, although note that no other immunosuppressant other than tacrolimus proved so far to be efficient for myoblast transplantation in monkeys.
In the muscle biopsy done in the cell-grafted site, large immune cell infiltrates are related to the reference suture and become confused with rejection.	The use of very reactive nonabsorbable sutures for reference points, especially silk, can produce important tissular reactions. Otherwise, the reference suture points were placed too close to the graft.	Use lowest reactive sutures (polypropylene) for reference points, and place them a few millimeters away from the cell-grafted region.

results suggest that refinements in the immunosuppression protocol would be potentially needed to optimize the control of acute rejection in future clinical applications of myoblast allotransplantation.

14.7 Application Notes

The method outlined in this chapter resumes the protocol that we have been using for years for preclinical studies of myoblast transplantation in monkeys. It is the basis and point of reference for any test conducted in our laboratory to analyze the potential approaches to improve the clinical protocol of myoblast transplantation. The protocol described here could be used for studies of myoblast transplantation in other large mammals if their myoblasts can be proliferated similarly. This is not always the case, since our experience in dogs showed that myoblasts can be proliferated quite easily from puppies but not from adults [42], in contrast to monkeys and humans, in which muscle biopsies from adults provide good myoblast cultures [12, 14–18, 43]. Dogs present additional specific problems for immunosuppression, since, as an example, they do not tolerate tacrolimus administration well [44, 45]. Possibly for these reasons, the results of myoblast transplantation in dogs [42] were not as good as those obtained in monkeys. Otherwise, the technique of cell implantation outlined in this chapter can be used in other large mammals, although in animals such as dogs, in which the skin is quite thicker than in monkeys, the distance from the skin surface to the muscle may be ideally determined for percutaneous injections. However, there could be differences in the muscle behavior after myoblast transplantation between different taxonomic groups of large mammals. For example, in monkeys muscle regeneration is excellent following our protocol of repetitive and concentrated cell injections throughout skeletal muscles, but other research teams reported less good muscle regeneration following a similar protocol of cell injections in rabbits [46].

With the indispensable variants that we indicated above, the method outlined here can be extrapolated to studies of myoblast transplantation in myopathic patients.

14.8 Summary Points

1. Myoblasts can be easily isolated by standard cell-culture techniques from samples of skeletal muscles obtained from adult monkeys, and largely proliferated in culture medium composed of modified MCDB 120 or MB-1 supplemented with 15% FBS, basic fibroblast growth factor, bovine serum albumin, dexamethasone and insulin. Sufficient amounts of myoblasts can be produced for transplantation in several muscles.
2. To be integrated optimally into the recipient's skeletal muscles, the transplanted myoblasts need to be delivered intramuscularly with a technique implicating a matrix of close cell injections throughout the muscle, delivering the cells homogeneously in each injection trajectory.
3. This protocol of cell injections can be performed with precision syringes operated manually, but the use of repeating dispensers considerably facilitates the task.
4. Adequate immunosuppression, preferably with tacrolimus, is necessary in allogeneic conditions.
5. With minor variations, the same protocol used for experiments in nonhuman primates can be used in humans.

References

[1] Skuk, D., "Myoblast Transplantation for Inherited Myopathies: A Clinical Approach," *Expert. Opin. Biol. Ther.*, Vol. 4, No. 12, 2004, pp. 1871–85.

[2] Skuk, D., and J. P. Tremblay, "Myoblast Transplantation: The Current Status of a Potential Therapeutic Tool for Myopathies," *J. Muscle. Res. Cell. Motil.*, Vol. 24, No. 4–6, 2003, pp. 285–300.

[3] Rose, S. M., N. Blustein, and D. Rotrosen, "Recommendations of the Expert Panel on Ethical Issues in Clinical Trials of Transplant Tolerance. National Institute of Allergy and Infectious Diseases of the National Institutes of Health," *Transplantation*, Vol. 66, No. 9, 1998, pp. 1123–25.

[4] Kirk, A. D., "Transplantation Tolerance: A Look At The Nonhuman Primate Literature In The Light Of Modern Tolerance Theories," *Crit. Rev. Immunol.*, Vol. 19, No. 5–6, 1999, pp. 349–88.

[5] Kirk, A. D., "Crossing the Bridge: Large Animal Models in Translational Transplantation Research," *Immunol. Rev.*, Vol. 196, 2003, pp. 176–96.

[6] Slierendregt, B. L., et al., "Evolutionary Stability of Transspecies Major Histocompatibility Complex Class II DRB Lineages in Humans and Rhesus Monkeys," *Hum. Immunol.*, Vol. 35, No. 1, 1992, pp. 29–39.

[7] Geluk, A., et al., "Evolutionary Conservation of Major Histocompatibility Complex- DR/Peptide/T Cell Interactions in Primates," *J. Exp. Med.*, Vol. 177, No. 4, 1993, pp. 979–87.

[8] Jaeger, E. E., R. E. Bontrop, and J. S. Lanchbury, "Structure, Diversity, and Evolution of the T-Cell Receptor VB Gene Repertoire in Primates," *Immunogenetics*, Vol. 40, No. 3, 1994, pp. 184–91.

[9] Levinson, G., A. L. Hughes, and N. L. Letvin, "Sequence and diversity of Rhesus Monkey T-Cell Receptor Beta Chain Genes," *Immunogenetics*, Vol. 35, No. 2, 1992, pp. 75–88.

[10] Kinoshita, I., et al., "Myoblast Transplantation in Monkeys: Control of Immune Response by FK506," *J. Neuropathol. Exp. Neurol.*, Vol. 55, No. 6, 1996, pp. 687–97.
This study shows for the first time the potential of tacrolimus as a key immunosuppressant for myoblast transplantation in nonhuman primates.

[11] Kinoshita, I., et al., "Myoblast Allotransplantation in Primates," *Muscle Nerve*, Vol. 18, No. 10, 1995, pp. 1217–18.

[12] Skuk, D., et al., "Successful Myoblast Transplantation in Primates Depends on Appropriate Cell Delivery and Induction of Regeneration in the Host Muscle," *Exp. Neurol.*, Vol. 155, No. 1, 1999, pp. 22–30.

[13] Skuk, D., M. Goulet, and J. P. Tremblay, "Use of Repeating Dispensers to Increase the Efficiency of the Intramuscular Myogenic Cell Injection Procedure," *Cell Transplant.*, Vol. 15, No. 7, 2006, pp. 659–63.

[14] Skuk, D., et al., "Efficacy of Myoblast Transplantation in Nonhuman Primates Following Simple Intramuscular Cell Injections: Toward Defining Strategies Applicable to Humans," *Exp. Neurol.*, Vol. 175, No. 1, 2002, pp. 112–26.
This study analyzes the relationship between the interinjection distance and the graft success for the intramuscular transplantation of myoblasts in nonhuman primates. It also gives the first correlation between tacrolimus blood levels and graft outcome.

[15] Skuk, D., et al., "Myoblast Transplantation in Whole Muscle of Nonhuman Primates," *J. Neuropathol. Exp. Neurol.*, Vol. 59, No. 3, 2000, pp. 197–206.

[16] Skuk, D., et al., "Dystrophin Expression in Muscles of Duchenne Muscular Dystrophy Patients after High-Density Injections of Normal Myogenic Cells," *J. Neuropathol. Exp. Neurol.*, Vol. 65, No. 4, 2006, pp. 371–86.
These are the first consistent results obtained by normal myoblast allotransplantation in a clinical context. The study, conducted on nine Duchenne muscular dystrophy patients, resumes several important issues of clinical myoblast transplantation, including analyses of donor-derived dystrophin expression and of immune responses.

[17] Skuk, D., et al., "First Test of a 'High-Density Injection' Protocol for Myogenic Cell Transplantation Throughout Large Volumes of Muscles in a Duchenne Muscular Dystrophy Patient: Eighteen Months Follow-Up," *Neuromuscul. Disord.*, Vol. 17, No. 1, 2007, pp. 38–46.
This is the first demonstration of long-term successful graft survival following normal myoblast allotransplantation in a patient with Duchenne muscular dystrophy immunosuppressed with tacrolimus, and the first case in which our technique of intramuscular cell implantation was applied in the clinics to a large skeletal muscle.

[18] Skuk, D., et al., "Dystrophin Expression in Myofibers of Duchenne Muscular Dystrophy Patients Following Intramuscular Injections of Normal Myogenic Cells," *Mol. Ther.*, Vol. 9, No. 3, 2004, pp. 475–82.

[19] Gussoni, E., et al., "Normal Dystrophin Transcripts Detected in Duchenne Muscular Dystrophy Patients after Myoblast Transplantation," *Nature*, Vol. 356, No. 6368, 1992, pp. 435–38.

[20] Karpati, G., et al., "Myoblast Transfer (MT) in McArdle's Disease (McD)," *Neurology*, Vol. 42 (Suppl. 3), No. 1992, p. 387.

[21] Karpati, G., et al., "Myoblast Transfer in Duchenne Muscular Dystrophy," *Ann. Neurol.*, Vol. 34, No. 1993, pp. 8-17.

[22] Tremblay, J. P., et al., "Myoblast Transplantation between Monozygotic Twin Girl Carriers of Duchenne Muscular Dystrophy," *Neuromuscul. Disord.*, Vol. 3, No. 5–6, 1993, pp. 583–92.

[23] Morandi, L., et al., "Lack of mRNA and Dystrophin Expression in DMD Patients Three Months after Myoblast Transfer," *Neuromuscul. Disord.*, Vol. 5, No. 4, 1995, pp. 291–95.

[24] Mendell, J. R., et al., "Myoblast Transfer in the Treatment of Duchenne's Muscular Dystrophy," *N. Engl. J. Med.*, Vol. 333, No. 1995, pp. 832–38.

[25] Miller, R. G., et al., "Myoblast Implantation in Duchenne Muscular Dystrophy: The San Francisco Study," *Muscle Nerve*, Vol. 20, No. 4, 1997, pp. 469–78.

[26] Neumeyer, A. M., et al., "Pilot Study of Myoblast Transfer in the Treatment of Becker Muscular Dystrophy," *Neurology*, Vol. 51, No. 2, 1998, pp. 589–92.

[27] Partridge, T. A., et al., "Conversion of MDX Myofibres from Dystrophin-Negative to -Positive by Injection of Normal Myoblasts," *Nature*, Vol. 337, No. 6203, 1989, pp. 176–79.

[28] Vilquin, J. T., et al., "Partial Laminin Alpha2 Chain Restoration in Alpha2 Chain-Deficient DY/DY Mouse by Primary Muscle Cell Culture Transplantation," *J. Cell. Biol.*, Vol. 133, No. 1, 1996, pp. 185–97.

[29] Kinoshita, I., et al., "Immunosuppression with FK 506 Insures Good Success of Myoblast Transplantation in MDX Mice," *Transplant. Proc.*, Vol. 26, No. 6, 1994, p. 3518.

[30] Ito, H., et al., "Myoblast Transplantation in Non-Dystrophic Dog," *Neuromuscul. Disord.*, Vol. 8, No. 2, 1998, pp. 95–110.

[31] Wernig, A., A. Irintchev, and G. Lange, "Functional effects of myoblast implantation into histoincompatible mice with or without immunosuppression," *J Physiol (Lond)*, Vol. 484, No. Pt 2 1995, pp. 493-504.

[32] Irintchev, A., M. Langer, M. Zweyer, R. Theisen, and A. Wernig, "Functional improvement of damaged adult mouse muscle by implantation of primary myoblasts," *J Physiol (Lond)*, Vol. 500, No. Pt 3 1997, pp. 775-785.

[33] Alameddine, H. S., et al., "Functional Recovery Induced by Satellite Cell Grafts in Irreversibly Injured Muscles," *Cell Transplant.*, Vol. 3, No. 1, 1994, pp. 3–14.

[34] Wernig, A., M. Zweyer, and A. Irintchev, "Function of Skeletal Muscle Tissue Formed After Myoblast Transplantation into Irradiated Mouse Muscles," *J. Physiol. (Lond.)*, Vol. 522,. Pt. 2, 2000, pp. 333–45.

[35] Watt, D. J., et al., "Incorporation of Donor Muscle Precursor Cells Into an Area of Muscle Regeneration in the Host Mouse," *J. Neurol. Sci.*, Vol. 57, No. 2–3 1982, pp. 319–31.

[36] Watt, D. J., "Factors which Affect the Fusion of Allogeneic Muscle Precursor Cells In Vivo," *Neuropathol. Appl. Neurobiol.*, Vol. 8, No. 2, 1982, pp. 135–47.

[37] Pavlath, G. K., et al., "Localization of Muscle Gene Products in Nuclear Domains," *Nature*, Vol. 337, No. 6207, 1989, pp. 570–73.

[38] Ham, R. G., et al., "Improved Media for Normal Human Muscle Satellite Cells: Serum-Free Clonal Growth and Enhanced Growth with Low Serum," *In Vitro Cell Dev. Biol.*, Vol. 24, No. 8, 1988, pp. 833–44.

[39] Cepko, C., "Preparation of a Specific Retrovirus Producer Cell Line," *Curr. Protoc. Mol. Biol.*, Chapter 9, 2001, Unit 9–10.
A detailed description of protocols to establish a cell line producing specific retroviral vectors.

[40] Yaffe, D., "Retention of Differentiation Potentialities During Prolonged Cultivation of Myogenic Cells," *Proc. Natl. Acad. Sci. USA*, Vol. 61, No. 2, 1968, pp. 477–83.

[41] Polinger, I. S., "Separation of Cell Types in Embryonic Heart Cell Cultures," *Exp. Cell Res.*, Vol. 63, No. 1, 1970, pp. 78–82.

[42] Ito, H., et al., "Myoblast Transplantation in Non-Dystrophic Dog," *Neuromuscul. Disord.*, Vol. 8, No. 2, 1998, pp. 95–110.

[43] Skuk, D., et al., "Ischemic Central Necrosis in Pockets of Transplanted Myoblasts in Nonhuman Primates: Implications for Cell-Transplantation Strategies," *Transplantation*, Vol. 84, No. 10, 2007, pp. 1307–15.

[44] Ueda, Y., et al., "Induction of Graft Acceptance after Dog Kidney or Liver Transplantation," *Transplant. Proc.*, Vol. 22, No. 1, 1990, pp. 80–82.

[45] Thomson, A. W., "Interspecies Comparison of the Immunosuppressive Efficacy and Safety of FK 506," *Transplant. Proc.*, Vol. 22, No. 1, 1990, pp. 100–105.

[46] Cesar, M., et al., "Transplantation of Adult Myoblasts or Adipose Tissue Precursor Cells by High-Density Injection Failed to Improve Reinnervated Skeletal Muscles," *Muscle Nerve*, Vol. 37, No. 2, 2008, pp. 219–30.

Acronyms and Abbreviations

3D-HA	three-dimensional cultures with a porous hyaluronate
6-OHDA	6-hydroxydopamine
AA	ascorbic acid
AAV	adeno-associated virus
ADASs	adipose-derived adherent stromal cells/adipose-derived adult stem cells
ADSCs	adipose-derived stromal cells
AE	amniotic epithelium
AFP	α-fetoprotein
ALB	albumin antibody
ALT	alanine aminotransferase
AM	amnion
ASF	asialofetuin
ASGPR	asialoglycoprotein receptors
ASGPs	asialoglycoproteins
AST	aspartate aminotransferase
AT-MSCs	adipose tissue-derived mesenchymal stem cells
ATSCs	adipose-tissue-derived stromal cells
BCA	bicinchoninic acid
BCG	bromocresol green
BD	Becton Dickenson
BDD	brain-dead donor
BDNF	brain-derived neurotrophic factor
BHK	baby hamster kidney
BIIM	Baylor islet isolation method
BMCs	bone marrow cells
BM-MSCs	bone marrow mesenchymal stem cells
BMS	Basso mouse scale
BrdU	bromodeoxyuridine
BSA	bovine serum albumin
$CaCl_2$	calcium chloride
CDI	N,N0-carbonyldiimidazole
CEFCM	collagen embedded floating culture method
CFU-F	colony-forming unit-fibroblasts
CK	cytokeratin
CK-18	mouse anti-human cytokeratin-18 antibody
CM	chorion
CMF-HBSS	calcium and magnesium free Hanks' balanced salt solution
CMV	cytomegalovirus

CNS	central nervous system
ConA	concanavalin A
CPA	carboxypeptidase A
CPR	cytochrome P-450 reductase
DAPI	6-diamino-2-phenylindole
DIC	differential interference contrast
Dil-Ac-LDL	1,1-dioctadecyl-3,3,3,3-tetramethylindocarbocyanine perchlorate
DLS	dynamic light scattering
DM	differentiation medium
DMEM	Dulbecco's Modified Eagle Medium
DMF	dimethyl-formamide
DMSO	dimethyl sulfoxid
DPP	depeptidyl peptidase
DTZ	dithizone
EBs	embryoid bodies
ECM	extracellular matix
ECMs	extracellular matrices
ED	embryonic day
EGF	epidermal growth factor
EGTA	ethylene glycol tetraacetic acid
EHC	enzymehistochemistry
ELS	electrophoretic light scattering
EMEM	Eagle's minimum essential medium
ES	embryonic stem
FACS	fluorescent-activated cell sorting
FAK	focal adhesion kinase
FBS	fetal bovine serum
FGF	fibroblast growth factor
FITC	fluorescein isothiocyanate
FLSPC	fetal liver stem/progenitor cells
FN	fibronectin
GDA	gastroduodenal artery
GDNF	glial cell line-derived neurotrophic factor
GFP	green fluorescent protein
GM	growth medium
GPMN	glycyl-proline-4-methoxy-α-naphtylamide
GTT	glucose tolerance test
H&E	hematoxylin and eosin
HBSS	Hanks' balanced salt solution
HBG	mM HEPES
HCM	hepatic culture medium
hES	human embryonic stem
HGF	hepatocyte growth factor
HIFC	hepatocyte induction factor cocktail
hIPS	human induced pluripotent stem
HNF	hepatocyte nuclear factor
HPGM	hematopoietic progenitor growth medium
HSP	heat shock protein

HSV	herpes simplex virus
HuNu	human nuclei
IBDU	intrahepatic bile duct unit
ICC	islet cell clusters
ICM	inner cell mass
ID	inner diameter
IDO	indoleamine 2,3-dioxygenase
IEQ	islet equivalent
IEs	islet equivalents
IHBECs	intrahepatic biliary epithelial cells
IHC	immunohistochemistry
IMV	inferior mesenteric vein
iPS	induced pluripotent stem
ITS	insulin-transferrin-selenium
IVC	inferior vena cava
KRB	Krebs ringer buffer
KRBB	Kreb's ringer bicarconate buffer
LB	Langmuir-Blodgett
LcH	lens culinaris
LCST	lower critical solution temperature
MACS	magnetic-activated cell sorting
MEF	mouse embryonic fibroblast
$MgCl_2$	magnesium chloride
MI	myocardial infarction
MK	modified Kyoto
MMI	Miami-defined media #1
MOI	multiplicity of infection
MPTP	1-methyl-4 phenyl-1,2,3,6-tetrahydropyridine
MSC	mesenchymal stromal cells
MSC-GM	mesenchymal stem cell-growth medium
MTT	3-(4,5-dimethylthiazol-2-yl)-2,5-diphenyl tetrazolium bromide
N:P	nitrogen to phosphate
NHBD	non-heart-beating donor
NOD	nonobese diabetic
NYU	New York University
OD	out diameter
OsM	Oncostatin M
OSU	Ohio State University
PAAm	polyacrylamide
PAS	periodic acid-schiff
PBS	phosphate-buffered saline
PC1	prohormone convertases 1
Pdx-1	pancreatic and deudenal homeobox-1
PET	poly(ethylene terephthalate)
PFA	paraformaldehyde
PFU	plaque forming units
PH	partial hepatectomy
PHA	proper hepatic artery

PI	propidium iodide
PIPAAM	N-isopropylacrylamide
PLA	processed lipoaspirate
PLGA	poly(D,L-lactic-co-glycolic acid)
PLL	poly(L-lysine)
PS	polystyrene
PVDF	polyvinylidine difluoride
PVLA	poly(N-P-vinylbenzyl-4-O-β-D-galactopyranosyl-D-gluconamide)
QC	quality control
RLU	relative light units
ROM	range of motion
RT-PCR	reverse transcription polymerase chain reaction
SCID	severe combined immunodeficient
SDS	sodium dodecyl sulfate
SHVC	superior hepatic vena cava
SI	stimulation index
SMA	superior mesenteric artery
SNL	spinal nerve ligation
SPION	superparamagnetic iron oxide nanoparticles
SSEA	stage-specific embryonic antigens
STZ	streptozotocin
TGF	transforming growth factor
TMS	Trisma® maleate/0.1 M NaCl solution
TTR	transthyretin antibody
TUNEL	TdT-mediated dUTP-biotin nick end labeling
UA	uric acid
UW	University of Wisconsin
VEGF	vascular endothelial growth factor
WE	Williams' Medium E
WST-8	water-soluble tetrazolium
wt	wild type

About the Editors

Alejandro Soto-Gutierrez is an assistant professor in the Department of Surgery, Transplantation Section at the Children's Hospital of Pittsburgh. He is also a faculty member in the McGowan Institute for Regenerative Medicine, University of Pittsburgh. Before his move to Pittsburgh, he was a surgery research fellow in the Department of Surgery at Massachusetts General Hospital, Harvard Medical School, and the Shriners Hospitals for Children in Boston. Dr. Soto-Gutiérrez earned his medical degree at the University of Guadalajara, Mexico. While working on his doctorate, he completed a surgical fellowship in the Department of Surgery of Okayama University Graduate School of Medicine and Dentistry, Japan, where he wrote his dissertation on liver tissue engineering. His major research interests focus on the development of experimental technologies to treat hepatic diseases using bioengineering techniques, cell transplantation, and organ engineering. He currently works on engineering and transplantation of hepatic grafts using decellularized native matrices and stem cell differentiation. He won the 2008 Clinical Research Award in Transplantation from the New England Organ Bank and the 2009 Thomas E. Starzl, M.D. Postdoctoral Research Award from the American Liver Foundation. He was recently awarded the National Institutes of Health Pathway to Independence Award (K99/R00). He is the author of more than 50 peer-reviewed publications in leading journals and six book chapters.

Nalu Navarro-Alvarez is a research scientist in the Columbia Center for Translational Immunology (CCTI) at Columbia University. Prior to her move to New York she was a transplant immunology research fellow at the Transplantation Biology Research Center (TBRC) in the Department of Surgery at Massachusetts General Hospital, Harvard Medical School in Boston. Dr. Navarro-Alvarez received her medical degree at the University of Guadalajara, Mexico. While working on her doctorate, she also completed a surgical fellowship in the Department of Surgery of Okayama University Graduate School of Medicine and Dentistry, Japan. Her main research interest involves the development of innovative therapies for liver disease using different cell transplant techniques as well as the study of the immunological barriers to liver repopulation through hepatocyte transplantation. Her current research involves the use of transgenic animal models to study the role of innate immunity in hepatocyte transplantation. She was awarded the Basic Science Research Award from the American Society of Transplantation (AST) in 2009. She is author of more than 40 peer-reviewed publications in leading journals and four book chapters.

Ira J. Fox is the director of the Center for Regenerative Therapies at Children's Hospital of Pittsburgh and the McGowan Institute for Regenerative Medicine. He is also a professor of surgery and pediatric transplantation at the University of Pittsburgh School of

Medicine. Prior to moving to Pittsburgh in June 2008, Dr. Fox was senior associate dean for Research and M.D–Ph.D. Program Director at the University of Nebraska Medical Center. His major research interests involve the study of experimental therapies for liver disease, including auxiliary liver transplantation, liver assist devices, liver cell transplantation, and liver stem cells. His laboratory examines the biologic and immunologic barriers to successful transplantation of primary allogeneic and xenogeneic hepatocytes in rodents and primates. He also studies differentiation of mouse and human embryonic stem cells into liver cells. He has initiated clinical trials involving extracorporeal human and pig liver perfusion to treat patients with acute liver failure, and has demonstrated successful correction of liver-based metabolic deficiencies in newborns and young children by transplantation of isolated liver cells. Dr. Fox is recognized as an authority on innovative therapies for the treatment of liver disease and as a leader in stem cell transplantation. His work has been continuously supported by grants from the NIH and DOD for over 18 years.

Contributors

Toshihiro Akaike
Graduate School of Bioscience and Biotechnology
Tokyo Institute of Technology
Yokohama, Japan

Suzanne Bertera
Rangos Research Center
Children's Hospital of Pittsburgh of UPMC
Pittsburgh, PA, U.S.A.

Rita Bottino
Rangos Research Center
Children's Hospital of Pittsburgh of UPMC
Pittsburgh, PA, U.S.A.

Victoria L. Browning
Division of Transplantation
Department of Surgery
University of Wisconsin-Madison School of Medicine
and Public Health
and WiCell Research Institute
Madison, WI, U.S.A.

Chong-Su Cho
Department of Agricultural Biotechnology
Research Institute for Agriculture and Life Sciences
Seoul National University
Seoul, Korea

Isao Date
Department of Neurological Surgery
Okayama University Graduate School of Medicine,
Dentistry, and Pharmaceutical Sciences
Okayama, Japan

Julio C. Davila
Pfizer, Inc.
St. Louis, MO, U.S.A.

Kenneth Dorko
Department of Pathology
School of Medicine
University of Pittsburgh
Pittsburgh, PA, U.S.A.

Shimon Efrat
Department of Human Molecular Genetics
and Biochemistry
Sackler School of Medicine
Tel Aviv University
Tel Aviv, Israel

Ewa C.S. Ellis
Department of Clinical Science, Intervention,
and Technology (CLINTEC)
Division of Transplantation Surgery
Liver Cell Lab
Karolinska University Hospital Huddinge
Stockholm, Sweden

Laura Gómez-Santos
Division of Molecular and Cellular Medicine
National Cancer Center Research Institute
Tokyo, Japan

Marlyne Goulet
Neurosciences Division, Human Genetics
CHUQ Research Center
CHUL
Quebec, Canada

Roberto Gramignoli
Department of Pathology
School of Medicine
University of Pittsburgh
Pittsburgh, PA, U.S.A.

Keitaro Hagiwara
Division of Molecular and Cellular Medicine
National Cancer Center Research Institute
Tokyo, Japan

Marc C. Hansel
Department of Pathology
School of Medicine
University of Pittsburgh
Pittsburgh, PA, U.S.A.

Takashi Hoshiba
Biomaterials Center
National Institute for Materials Science
Tsukuba, Japan

Tetsuya Ishikawa
Division of Molecular and Cellular Medicine
National Cancer Center Research Institute
Tokyo, Japan

Hideki Iwaguro
Division of Molecular and Cellular Medicine
National Cancer Center Research Institute
Tokyo, Japan

Akio Iwanami
Department of Orthopedic Surgery
Keio University School of Medicine
Tokyo, Japan

Takanori Iwata
Institute of Advanced Biomedical Engineering
and Science
Tokyo Woman's Medical University
Tokyo, Japan

Hu-Lin Jiang
College of Veterinary Medicine
Seoul National University
Seoul, Korea

Jun-Ichiro Jo
Institute for Frontier Medical Sciences
Kyoto University
Kyoto, Japan

Sang-Hyon Kim
Biomaterials Research Center
Korea Institute of Science and Technology
Seoul, Korea

Naoya Kobayashi
Department of Surgery
Okayama University Graduate School of Medicine and Dentistry
Okayama, Japan

Marco Marigliano
Rangos Research Center
Children's Hospital of Pittsburgh of UPMC
Pittsburgh, PA, U.S.A.

Fabio Marongiu
Department of Biomedical Sciences and Technologies
University of Cagliari
Cagliari, Italy

Shinichi Matsumoto
Baylor All Saints Medical Center
Baylor Research Institute
Fort Worth, TX, U.S.A.

Toshio Miki
Department of Developmental Biology
School of Medicine
University of Pittsburgh
Pittsburgh, PA, U.S.A.

Kentaro Nagane
Institute for Frontier Medical Sciences
Kyoto University
Kyoto, Japan

Masaki Nagaya
Research Division, Section on Islet Transplantation and Cell Biology
Joslin Diabetes Center
Boston, MA, U.S.A.

Narihito Nagoshi
Department of Orthopedic Surgery
Keio University School of Medicine
Tokyo, Japan

Masaya Nakamura
Department of Orthopedic Surgery
Keio University School of Medicine
Tokyo, Japan

Hirofumi Noguchi
Baylor All Saints Medical Center
Baylor Research Institute
Fort Worth, TX, U.S.A
and Department of Surgery
Okayama University Graduate School of Medicine and Dentistry
Okayama, Japan

Hideaki Obata
Department of Anesthesiology
Gunma University
Japan

Takahiro Ochiya
Division of Molecular and Cellular Medicine
National Cancer Center Research Institute
Tokyo, Japan

Jon S. Odorico
Division of Transplantation
Department of Surgery
University of Wisconsin-Madison School of Medicine and Public Health
and WiCell Research Institute
Madison, WI, U.S.A.

Michael Oertel
Marion Bessin Liver Research Center
and Division of Gastroenterology and Liver Diseases
Department of Medicine
Albert Einstein College of Medicine of Yeshiva University
Bronx, NY, U.S.A.

Hideyuki Okano
Department of Physiology
Keio University School of Medicine
Tokyo, Japan

Teruo Okano
Institute of Advanced Biomedical Engineering and Science
Tokyo Woman's Medical University
Tokyo, Japan

Martin Paradis
Neurosciences Division, Human Genetics
CHUQ Research Center
CHUL
Quebec, Canada

Holger A. Russ
Department of Human Molecular Genetics and Biochemistry
Sackler School of Medicine
Tel Aviv University
Tel Aviv, Israel

Shigeru Saito
Department of Anesthesiology
Gunma University
Japan

David A. Shafritz
Marion Bessin Liver Research Center
and Division of Gastroenterology and Liver Diseases
Department of Medicine
Albert Einstein College of Medicine of Yeshiva University
Bronx, NY, U.S.A.

Daniel Skuk
Neurosciences Division, Human Genetics
CHUQ Research Center
CHUL
Quebec, Canada

Kristen J. Skvorak
Department of Pathology
School of Medicine
University of Pittsburgh
Pittsburgh, PA, U.S.A.

Stephen C. Strom
Department of Pathology
School of Medicine
University of Pittsburgh
Pittsburgh, PA, U.S.A.

Yasuhiko Tabata
Institute for Frontier Medical Sciences
Kyoto University
Kyoto, Japan

Veysel Tahan
Department of Pathology
School of Medicine
University of Pittsburgh
Pittsburgh, PA, U.S.A.

Yang D. Teng
Departments of Neurosurgery, and Physical Medicine
and Rehabilitation
Harvard Medical School
Boston, MA, U.S.A.
and Division of Spinal Cord Injury Research
Veterans Affairs Boston Healthcare System
Boston, MA, U.S.A.

Devang K, Thakor
Departments of Neurosurgery, and
Physical Medicine and Rehabilitation
Harvard Medical School
Boston, MA, U.S.A.
and Division of Spinal Cord Injury Research
Veterans Affairs Boston Healthcare System
Boston, MA, U.S.A.

Yoshiaki Toyama
Department of Orthopedic Surgery
Keio University School of Medicine
Tokyo, Japan

Jacques P. Tremblay
Neurosciences Division, Human Genetics
CHUQ Research Center
CHUL
Quebec, Canada

Massimo Trucco
Rangos Research Center
Children's Hospital of Pittsburgh of UPMC
Pittsburgh, PA, U.S.A.

Gordon C. Weir
Research Division, Section on Islet Transplantation
and Cell Biology
Joslin Diabetes Center
Boston, MA, U.S.A.

Xiaofang Xu
Division of Transplantation
Department of Surgery
University of Wisconsin-Madison School
of Medicine and Public Health
and WiCell Research Institute
Madison, WI, U.S.A.

Masaya Yamamoto
Institute for Frontier Medical Sciences
Kyoto University
Kyoto, Japan

Masayuki Yamato
Institute of Advanced Biomedical Engineering
and Science
Tokyo Woman's Medical University
Tokyo, Japan

Takao Yasuhara
Department of Neurological Surgery
Okayama University Graduate School of Medicine,
Dentistry, and Pharmaceutical Sciences
Okayama, Japan

Index